T0281323

# Lecture Notes in Mathematics 1907

**Editors:**
J.-M. Morel, Cachan
F. Takens, Groningen
B. Teissier, Paris

Martin Rasmussen

# Attractivity and Bifurcation for Nonautonomous Dynamical Systems

 Springer

Author

Martin Rasmussen
Institut für Mathematik
Lehrstuhl für Angewandte Analysis
Universität Augsburg
86135 Augsburg
Germany
*e-mail: martin.rasmussen@math.uni-augsburg.de*

Library of Congress Control Number: 2007925370

Mathematics Subject Classification (2000): 34D05, 37B25, 37B55, 37D10, 37G35

ISSN print edition: 0075-8434
ISSN electronic edition: 1617-9692
ISBN-10   3-540-71224-0 Springer Berlin Heidelberg New York
ISBN-13   978-3-540-71224-4 Springer Berlin Heidelberg New York

DOI 10.1007/978-3-540-71225-1

This work is subject to copyright. All rights are reserved, whether the whole or part of the material is concerned, specifically the rights of translation, reprinting, reuse of illustrations, recitation, broadcasting, reproduction on microfilm or in any other way, and storage in data banks. Duplication of this publication or parts thereof is permitted only under the provisions of the German Copyright Law of September 9, 1965, in its current version, and permission for use must always be obtained from Springer. Violations are liable for prosecution under the German Copyright Law.

Springer is a part of Springer Science+Business Media
springer.com
© Springer-Verlag Berlin Heidelberg 2007

The use of general descriptive names, registered names, trademarks, etc. in this publication does not imply, even in the absence of a specific statement, that such names are exempt from the relevant protective laws and regulations and therefore free for general use.

Typesetting by the authors and SPi using a Springer LaTeX macro package

Cover design: *design & production* GmbH, Heidelberg

Printed on acid-free paper     SPIN: 12027767     VA41/3100/SPi   5 4 3 2 1 0

To Professor Bernd Aulbach
and my parents

# Preface

This book has been developed from my dissertation, which I wrote at the University of Augsburg from 2002 to 2005. I first became acquainted with several definitions of attractor for nonautonomous dynamical systems when I was preparing my diploma thesis, and the question arose whether a nonautonomous bifurcation theory can be founded based on suitable notions of nonautonomous attractor (and repeller).

At the beginning of my time as a Ph. D. student, I developed local notions of attractor and repeller for several time domains (the past, the future, the whole time and finite time intervals), and I distinguished between two bifurcation scenarios. The first scenario describes the loss of attractivity and repulsivity, and the second one deals with transitions of attractors and repellers. All definitions are introduced in Chapter 2 of this book. As a test for the new definitions, I then considered asymptotically autonomous differential equations; these are systems whose behavior becomes autonomous when time tends to the past or the future. I found conditions for the occurrence of a nonautonomous bifurcation in case the underlying autonomous system admits a bifurcation (see Chapter 7). Moreover, I developed nonautonomous counterparts for classical one-dimensional bifurcation patterns (see Chapter 6).

The remaining part of my work was focussed on the study of qualitative properties of the local notions of attractivity and repulsivity. I showed that these are suitable to describe the *global* asymptotic behavior via Morse decompositions (see Chapter 3), and for linear systems, I introduced notions of dichotomy and dichotomy spectra for the four different time domains (see Chapter 4). Furthermore, I constructed invariant manifolds of nonlinear systems for the different time domains in order to obtain attractivity and repulsivity from the linearization (see Chapter 5).

Writing this book would not have been possible without the aid of many people to whom I would like to express my gratitude. First of all, I would like to thank my supervisor Professor Bernd Aulbach, who unfortunately suddenly and unexpectedly passed away on January 14, 2005, at the age of 57 years. I am grateful for his longstanding support while writing my diploma thesis and dissertation. I benefited from his great ability to explain complicated facts very clearly and lucidly, and I am thankful to him for many fruitful discussions. Moreover, I am greatly indebted to Professor Fritz Colonius who became my advisor after the death of Professor Aulbach. He was very interested in the details of my work, and I was very encouraged by his positive attitude to my ideas and suggestions. Furthermore, I am grateful to Professor Lars Grüne for his interest in my work and for being a referee for my dissertation. I would also like to thank Dr. Stefan Siegmund for many useful discussions and remarks, especially in the first year of my work. Special thanks go to my friends and colleagues Dr. Christian Pötzsche and Dr. Ludwig Neidhart for reading the manuscript and making useful comments. I also thank the *Deutsche Forschungsgemeinschaft* for the financial support I received from them, when I was a member of the *Graduiertenkolleg "Nichtlineare Probleme in Analysis, Geometrie und Physik"* in the department for mathematics and physics at the University of Augsburg. Finally, I would like to thank my parents for making it possible for me to study mathematics and for their support during all these years.

Augsburg, February 2007                                    *Martin Rasmussen*

# Contents

# 1

# Introduction

The mathematical concept of dynamical system is founded on the fact that motions of many application processes are subjected to certain rules. In Newtonian mechanics, in other natural sciences and even in an economical and social context, these laws are given implicitly by a relation that determines the state of a system for all future times just by the knowledge of the present state. A dynamical system therefore consists of the following two components: the space of states and the rule which, given an initial state, allows the projection of the state of the system in the future.

Historically, the notion of dynamical system was derived as an abstraction and generalization of ordinary differential equations. It was first used in 1927 by the American mathematician George D. Birkhoff (1884–1944) in his homonymous book [31]. Birkhoff was strongly influenced by the French mathematician Henri Poincaré (1854–1912), who is regarded—together with the Russian mathematician and engineer Aleksandr M. Lyapunov (1857–1918)—as the father of the so-called qualitative theory of dynamical systems. The goal of the qualitative theory is to understand the behavior of solutions from a more geometrical and topological point of view. In this book, we mainly address two aspects of this theory: the theory of attractivity and the theory of bifurcation. These fields are strongly related, since bifurcations from a dynamical viewpoint are associated with loss or gain of attractivity.

The theory of attractivity has its origin in the thesis *The General Problem of the Stability of Motion* [110, 112, 113], where Lyapunov introduced several definitions and methods to analyze the dynamical behavior in the vicinity of an equilibrium or—more generally—an arbitrary solution of an ordinary differential equation. The term *attractor* was first used by Coddington and Levinson [48] and Mendelson [119]. In the article *Attractors in Dynamical Systems* [23], Auslander, Bhatia and Seibert considered attractors consisting of more than one point. In 1967, Stephen Smale introduced in *Differential Dynamical Systems* [175] a new type of attractor, the axiom A attractor.

A new highlight in the theory of attractor was reached in 1971, when Ruelle and Takens regarded so-called strange attractors as a reason for the turbulent behavior in fluids (*On the Nature of Turbulence* [148]). This notion of attractor allowed the connection of the attractor theory and the upcoming chaos theory. Similar ideas have formerly been used by Edward N. Lorenz in *Deterministic Nonperiodic Flow* [108]. In *Isolated Invariant Sets and the Morse Index* [53], Charles C. Conley introduced in 1978 a very natural notion of local attractor which allowed the construction of so-called attractor-repeller pairs and Morse decompositions, and Ruelle modified this concept by considering so-called pseudoorbits in *Small Random Perturbations of Dynamical Systems and the Definition of Attractors* [147].

The fundamental ideas and elements of bifurcation theory go back to Poincaré [137] and Lyapunov [111]. Poincaré first used the term *bifurcation* to describe the splitting of asymptotic states of a dynamical system in his article *Sur l'equilibre d'une masse fluids animes d'un mouvement de rotation* [135, §2 Equilibre de bifurcation, p. 261]. In 1937, a great step towards a formalization of bifurcation theory was undertaken by the definition of structural stability by Andronov and Pontryagin (*Systemes grossiers* [4]). Since the 1960s, the bifurcation theory was fast-paced. One reason for this development was the introduction of the center manifold theory by Pliss [132] and Kelley [92], which allowed systems of high dimension amenable to a low-dimensional bifurcation analysis. Moreover, the normal form theory, which dates back to the thesis of Poincaré [134] and Birkhoff [31], became a field of intensive research.

In many cases, the notion of dynamical system is not general enough to model real world phenomena, since it is often indicated to assume that the underlying rules are time-dependent. For biological processes, for instance, it is more realistic to take evolutionary adaptations into account, and sometimes it is unavoidable to consider random perturbations such as white noise or to model the control of a process by a human being. The appropriate class to treat such problems are the so-called nonautonomous dynamical systems. Another reason to consider nonautonomous dynamical systems is given by the fact that the investigation of states of dynamical systems which are nonconstant in time leads to nonautonomous problems in form of the equation of perturbed motion. The notion of nonautonomous dynamical system was created in the 1990s from the studies of both topological skew product flows and random dynamical systems. The theory of topological skew product flows was founded in the late 1960s by George R. Sell and Richard K. Miller (see [120, 165, 166, 167]), and the notion of the random dynamical system is based on research by Baxendale, Bismut, Elworthy, Ikeda, Kunita, Watanabe and many others (see [25, 32, 65, 83, 100]). Further progress in this field was achieved by Ludwig Arnold and his "Bremen Group".

The nonautonomous theory of attractivity has been stimulated in the last fifteen years by the introduction of the notions of pullback attractor, forward attractor, random attractor and weak random attractor. In particular, questions of existence, uniqueness, perturbation and discretization have been addressed. These contributions were made by Cheban, Crauel, Flandoli, Kloeden, Ochs, Schmalfuß and others (see [40, 41, 58, 96, 123]; cf. also Subsection 2.4.3). Nonautonomous bifurcation theory is a new branch which has been developed quite independently for topological skew product flows (see Fabbri, Johnson, Kloeden, Mantellini [67, 85, 86, 87]; cf. also Subsection 2.6.2) and random dynamical systems (see Arnold, Sri Namachchivaya, Schenk-Hoppé [6, 8, 156, 176]; cf. also Subsection 2.6.3) so far.

The philosophy behind the present bifurcation theory of nonautonomous dynamical systems is based on a given structure of nonautonomy such as quasi-periodicity or the existence of an invariant measure, and the question arises how to describe bifurcations in a more general nonautonomous context. Recently, Langa, Robinson and Suárez discussed an answer to this question by defining a bifurcation of a nonautonomous differential equation as a merging process of two distinct solutions with different stability behavior (see [103, 105]; cf. also

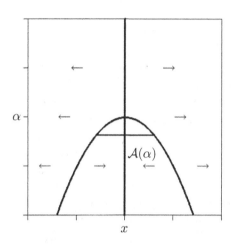

**Fig. 1.1.** Pitchfork bifurcation

Subsection 2.6.4). In this book, other possible approaches are pursued, which are explained demonstratively in the following.

Since the basic understanding of nonautonomous bifurcations in this book is based on phenomenological observations from the autonomous bifurcation theory, it is useful to look exemplarily at an autonomous bifurcation. For a real parameter $\alpha$, consider the ordinary differential equation $\dot{x} = x(\alpha + x^2)$, which is a prototype of a pitchfork bifurcation as indicated in Figure 1.1. For $\alpha \geq 0$, there is only one equilibrium, which is given by zero and which is repulsive. By letting the parameter $\alpha$ pass through zero in negative direction, this equilibrium becomes attractive, and two other repulsive equilibria, given by $\pm\sqrt{-\alpha}$, are bifurcating.

In order to establish a nonautonomous bifurcation theory, consider this scenario in the following way: For $\alpha < 0$, the trivial solution is attractive, and the domain of attraction $\mathcal{A}(\alpha)$ is given by the open interval between the two other equilibria. Now, the main point is that this domain of attraction undergoes a qualitative change from a nontrivial to a trivial object in the limit $\alpha \nearrow 0$. Moreover, $\mathcal{A}(\alpha)$ is also a repeller, and therefore, also a repeller changes qualitatively for $\alpha \nearrow 0$. We call the shrinking of a domain of attraction (repulsion, respectively) a bifurcation, whereas the case of a changing repeller (attractor, respectively) is denoted as a transition.

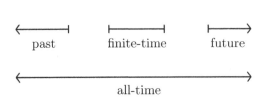

**Fig. 1.2.** Time domains

To implement this idea in the nonautonomous context, locally defined notions of attractive and repulsive solutions, domains of attractivity and repulsivity, as well as attractor and repeller are needed. This book distinguishes between four points of view concerning different time domains. The new concepts are introduced for the past (past attractivity, repulsivity, bifurcation and transition), the future (future attractivity, repulsivity, bifurcation and transition), the entire time (all-time attractivity, repulsivity, bifurcation and transition) and compact time intervals (finite-time attractivity, repulsivity, bifurcation and transition) (see Figure 1.2).

The second chapter of this book is devoted to notational preparations and the introduction of nonautonomous dynamical systems, and it contains all relevant notions of attractivity, repulsivity, bifurcation and transition. Several examples illustrate these definitions, and fundamental questions such as existence and uniqueness are discussed. Moreover, the relationship to other notions of attractivity and bifurcation is examined.

Chapter 3 is devoted to Morse decompositions, which were introduced by Charles C. Conley in 1978 to describe the global asymptotic behavior of (autonomous) dynamical systems on compact metric spaces (see [53]). Their components, the so-called Morse sets, are obtained as intersections of attractors and repellers. It is shown that the notions of past and future attractivity and repulsivity are designed to establish nonautonomous generalizations of the Morse decomposition. The dynamical properties of these decompositions are discussed and nonautonomous Lyapunov functions which are constant on the Morse sets are constructed explicitly. Moreover, Morse decompositions of linear systems on the projective space are examined, and a nonautonomous analogon to the Theorem of Selgrade (see [164]) is proved.

In Chapter 4, methods for the analysis of linear systems with respect to the notions of attractivity and repulsivity are introduced. First, several notions of dichotomy are defined, and it is shown that the ranges and null spaces of the corresponding invariant projectors form repellers and attractors of the linear system on the projective space. Furthermore, for the different time domains, dichotomy spectra are introduced which are based on the analysis of the entire time by Sacker and Sell (see [154]) and Siegmund and Aulbach (see [172, 171, 19]). It is also shown that the so-called spectral manifolds give rise to a Morse decomposition on the projective space. This chapter is concluded with a discussion of the relationship to the Lyapunov spectra and some roughness results.

Chapter 5 is devoted to the development of the qualitative theory with respect to the notions of attractivity and repulsivity for nonlinear systems. First, nonautonomous invariant manifolds are constructed, and methods are derived to obtain attractivity and repulsivity from the linearization. Moreover, as an application to bifurcation theory, it is shown that the zero is contained in the dichotomy spectrum of a bifurcating solution, and the relationship of the concept of finite-time bifurcation to the bifurcation theory of adiabatic systems is discussed.

The aim of Chapter 6 is to develop counterparts for the classical one-dimensional transcritical and pitchfork bifurcation patterns in the context of nonautonomous bifurcations and transitions. The sufficient conditions are formulated in terms of Taylor coefficients for the right hand side of ordinary differential equations. It is shown that the results are proper generalizations of the autonomous bifurcation scenarios.

In the last chapter of this book, asymptotically autonomous systems are discussed. It is supposed that the underlying autonomous system admits a one-dimensional bifurcation of saddle node, pitchfork or transcritical type or a two-dimensional Hopf bifurcation. Sufficient conditions are obtained for the transfer of this bifurcation behavior to the asymptotically autonomous system.

In order to keep this book self-contained, some basic facts about ordinary differential equations and projective spaces are noted in the Appendix. Furthermore, the definitions and results are formulated—whenever it was possible—in a very general form. However, to provide reading fluency, attention is restricted to continuous time in Chapter 5, 6 and 7. Extensions for the discrete time can be obtained similarly. Please note that for future reference, the definitions in Chapter 2 are also formulated for noninvertible systems, although invertibility is supposed in all other chapters.

Chapter 2 is necessary for the understanding of all the other chapters, since it contains both basic facts and the notions of attractivity and bifurcation. The other chapters can be read quite independently of each other. Please note that in Chapter 4, the assertions concerning the Morse decomposition require Chapter 3, and in Chapter 6, results concerning linearized attractivity and repulsivity from Chapter 5 are used.

Finally, please note that—although the applications in this book are mainly of low dimension—the concepts of bifurcation and transition also apply in a higher dimensional setting, since the definitions of attractivity and repulsivity are given in a very general form. The main tool for the analysis of such systems is the method of center manifold reduction (cf. Example 7.14). The basic idea is to detect a bifurcation of the system restricted to a center manifold. For instance, consider again the motivating example $\dot{x} = x(\alpha + x^2)$ with an additional second equation, given by $\dot{y} = \lambda y$. In case $\lambda > 0$, the trivial solution is not attractive for $\alpha < 0$, in contrast to the one-dimensional system, and therefore, we have no bifurcation of attraction areas but only a transition of repellers. For $\lambda < 0$, the trivial solution is attractive, and thus the two-dimensional system admits a bifurcation of attraction areas, but no longer a repeller transition. Restricting the attention to the lower dimensional invariant manifold $\mathbb{R} \times \{0\}$, however, yields the original one-dimensional system, and for this system we obtain both a bifurcation and a transition.

# 2

# Notions of Attractivity and Bifurcation

In this chapter, new concepts of (local) attractivity and repulsivity (in Section 2.3) and bifurcation and transition (in Section 2.5) are introduced for nonautonomous dynamical systems. By a bifurcation and transition, a qualitative change of attractivity or repulsivity is meant. Due to the nonautonomous framework, it is distinguished between four distinct points of view concerning different time domains. The notions of attractivity and repulsivity—and for this reason also the notions of bifurcation and transition—are introduced for the past (past attractivity and repulsivity), the future (future attractivity and repulsivity), the entire time (all-time attractivity and repulsivity) and the present (finite-time attractivity and repulsivity) of the system.

Since the definitions in this chapter are new to a broad extent, the relationship to well-known concepts is discussed in Section 2.4 (in case of attractivity and repulsivity) and Section 2.6 (in case of bifurcation and transition).

Before introducing the concepts of attractivity and bifurcation, the first section of this chapter is devoted to elementary definitions and notational preparations, and in Section 2.2, nonautonomous dynamical systems are treated.

## 2.1 Preliminary Definitions

As usual, we denote by $\mathbb{Z}$ and $\mathbb{R}$ the sets of all integers and reals, respectively, and we define $\overline{\mathbb{R}} := \mathbb{R} \cup \{-\infty, \infty\}$. $\mathbb{R}^{M \times N}$ is the set of all real $M \times N$ matrices, and we write $\mathbb{1}$ for the unit matrix and $0$ for the zero matrix. Given an arbitrary set $A \subset \mathbb{R}$ and $\kappa \in \mathbb{R}$, we define $A^{\pm} := \{x \in \mathbb{R} : x \in A \text{ or } -x \in A\}$, $A^+ := A \cap (0, \infty)$, $A_\kappa^+ := A \cap [\kappa, \infty)$, $A^- := A \cap (-\infty, 0)$ and $A_\kappa^- := A \cap (-\infty, \kappa]$. Moreover, we set $\mathbb{N} := \mathbb{Z}^+$. For $\mathbb{T} = \mathbb{R}$ or $\mathbb{T} = \mathbb{Z}$, a $\mathbb{T}$-interval is given by the intersection of a real interval with $\mathbb{T}$.

Let $f : X \to Y$ be a function from a set $X$ to a set $Y$. Then the *graph* of $f$ is defined by $\operatorname{graph} f := \{(x, y) \in X \times Y : y = f(x)\}$.

Given a metric space $(X, d)$ and $\varepsilon > 0$, let $U_\varepsilon(x_0) = \{x \in X : d(x, x_0) < \varepsilon\}$ be the $\varepsilon$-neighborhood of a point $x_0 \in X$, and we write $U_\varepsilon(A) = \cup_{x \in A} U_\varepsilon(x)$ for the $\varepsilon$-neighborhood of a set $A \subset X$. The set of all inner points of a nonempty set $A \subset X$ is denoted by int $A$; we write cls $A$ for the closure of $A$ and $\partial A$ for the boundary of $A$. We define the *distance* of a point $x \in X$ to a nonempty set $A \subset X$ by $d(x, A) := \inf_{y \in A} d(x, y)$ and the *Hausdorff semi-distance* of two nonempty sets $A, B \subset X$ by

$$d(A|B) := \sup_{x \in A} d(x, B).$$

In addition, if both $A$ and $B$ are empty, we set $d(A|B) := 0$. The *Hausdorff distance* of $A$ and $B$ is defined by

$$d_H(A, B) := \max \{d(A|B), d(B|A)\}.$$

Moreover, for $A, B \subset X$ with $B \subset$ int $A$, we define

$$\hat{d}(A|B) := \sup \{r > 0 : U_r(B) \subset A\}.$$

By $\operatorname{diam}(A) := \sup \{d(x, y) : x, y \in A\}$, the *diameter* of a nonempty set $A \subset X$ is given, and additionally, we define $\operatorname{diam}(\emptyset) := 0$.

If $X$ is a vector space, $A, B \subset X$ and $x \in X$, the following notations will be used:

$$x + A := \{x + a : a \in A\} \quad \text{and} \quad A + B := \{a + b : a \in A, b \in B\}.$$

With the Euclidean norm

$$\|(x_1, \ldots, x_N)\| := \sqrt{\sum_{i=1}^{N} x_i^2} \quad \text{for all } (x_1, \ldots, x_N) \in \mathbb{R}^N,$$

induced by the Euclidean scalar product $\langle \cdot, \cdot \rangle$, defined by

$$\langle x, y \rangle := \sum_{i=1}^{N} x_i y_i \quad \text{for all } x = (x_1, \ldots, x_N), y = (y_1, \ldots, y_N) \in \mathbb{R}^N,$$

the $\mathbb{R}^N$ is a normed vector space.

Let $\mathbb{I}$ be a $\mathbb{T}$-interval $(\mathbb{T} = \mathbb{R}, \mathbb{Z})$ and $\gamma \in \mathbb{R}$. We call a function $g : \mathbb{I} \to \mathbb{R}^N$ $\gamma^+$-*quasibounded* if $\mathbb{I}$ is unbounded above and $\sup_{t \in \mathbb{I}^+} \|g(t)\| e^{-\gamma t} < \infty$. Accordingly, we say, a function $g : \mathbb{I} \to \mathbb{R}^N$ is $\gamma^-$-*quasibounded* if $\mathbb{I}$ is unbounded below and $\sup_{t \in \mathbb{I}^-} \|g(t)\| e^{-\gamma t} < \infty$. The $(N-1)$-*sphere* of the $\mathbb{R}^N$ is defined by $\mathbb{S}^{N-1} := \{x \in \mathbb{R}^N : \|x\| = 1\}$.

Given a differentiable function $f : X \subset \mathbb{R}^N \to \mathbb{R}^M$, we write $Df : X \to \mathbb{R}^{M \times N}$ for its derivative and $D_i f : X \to \mathbb{R}^M$ for its partial derivative with respect to the $i$-th variable, $i \in \{1, \ldots, N\}$. Higher order derivatives $D^n f$ or $D_i^n f$ are defined inductively.

## 2.2 Nonautonomous Dynamical Systems

The notion of nonautonomous dynamical system emerged in the late 1990s as an abstraction of both continuous skew product flows (see, e.g., MILLER [120] and SELL [165, 166, 167]) and random dynamical systems (see, e.g., the monograph ARNOLD [5]). The definition is given as follows (see also the conference proceedings COLONIUS & KLOEDEN & SIEGMUND [52]).

**Definition 2.1 (Nonautonomous dynamical system).** *A (local) nonautonomous dynamical system (NDS for short) on a metric space $X$ with a time $\mathbb{T}$ $(=\mathbb{R}, \mathbb{R}_0^+, \mathbb{Z}, \mathbb{Z}_0^+)$ and base set $P$ is a pair of mappings*

$$\left(\theta : \mathbb{T}^{\pm} \times P \to P, \, \varphi : D \subset \mathbb{T} \times P \times X \to X\right)$$

*with the following properties:*

(i) *The so-called* base flow *or* driving system *$\theta$ is a dynamical system, i.e., we have the relations*

$$\theta(0, p) = p \text{ and } \theta(t + s, p) = \theta(t, \theta(s, p)) \quad \text{for all } p \in P \text{ and } t, s \in \mathbb{T}^{\pm}.$$

(ii) *The* maximal interval of existence *$D_{max}(p, x) := \{t \in \mathbb{T} : (t, p, x) \in D\}$ for $p \in P$ and $x \in X$ is either empty or an open $\mathbb{T}$-interval which contains $0 \in \mathbb{T}$.*

(iii) *$\varphi$ is a* cocycle *over $\theta$, i.e., for all $t, s \in \mathbb{T}$ and $(p, x) \in P \times X$ fulfilling both $s \in D_{max}(p, x)$ and $t + s \in D_{max}(p, x)$, we have the relations $t \in D_{max}(\theta(s, p), \varphi(s, p, x))$,*

$$\varphi(0, p, x) = x \quad \text{and} \quad \varphi(t + s, p, x) = \varphi(t, \theta(s, p), \varphi(s, p, x)).$$

(iv) *$\varphi$ is continuous with respect to $t \in \mathbb{T}$ and $x \in X$.*

$X$ is called *phase space*, and $P \times X$ is called *extended phase space*. We say, a NDS $(\theta, \varphi)$ is *invertible* if $\mathbb{T} = \mathbb{R}$ or $\mathbb{T} = \mathbb{Z}$ and $t \in D_{max}(p, x)$ for some $p \in P$ and $x \in X$ implies $-t \in D_{max}(\theta(t, p), \varphi(t, p, x))$. Throughout this book, we will only consider invertible nonautonomous dynamical systems, except in Subsection 2.3.2, where the relevant definitions of attractivity and repulsivity are stated for the noninvertible case.

For simplicity in notation, we also write $\theta_t p$ instead of $\theta(t, p)$ and $\varphi(t, p)x$ for $\varphi(t, p, x)$.

A standard example of a nonautonomous dynamical system, which is of main interest in this book, is provided by a nonautonomous ordinary differential equation

$$\dot{x} = f(t, x) \tag{2.1}$$

with $f : D \subset \mathbb{R} \times \mathbb{R}^N \to \mathbb{R}^N$ (see Appendix A.1). Here $\mathbb{T} = \mathbb{R}$, and the base set $P$ can simply be chosen to be $\mathbb{R}$ with base flow $(t, s) \mapsto t + s$. In case

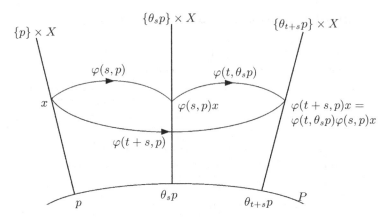

**Fig. 2.1.** The cocycle property of Definition 2.1

the function $f$ is fulfilling special conditions, the nonautonomous differential equation (2.1) gives rise to a general solution $\lambda : \Omega \subset \mathbb{R} \times \mathbb{R} \times \mathbb{R}^N \to \mathbb{R}^N$ (see Proposition A.3), and $\varphi$ can then be defined by

$$\varphi(t,s)x := \lambda(t+s,s,x) \text{ for all } (t,s,x) \in \mathbb{R} \times \mathbb{R} \times \mathbb{R}^N \text{ such that } (t+s,s,x) \in \Omega \,.$$

Without further notice, we assume that all ordinary differential equations considered in this book fulfill conditions of local existence and uniqueness of solutions.

A similar construction is also possible for nonautonomous difference equations of the form $x_{n+1} = f(n, x_n)$, $\mathbb{T} = P = \mathbb{Z}$.

In both cases above, however, $P$ is noncompact, which may cause difficulties. This can be avoided for a special class of right hand sides $f$ by considering the Bebutov flow on the hull of $f$ (see, e.g., BEBUTOV [26] and SELL [167]).

Apart from deterministic also random and stochastic differential and difference equations (see, e.g., ARNOLD [5]) and some other types of equations such as functional differential equations or nonautonomous evolutionary equations generate nonautonomous dynamical systems.

*Remark 2.2.*

(i)  Normally, one has additional structures concerning the driving system $\theta$. In the deterministic case, the base set $P$ is a metric space and $\theta$ is continuous; in case of random dynamical systems, $\theta$ represents an ergodic dynamical system. For a discussion of the relationship of these two concepts, see BERGER & SIEGMUND [28].

(ii) In the literature, one usually considers *global* nonautonomous dynamical systems, i.e., $D = \mathbb{T} \times P \times X$. The above definition of a *local* nonautonomous dynamical system stems from AULBACH & SIEGMUND

& RASMUSSEN [16] (see also RASMUSSEN [144]; in case of random dynamical systems, see ARNOLD [5] and ARNOLD & NAMACHCHIVAYA & SCHENK-HOPPÉ [8]).

Let $\theta$ be a base flow on $P$. For an element $p \in P$, we define the *forward orbit* of $p$ by $\mathcal{O}^+(p) := \{\theta_t p : t \geq 0\}$, the *backward orbit* of $p$ by $\mathcal{O}^-(p) := \{\theta_t p : t \leq 0\}$, the *$T$-orbit* of $p$, $T \in \mathbb{T}^+$, by $\mathcal{O}^T(p) := \{\theta_t p : t \in [0, T] \cap \mathbb{T}\}$ and the *orbit* of $p$ by $\mathcal{O}(p) := \{\theta_t p : t \in \mathbb{T}^\pm\}$. Two elements $p_1, p_2 \in P$ are called *equivalent* ($p_1 \sim p_2$) if $p_1 \in \mathcal{O}(p_2)$. We denote the set of all equivalence classes $[p]$ by $P/\sim$.

**Definition 2.3 (Nonautonomous sets).** *We consider a nonautonomous dynamical system $(\theta, \varphi)$ on a metric space $X$ with a base set $P$. For an arbitrary set $M \subset P \times X$, we define the so-called $p$-fibre of $M$ by*

$$M(p) := \{x \in X : (p, x) \in M\} \quad \text{for all } p \in P,$$

*and we denote by $P^*(M) := \{p \in P : M(p) \neq \emptyset\}$ the set of all base elements leading to nonempty fibres. $M$ is called*

(i) *past nonautonomous set if $\mathcal{O}^-(p) \subset P^*(M)$ for all $p \in P^*(M)$,*

(ii) *future nonautonomous set if $\mathcal{O}^+(p) \subset P^*(M)$ for all $p \in P^*(M)$,*

(iii) *all-time nonautonomous set if $\mathcal{O}(p) \subset P^*(M)$ for all $p \in P^*(M)$,*

(iv) *$(p, T)$-nonautonomous set if $\mathcal{O}^T(p) \subset P^*(M)$.*

*We say that $M$ is*

(i) *invariant if $\varphi(t, p)M(p) = M(\theta_t p)$ for all $p \in P^*(M)$ and $t \in \mathbb{T}$ with $\theta_t p \in P^*(M)$,*

(ii) *closed if $M(p)$ is closed for all $p \in P^*(M)$,*

(iii) *compact if $M(p)$ is compact for all $p \in P^*(M)$.*

*Remark 2.4.*

(i) An all-time nonautonomous set is a past, as well as a future nonautonomous set. The reversal is certainly not true.

(ii) In the literature, an all-time nonautonomous set $M$ with $P^*(M) = P$ is called nonautonomous set.

The following definition is adapted from AUBIN & FRANKOWSKA [12] (see also AKIN [3, Exercise 1.5, p. 9] and ELSTRODT [64, p. 9]).

**Definition 2.5.** *For a past nonautonomous set $M \subset P \times X$ and $p \in P^*(M)$, we define*

$$\limsup_{t \to \infty} M(\theta_{-t} p) := \bigcap_{\tau \geq 0} \bigcup_{t \geq \tau} M(\theta_{-t} p)$$

*and*

$$\liminf_{t\to\infty} M(\theta_{-t}p) := \bigcup_{\tau\geq 0}\bigcap_{t\geq\tau} M(\theta_{-t}p).$$

*Given a future nonautonomous set $M \subset P \times X$ and $p \in P^*(M)$, we define*

$$\limsup_{t\to\infty} M(\theta_t p) := \bigcap_{\tau\geq 0}\bigcup_{t\geq\tau} M(\theta_t p)$$

*and*

$$\liminf_{t\to\infty} M(\theta_t p) := \bigcup_{\tau\geq 0}\bigcap_{t\geq\tau} M(\theta_t p).$$

It is easy to show that the following characterizations hold:

- $\limsup_{t\to\infty} M(\theta_{-t}p) = \left\{x \in X : \forall\,\tau \geq 0 : \exists\, t \geq \tau : x \in M(\theta_{-t}p)\right\}$,
- $\liminf_{t\to\infty} M(\theta_{-t}p) = \left\{x \in X : \exists\,\tau \geq 0 : \forall\, t \geq \tau : x \in M(\theta_{-t}p)\right\}$,
- $\limsup_{t\to\infty} M(\theta_t p) = \left\{x \in X : \forall\,\tau \geq 0 : \exists\, t \geq \tau : x \in M(\theta_t p)\right\}$,
- $\liminf_{t\to\infty} M(\theta_t p) = \left\{x \in X : \exists\,\tau \geq 0 : \forall\, t \geq \tau : x \in M(\theta_t p)\right\}$.

## 2.3 Attractivity and Repulsivity

This section is divided into six subsections. First, several notions of attractor and repeller are introduced for invertible (in Subsection 2.3.1) and non-invertible (in Subsection 2.3.2) nonautonomous dynamical systems. In Subsection 2.3.3 and 2.3.4, the theoretical background to analyze the strength of attractivity and repulsivity is established, and in Subsection 2.3.5, properties of the definitions under time reversal are studied. Finally, criteria for the existence of attractors and repellers are formulated in the last subsection, and the question of their uniqueness is discussed.

Throughout this section, let $\left(\theta : \mathbb{T} \times P \to P, \varphi : D \subset \mathbb{T} \times P \times X \to X\right)$ be an invertible nonautonomous dynamical system with an arbitrary base set $P$ and a metric space $(X, d)$.

### 2.3.1 Definitions

We begin with the definitions concerning the past of the system. In addition to the important notions of past attractor and past repeller, also $\mathcal{M}$-past attractors and repellers are introduced, which are generalizations of past attractors and repellers, respectively. Please note that past attractors are local forms of pullback attractors (see also Subsection 2.4.3).

**Definition 2.6 (Past attractivity and repulsivity).** *Let $A$ and $R$ be invariant and compact past nonautonomous sets and $\mathcal{M}$ be a collection of past nonautonomous sets.*

(i) *$A$ is called* past attractor *if there exists an $\eta > 0$ such that for all $p \in P^*(A)$, there exists a $\hat{p} \in [p] \cap P^*(A)$ with*

$$\lim_{t \to \infty} d\big(\varphi(t, \theta_{-\tau-t}\hat{p})U_\eta(A(\theta_{-\tau-t}\hat{p}))\big|A(\theta_{-\tau}\hat{p})\big) = 0 \quad \text{for all } \tau \geq 0.$$

(ii) *$R$ is called* past repeller *if there exists an $\eta > 0$ such that for all $p \in P^*(R)$, there exists a $\hat{p} \in [p] \cap P^*(R)$ with*

$$\lim_{t \to \infty} d\big(\varphi(-t, \theta_{-\tau}\hat{p})U_\eta(R(\theta_{-\tau}\hat{p}))\big|R(\theta_{-\tau-t}\hat{p})\big) = 0 \quad \text{for all } \tau \geq 0.$$

(iii) *$A$ is called $\mathcal{M}$-past attractor if for all $M \in \mathcal{M}$, we have $P^*(M) \subset P^*(A)$ and*

$$\lim_{t \to \infty} d\big(\varphi(t, \theta_{-t}p)M(\theta_{-t}p)\big|A(p)\big) = 0 \quad \text{for all } p \in P^*(M).$$

(iv) *$R$ is called $\mathcal{M}$-past repeller if for all $M \in \mathcal{M}$, we have $P^*(M) \subset P^*(R)$ and*

$$\lim_{t \to \infty} d\big(\varphi(-t, p)M(p)\big|R(\theta_{-t}p)\big) = 0 \quad \text{for all } p \in P^*(M).$$

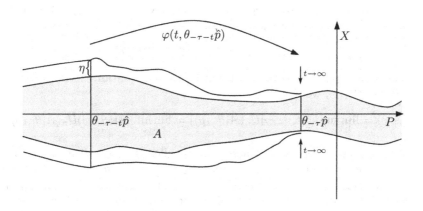

**Fig. 2.2.** Past attractor

It follows directly from the definitions that the empty set is both a past attractor and a past repeller. If $X$ is compact and $D = \mathbb{T} \times P \times X$, then $P \times X$ is also both a past attractor and a past repeller.

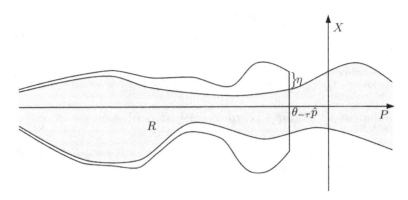

**Fig. 2.3.** Past repeller

*Remark 2.7.*

(i) The notions of Definition 2.6 represent the behavior of $(\theta, \varphi)$ in the past. This can be seen by considering another nonautonomous dynamical system $(\theta, \bar{\varphi} : D \subset \mathbb{T} \times P \times X \to X)$ with the following property: For all $\bar{p} \in P$, there exists a $\hat{p} \in [\bar{p}]$ with

$$\varphi(t,p)x = \bar{\varphi}(t,p)x \quad \text{for all } (t,p,x) \in D \text{ fulfilling } p, \theta_t p \in \mathcal{O}^-(\hat{p}).$$

Then for any past attractor (repeller, respectively) $A$ of $(\theta, \varphi)$, there exists a past attractor (repeller, respectively) $\bar{A}$ of $(\theta, \bar{\varphi})$ such that for all $\bar{p} \in P$, there exists a $\hat{p} \in [\bar{p}]$ with

$$A(p) = \bar{A}(p) \quad \text{for all } p \in \mathcal{O}^-(\hat{p}).$$

(ii) Let $A$ be a past attractor with $\eta$ given as in Definition 2.6 (i). Then for all $p \in P^*(A)$, we have

$$A(p) = \limsup_{t \to \infty} \varphi(t, \theta_{-t}p)U_\eta(A(\theta_{-t}p)) = \liminf_{t \to \infty} \varphi(t, \theta_{-t}p)U_\eta(A(\theta_{-t}p)).$$

(iii) The notions of $\mathcal{M}$-past attractor and repeller are generalizations of past attractors and repellers, since a past attractor $A$ is an $\{M\}$-past attractor for some past nonautonomous set $M$ fulfilling the following property: There exists an $\eta > 0$ such that for all $p \in P^*(A)$, there exists a $\hat{p} \in [p] \cap P^*(A)$ with

$$U_\eta(A(\theta_{-t}\hat{p})) \subset M(\theta_{-t}\hat{p}) \quad \text{for all } t \geq 0.$$

Moreover, a past repeller $R$ is an $\{M\}$-past repeller for some past nonautonomous set $M$ fulfilling the following property: There exists an $\eta > 0$ such that for all $p \in P^*(R)$, there exists a $\hat{p} \in [p] \cap P^*(R)$ with

$$U_\eta(R(\theta_{-t}\hat{p})) \subset M(\theta_{-t}\hat{p}) \quad \text{for all } t \geq 0.$$

(iv)  Due to the continuity of $\varphi$, one can derive the following equivalent characterization: An invariant and compact nonautonomous set $A$ is a past attractor if and only if there exists an $\eta > 0$ such that for all $p \in P^*(A)$, we have

$$\lim_{t \to \infty} d\big(\varphi(t, \theta_{-t}p)U_\eta(A(\theta_{-t}p))\big|A(p)\big) = 0.$$

Such a reduction is not possible for past repellers.

Before proceeding with the case of future attractivity and repulsivity, the definitions for the past are illustrated by means of the following two examples.

*Example 2.8.* We consider the linear nonautonomous differential equation

$$\dot{x} = a(t)x \tag{2.2}$$

with a continuous function $a : \mathbb{R} \to \mathbb{R}$, which generates a nonautonomous dynamical system with $\mathbb{T} = P = \mathbb{R}$ (see Section 2.2). It is easy to see that every invariant and compact past nonautonomous set is a past attractor if and only if

$$\lim_{t \to -\infty} \int_t^0 a(s)\, ds = -\infty$$

and a past repeller if and only if

$$\lim_{t \to -\infty} \int_t^0 a(s)\, ds = \infty.$$

*Example 2.9.* The nonautonomous differential equation

$$\dot{x} = a(t)x + b(t)x^3 = x\big(a(t) + b(t)x^2\big) \tag{2.3}$$

with continuous functions $a : \mathbb{R} \to \mathbb{R}$ and $b : \mathbb{R} \to \mathbb{R}_\kappa^+$ for some $\kappa > 0$ generates a nonautonomous dynamical system with $\mathbb{T} = P = \mathbb{R}$ (see Section 2.2). For simplicity, we define

$$w(t) := \sqrt{-\frac{a(t)}{b(t)}} \quad \text{for all } t \in \mathbb{R} \text{ with } a(t) < 0.$$

Then, for fixed $t \in \mathbb{R}$ with $a(t) < 0$, the zero set of the right hand side is $\{0, \pm w(t)\}$; for all $t \in \mathbb{R}$ with $a(t) \geq 0$, this zero set is the singleton $\{0\}$. An elementary discussion of the sign of the right hand side of (2.3) yields that $\mathbb{R} \times \{0\}$ is a past attractor if $\liminf_{t \to -\infty} -a(t)/b(t) > 0$, and $\mathbb{R} \times \{0\}$ is a past repeller if $\limsup_{t \to -\infty} -a(t)/b(t) \leq 0$. These conditions are only sufficient for attractivity or repulsivity of $\mathbb{R} \times \{0\}$ but not necessary.

In the following definition, the notions of future attractivity and repulsivity are explained. Please note that future attractors are local forms of forward attractors (see Subsection 2.4.3 for further information).

**Definition 2.10 (Future attractivity and repulsivity).** *Let $A$ and $R$ be invariant and compact future nonautonomous sets and $\mathcal{M}$ be a collection of future nonautonomous sets.*

(i) *$A$ is called* future attractor *if there exists an $\eta > 0$ such that for all $p \in P^*(A)$, there exists a $\hat{p} \in [p] \cap P^*(A)$ with*

$$\lim_{t \to \infty} d\big(\varphi(t, \theta_\tau \hat{p}) U_\eta(A(\theta_\tau \hat{p})) \big| A(\theta_{\tau+t}\hat{p})\big) = 0 \quad \text{for all } \tau \geq 0.$$

(ii) *$R$ is called* future repeller *if and only if there exists an $\eta > 0$ such that for all $p \in P^*(R)$, there exists a $\hat{p} \in [p] \cap P^*(R)$ with*

$$\lim_{t \to \infty} d\big(\varphi(-t, \theta_{\tau+t}\hat{p}) U_\eta(R(\theta_{\tau+t}\hat{p})) \big| R(\theta_\tau \hat{p})\big) = 0 \quad \text{for all } \tau \geq 0.$$

(iii) *$A$ is called* $\mathcal{M}$-future attractor *if for all $M \in \mathcal{M}$, we have $P^*(M) \subset P^*(A)$ and*

$$\lim_{t \to \infty} d\big(\varphi(t, p) M(p) \big| A(\theta_t p)\big) = 0 \quad \text{for all } p \in P^*(M).$$

(iv) *$R$ is called* $\mathcal{M}$-future repeller *if for all $M \in \mathcal{M}$, we have $P^*(M) \subset P^*(R)$ and*

$$\lim_{t \to \infty} d\big(\varphi(-t, \theta_t p) M(\theta_t p) \big| R(p)\big) = 0 \quad \text{for all } p \in P^*(M).$$

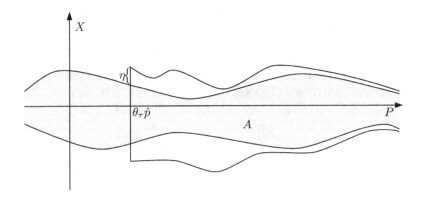

**Fig. 2.4.** Future attractor

It follows directly from the definitions that the empty set is both a future attractor and future repeller. If $X$ is compact and $D = \mathbb{T} \times P \times X$, then $P \times X$ is also a future attractor and future repeller.

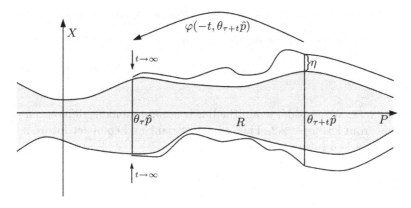

**Fig. 2.5.** Future repeller

*Remark 2.11.*

(i) As seen in Remark 2.7 (i) in case of past attractivity and repulsivity, the notions of Definition 2.10 represent the behavior of $(\theta, \varphi)$ in the future.

(ii) There are important analogies in the concepts of past and future attractivity and repulsivity. This question is treated in Section 2.3.5. It is shown that a past attractor corresponds to a future repeller, and a past repeller is related to a future attractor.

(iii) Let $R$ be a future repeller with $\eta$ given as in Definition 2.10 (ii). Then for all $p \in P^*(R)$, we have

$$R(p) = \limsup_{t \to \infty} \varphi(-t, \theta_t p) U_\eta(R(\theta_t p)) = \liminf_{t \to \infty} \varphi(-t, \theta_t p) U_\eta(R(\theta_t p)).$$

(iv) The notions of $\mathcal{M}$-future attractor and repeller are generalizations of future attractors and repellers, since a future attractor $A$ is an $\{M\}$-future attractor for some future nonautonomous set $M$ fulfilling the following property: There exists an $\eta > 0$ such that for all $p \in P^*(A)$, there exists a $\hat{p} \in [p] \cap P^*(A)$ with

$$U_\eta(A(\theta_t \hat{p})) \subset M(\theta_t \hat{p}) \quad \text{for all } t \geq 0.$$

Moreover, a future repeller $R$ is an $\{M\}$-future repeller for some future nonautonomous set $M$ fulfilling the following property: There exists an $\eta > 0$ such that for all $p \in P^*(R)$, there exists a $\hat{p} \in [p] \cap P^*(R)$ with

$$U_\eta(R(\theta_t \hat{p})) \subset M(\theta_t \hat{p}) \quad \text{for all } t \geq 0.$$

(v) Due to the continuity of $\varphi$, one can derive the following equivalent characterization: An invariant and compact nonautonomous set $R$ is a future repeller if and only if there exists an $\eta > 0$ such that for all $p \in P^*(R)$, we have

$$\lim_{t\to\infty} d\big(\varphi(-t,\theta_t p)U_\eta(R(\theta_t p))\big|R(p)\big) = 0 \, .$$

Such a reduction is not possible for future attractors.

The following two examples illustrate the notions of future attractivity and repulsivity.

*Example 2.12.* We consider again the linear nonautonomous differential equation (2.2) from Example 2.8. Then each invariant and compact future nonautonomous set is a future attractor if and only if

$$\lim_{t\to\infty} \int_0^t a(s)\,\mathrm{d}s = -\infty$$

and a future repeller if and only if

$$\lim_{t\to\infty} \int_0^t a(s)\,\mathrm{d}s = \infty \, .$$

*Example 2.13.* Let (2.3) be the scalar nonautonomous differential equation from Example 2.9. Analogously to the observations in this example, one can see that $\mathbb{R}\times\{0\}$ is a future attractor if $\liminf_{t\to\infty} -a(t)/b(t) > 0$ and a future repeller if $\limsup_{t\to\infty} -a(t)/b(t) \leq 0$.

In the following definition, the notions of all-time attractivity and repulsivity are explained. First, note that an all-time attractor is a local form of a uniform attractor as discussed, e.g., in CHEPYZHOV & VISHIK [44] (see also Subsection 2.4.3).

**Definition 2.14 (All-time attractivity and repulsivity).** *Let $A$ and $R$ be invariant and compact all-time nonautonomous sets.*

(i)  *$A$ is called* all-time attractor *if there exists an $\eta > 0$ with*

$$\lim_{t\to\infty} \sup_{p\in P^*(A)} d\big(\varphi(t,p)U_\eta(A(p))\big|A(\theta_t p)\big) = 0 \, .$$

(ii)  *$R$ is called* all-time repeller *if there exists an $\eta > 0$ with*

$$\lim_{t\to\infty} \sup_{p\in P^*(R)} d\big(\varphi(-t,p)U_\eta(R(p))\big|R(\theta_{-t} p)\big) = 0 \, .$$

It follows directly from the definitions that the empty set is both an all-time attractor and an all-time repeller. If $X$ is compact and $D = \mathbb{T}\times P\times X$, then $P\times X$ is also an all-time attractor and all-time repeller. An all-time attractor (all-time repeller, respectively) is also both a past attractor (past repeller, respectively) and a future attractor (future repeller, respectively), since in the above definition, $p$ can be replaced by $\theta_{-t}p$.

We look again at the two examples for the attractivity and repulsivity of a linear and nonlinear scalar differential equation.

*Example 2.15.* Consider the nonautonomous differential equation (2.2) from Example 2.8. Then each invariant and compact all-time nonautonomous set is an all-time attractor if and only if

$$\lim_{t \to \infty} \sup_{\tau \in \mathbb{R}} \int_{\tau}^{\tau+t} a(s)\,\mathrm{d}s = -\infty$$

and an all-time repeller if and only if

$$\lim_{t \to \infty} \sup_{\tau \in \mathbb{R}} \int_{\tau}^{\tau+t} a(s)\,\mathrm{d}s = \infty \,.$$

*Example 2.16.* Consider the nonautonomous differential equation (2.3) from Example 2.9. Analogously to the observations in this example, one can see that $\mathbb{R} \times \{0\}$ is an all-time attractor if $\inf_{t \in \mathbb{R}} -a(t)/b(t) > 0$ and an all-time repeller if $-a(t)/b(t) \leq 0$ for all $t \in \mathbb{R}$.

Finally, the definitions of finite-time attractivity and repulsivity are introduced.

**Definition 2.17 (Finite-time attractivity and repulsivity).** *For $p \in P$ and $T > 0$, let $A$ and $R$ be invariant and compact $(p, T)$-nonautonomous sets.*

(i)  *$A$ is called $(p, T)$-attractor if*

$$\limsup_{\eta \searrow 0} \frac{1}{\eta} d\big(\varphi(T, p)U_{\eta}(A(p))\big|A(\theta_T p)\big) < 1 \,.$$

(ii)  *$R$ is called $(p, T)$-repeller if*

$$\limsup_{\eta \searrow 0} \frac{1}{\eta} d\big(\varphi(-T, \theta_T p)U_{\eta}(R(\theta_T p))\big|R(p)\big) < 1 \,.$$

*Remark 2.18.* In contrast to the above definitions in case of past, future and all-time attractivity and repulsivity, the notions of finite-time attractivity and repulsivity are not invariant with respect to a change of the metric $d$ to an equivalent metric.

The following two examples illustrate the notions of finite-time attractivity and repulsivity.

*Example 2.19.* Consider again the nonautonomous differential equation (2.2) from Example 2.8, and let $p \in \mathbb{R}$ and $T > 0$. Then each invariant and compact $(p, T)$-nonautonomous set is a $(p, T)$-attractor if and only if

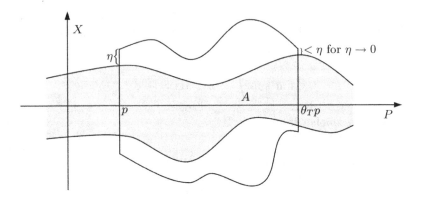

**Fig. 2.6.** $(p, T)$-attractor

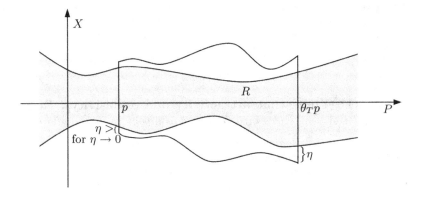

**Fig. 2.7.** $(p, T)$-repeller

$$\int_{p}^{p+T} a(s)\, \mathrm{d}s < 0$$

and a future repeller if and only if

$$\int_{p}^{p+T} a(s)\, \mathrm{d}s > 0\,.$$

*Example 2.20.* Let (2.3) be the scalar nonautonomous differential equation from Example 2.9, $p \in \mathbb{R}$ and $T > 0$. Analogously to the observations in this example, one can see that $\mathbb{R} \times \{0\}$ is a $(p, T)$-attractor if $-a(t)/b(t) > 0$ for all $t \in [p, p + T]$ and a $(p, T)$-repeller if $-a(t)/b(t) \leq 0$ for all $t \in [p, p + T]$.

We will often consider invariant nonautonomous sets which are solutions of the nonautonomous dynamical system $(\theta, \varphi)$.

**Definition 2.21 (Solution).** *Let* $\hat{P} \subset P$ *be nonempty with* $\hat{P} \subset \mathcal{O}(p)$ *for some* $p \in P$*. A function* $\mu : \hat{P} \to X$ *is called* solution *of* $(\theta, \varphi)$ *if* graph $\mu$ *is invariant.*

The attractivity and repulsivity of solutions are defined by considering the graph of the solution as an invariant nonautonomous set.

**Definition 2.22 (Attractivity and repulsivity of solutions).** *Let* $\mu$ *be a solution of* $(\theta, \varphi)$*.*

(i) $\mu$ *is called* past *(*future, all-time, $(p, T)$-, *respectively) attractive if* graph($\mu$) *is a past (future, all-time,* $(p, T)$-, *respectively) attractor.*

(ii) $\mu$ *is called* past *(*future, all-time $(p, T)$-, *respectively) repulsive if* graph($\mu$) *is a past (future, all-time,* $(p, T)$-, *respectively) repeller.*

Concluding this subsection, we state the following proposition, whose obvious proof will be omitted.

**Proposition 2.23.** *A past (future, all-time,* $(p, T)$-, *respectively) nonautonomous set can never be both a past (future, all-time,* $(p, T)$-, *respectively) attractor and a past (future, all-time,* $(p, T)$-, *respectively) repeller.*

### 2.3.2 The Noninvertible Case

Although, noninvertible systems are not considered in this book, the notions of repulsivity are introduced for not necessarily invertible nonautonomous dynamical systems $\left(\theta : \mathbb{T}^{\pm} \times P \to P, \varphi : D \subset \mathbb{T} \times P \times X \to X\right)$ in the following definition. In case of invertible systems, these definitions are equivalent to the definitions of the previous subsection.

**Definition 2.24 (Repulsivity in the noninvertible case).** *Given* $p \in P$ *and* $T > 0$*, and let* $A$ *and* $R$ *be invariant and compact past (future, all-time,* $(p, T)$-, *respectively) nonautonomous sets.*

(i) $R$ *is called* past repeller *if there exists an* $\eta > 0$ *such that for all* $p \in P^*(R)$*, there exists a* $\hat{p} \in [p] \cap P^*(R)$ *such that for all* $\varepsilon > 0$ *and* $\tau \geq 0$*, we have a* $t^* > 0$ *with*

$$\varphi(t, \theta_{-\tau-t}\hat{p})U_{\varepsilon}(R(\theta_{-\tau-t}\hat{p})) \supset U_{\eta}(R(\theta_{-\tau}\hat{p})) \quad \text{for all } t \geq t^*\,.$$

(ii) $R$ *is called* $\mathcal{M}$-past repeller *if for all* $M \in \mathcal{M}$*, we have* $P^*(M) \subset P^*(R)$*, and for all* $M \in \mathcal{M}$*,* $p \in P^*(M)$ *and* $\varepsilon > 0$*, there exists a* $t^* > 0$ *with*

$$\varphi(t, \theta_{-t}p)U_{\varepsilon}(R(\theta_{-t}p)) \supset M(p) \quad \text{for all } t \geq t^*\,.$$

(iii) $R$ *is called a* future repeller *if there exists an* $\eta > 0$ *such that for all* $p \in P^*(R)$, *there exists a* $\hat{p} \in [p] \cap P^*(R)$ *such that for all* $\varepsilon > 0$ *and* $\tau \geq 0$, *we have a* $t^* > 0$ *with*

$$\varphi(t, \theta_\tau \hat{p}) U_\varepsilon(R(\theta_\tau \hat{p})) \supset U_\eta(R(\theta_{\tau+t} \hat{p})) \quad \text{for all } t \geq t^* .$$

(iv) $R$ *is called* $\mathcal{M}$-future repeller *if for all* $M \in \mathcal{M}$, *we have* $P^*(M) \subset P^*(R)$, *and for all* $M \in \mathcal{M}$, $p \in P^*(M)$ *and* $\varepsilon > 0$, *there exists a* $t^* > 0$ *with*

$$\varphi(t, p) U_\varepsilon(R(p)) \supset M(\theta_t p) \quad \text{for all } t \geq t^* .$$

(v) $R$ *is called* all-time repeller *if there exists an* $\eta > 0$ *such that for all* $\varepsilon > 0$, *there exists a* $t^* > 0$ *with*

$$\varphi(t, p) U_\varepsilon(R(p)) \supset U_\eta(R(\theta_t p)) \quad \text{for all } p \in P^*(R) \text{ and } t \geq t^* .$$

(vi) $R$ *is called* $(p, T)$-repeller *if*

$$\liminf_{\eta \searrow 0} \frac{1}{\eta} \hat{d}\big(\varphi(T, p) U_\eta(R(p)) \big| R(\theta_T p)\big) > 1 .$$

### 2.3.3 Radii of Attraction and Repulsion

Since the *local* dynamical behavior of nonautonomous sets is studied in the definitions of the preceding subsections, it is useful to know something about the range of attractivity or repulsivity. In this subsection, notions of radii of attraction and repulsion are introduced.

**Definition 2.25 (Radii of attraction and repulsion).** *We define the* radius of past attraction *of a past attractor* $A$ *by*

$$\mathfrak{A}_A^\leftarrow := \sup \Big\{ \eta > 0 : \textit{For all } p \in P^*(A), \textit{there exists a } \hat{p} \in [p] \cap P^*(A) \textit{ with}$$
$$\lim_{t \to \infty} d\big(\varphi(t, \theta_{-\tau-t} \hat{p}) U_\eta(A(\theta_{-\tau-t} \hat{p})) \big| A(\theta_{-\tau} \hat{p})\big) = 0 \textit{ for all } \tau \geq 0 \Big\}$$

*and the* radius of past repulsion *of a past repeller* $R$ *by*

$$\mathfrak{R}_R^\leftarrow := \sup \Big\{ \eta > 0 : \textit{For all } p \in P^*(R), \textit{there exists a } \hat{p} \in [p] \cap P^*(R) \textit{ with}$$
$$\lim_{t \to \infty} d\big(\varphi(-t, \theta_{-\tau} \hat{p}) U_\eta(R(\theta_{-\tau} \hat{p})) \big| R(\theta_{-\tau-t} \hat{p})\big) = 0 \textit{ for all } \tau \geq 0 \Big\} .$$

*The* radius of future attraction *of a future attractor* $A$ *is defined by*

$$\mathfrak{R}_A^\rightarrow := \sup \Big\{ \eta > 0 : \textit{For all } p \in P^*(A), \textit{there exists a } \hat{p} \in [p] \cap P^*(A) \textit{ with}$$
$$\lim_{t \to \infty} d\big(\varphi(t, \theta_\tau \hat{p}) U_\eta(A(\theta_\tau \hat{p})) \big| A(\theta_{t+\tau} \hat{p})\big) = 0 \textit{ for all } \tau \geq 0 \Big\} ,$$

*and the* radius of future repulsion *of a future repeller* $R$ *is defined by*

$$\mathfrak{R}_{R}^{\rightarrow} := \sup \Big\{ \eta > 0 : \textit{For all } p \in P^*(R), \textit{ there exists a } \hat{p} \in [p] \cap P^*(R) \textit{ with}$$
$$\lim_{t \to \infty} d\big(\varphi(-t, \theta_{\tau+t}\hat{p})U_\eta(R(\theta_{\tau+t}\hat{p}))\big| R(\theta_\tau \hat{p})\big) = 0 \textit{ for all } \tau \geq 0 \Big\}.$$

*The* radius of all-time attraction *of an all-time attractor A is defined by*

$$\mathfrak{R}_{A}^{\leftrightarrow} := \sup \Big\{ \eta > 0 : \lim_{t \to \infty} \sup_{p \in P^*(A)} d\big(\varphi(t, p)U_\eta(A(p))\big| A(\theta_t p)\big) = 0 \Big\},$$

*and the* radius of all-time repulsion *of an all-time repeller R is defined by*

$$\mathfrak{R}_{R}^{\leftrightarrow} := \sup \Big\{ \eta > 0 : \lim_{t \to \infty} \sup_{p \in P^*(R)} d\big(\varphi(-t, p)U_\eta(R(p))\big| R(\theta_{-t}p)\big) = 0 \Big\}.$$

*The* radius of $(p, T)$-attraction *of a $(p, T)$-attractor A is defined by*

$$\mathfrak{A}_{A}^{(p,T)} := \sup \big\{ \eta > 0 : d\big(\varphi(T, p)U_{\hat{\eta}}(A(p))\big| A(\theta_T p)\big) < \hat{\eta} \textit{ for all } \hat{\eta} \in (0, \eta) \big\},$$

*and the* radius of $(p, T)$-repulsion *of a $(p, T)$-repeller R is defined by*

$$\mathfrak{R}_{R}^{(p,T)} := \sup \big\{ \eta > 0 : d\big(\varphi(-T, \theta_T p)U_{\hat{\eta}}(R(\theta_T p))\big| R(p)\big) < \hat{\eta}$$
$$\textit{for all } \hat{\eta} \in (0, \eta) \big\}.$$

When considering a solution $\mu$ of $(\theta, \varphi)$ which is either past (future, all-time, $(p, T)$-, respectively) attractive or repulsive, one of the above definitions applies for graph $\mu$. We write $\mathfrak{A}_\mu := \mathfrak{A}_{\mathrm{graph}\,\mu}$ or $\mathfrak{R}_\mu := \mathfrak{R}_{\mathrm{graph}\,\mu}$ and proceed similarly with further notation (concerning, e.g., the domains of attraction and repulsion introduced in the next subsection).

### 2.3.4 Domains of Attraction and Repulsion

The radii of attraction and repulsion defined in the last subsection are positive real numbers. However, if $X$ is a Banach space and in case of past, future and all-time attractivity and repulsivity, we will, in addition to the radii of attraction and repulsion, consider domains of attraction and repulsion as subsets of the phase space.

We begin with some auxiliary definitions for the *extended* phase space. Given a past attractor $A$, we define for all $p \in P^*(A)$,

$$\mathcal{A}_{A}^{\leftarrow}(p) := \Big\{ x \in X : \textit{There exists a neighborhood } U \textit{ of } x \textit{ such that}$$
$$\lim_{t \to \infty} d\big(\varphi(t, \theta_{-t}p)(A(\theta_{-t}p) + U)\big| A(p)\big) = 0 \Big\},$$

and for a past repeller $R$, we define for all $p \in P^*(R)$,

$$\mathcal{R}_R^{\leftarrow}(p) := \Big\{ x \in X : \text{There exists a neighborhood } U \text{ of } x \text{ such that}$$
$$\lim_{t \to \infty} d\big(\varphi(-t,p)(R(p) + U)\big| R(\theta_{-t}p)\big) = 0 \Big\}.$$

Given a future attractor $A$, we define for all $p \in P^*(A)$,

$$\mathcal{A}_A^{\rightarrow}(p) := \Big\{ x \in X : \text{There exists a neighborhood } U \text{ of } x \text{ such that}$$
$$\lim_{t \to \infty} d\big(\varphi(t,p)(A(p) + U)\big| A(\theta_t p)\big) = 0 \Big\},$$

and for a future repeller $R$, we define for all $p \in P^*(R)$,

$$\mathcal{R}_R^{\rightarrow}(p) := \Big\{ x \in X : \text{There exists a neighborhood } U \text{ of } x \text{ such that}$$
$$\lim_{t \to \infty} d\big(\varphi(-t,\theta_t p)(R(\theta_t p) + U)\big| R(p)\big) = 0 \Big\}.$$

Some properties of these sets are derived in the following proposition.

**Proposition 2.26.** *The following statements are fulfilled:*

(i) *Given a past attractor $A$, the set $\mathcal{A}_A^{\leftarrow}(p)$ is open for all $p \in P^*(A)$, and we have $\mathcal{A}_A^{\leftarrow}(p) = \mathcal{A}_A^{\leftarrow}(\hat{p})$ for all $\hat{p} \in [p] \cap P^*(A)$. Furthermore, for all $p \in P^*(A)$ and compact sets $C \subset \mathcal{A}_A^{\leftarrow}(p)$, the relation*

$$\lim_{t \to \infty} d\big(\varphi(t,\theta_{-t}p)(A(\theta_{-t}p) + C)\big| A(p)\big) = 0$$

*is fulfilled.*

(ii) *Given a past repeller $R$, the set $\mathcal{R}_R^{\leftarrow}(p)$ is open for all $p \in P^*(R)$. Furthermore, for all $p \in P^*(R)$ and compact sets $C \subset \mathcal{R}_R^{\leftarrow}(p)$, the relation*

$$\lim_{t \to \infty} d\big(\varphi(-t,p)(R(p) + C)\big| R(\theta_{-t}p)\big) = 0$$

*is fulfilled.*

(iii) *Given a future attractor $A$, the set $\mathcal{A}_A^{\rightarrow}(p)$ is open for all $p \in P^*(A)$. Furthermore, for all $p \in P^*(A)$ and compact sets $C \subset \mathcal{A}_A^{\rightarrow}(p)$, the relation*

$$\lim_{t \to \infty} d\big(\varphi(t,p)(A(p) + C)\big| A(\theta_t p)\big) = 0$$

*is fulfilled.*

(iv) *Given a future repeller $R$, the set $\mathcal{R}_R^{\rightarrow}(p)$ is open for all $p \in P^*(R)$, and we have $\mathcal{R}_R^{\rightarrow}(p) = \mathcal{R}_R^{\rightarrow}(\hat{p})$ for all $\hat{p} \in [p] \cap P^*(R)$. Furthermore, for all $p \in P^*(R)$ and compact sets $C \subset \mathcal{R}_R^{\rightarrow}(p)$, the relation*

$$\lim_{t \to \infty} d\big(\varphi(-t,\theta_t p)(R(\theta_t p) + C)\big| R(p)\big) = 0$$

*is fulfilled.*

*Proof.* (i) The openness of $\mathcal{A}_A^{\leftarrow}(p)$ is a direct consequence of its definition, and the second assertion follows from Remark 2.7 (iv). Let us now assume that there exist $p \in P^*(A)$, a compact set $C \subset \mathcal{A}_A^{\leftarrow}(p)$, an $\varepsilon > 0$ and sequences $\{x_n\}_{n \in \mathbb{N}}$ in $C$ and $\{t_n\}_{n \in \mathbb{N}}$ in $\mathbb{T}$ such that $\lim_{n \to \infty} t_n = \infty$ and

$$d\big(\varphi(t_n, \theta_{-t_n} p)(A(\theta_{-t_n} p) + x_n)\big| A(p)\big) \geq \varepsilon \quad \text{for all } n \in \mathbb{N}.$$

Since $C$ is compact, we assume w.l.o.g. that $\{x_n\}_{n \in \mathbb{N}}$ is convergent with $\lim_{n \to \infty} x_n = x_0$. Since $x_0 \in C \subset \mathcal{A}_A^{\leftarrow}(p)$, there exists a neighborhood $U$ of $x_0$ such that

$$\lim_{t \to \infty} d\big(\varphi(t, \theta_{-t} p)(A(\theta_{-t} p) + U)\big| A(p)\big) = 0.$$

This is a contradiction.

(ii) As in (i), the first assertion is clear. Suppose now, there exist a $p \in P^*(R)$, a compact set $C \subset \mathcal{R}_R^{\leftarrow}(p)$, an $\varepsilon > 0$ and sequences $\{x_n\}_{n \in \mathbb{N}}$ in $C$ and $\{t_n\}_{n \in \mathbb{N}}$ in $\mathbb{T}$ such that $\lim_{n \to \infty} t_n = \infty$ and

$$d\big(\varphi(-t_n, p)(R(p) + x_n)\big| R(\theta_{-t_n} p)\big) \geq \varepsilon \quad \text{for all } n \in \mathbb{N}.$$

Since $C$ is compact, we assume w.l.o.g. that $\{x_n\}_{n \in \mathbb{N}}$ is convergent with $\lim_{n \to \infty} x_n = x_0$. Since $x_0 \in C \subset \mathcal{R}_R^{\leftarrow}(p)$, there exists a neighborhood $U$ of $x_0$ such that

$$\lim_{t \to \infty} d\big(\varphi(-t, p)(R(p) + U)\big| R(\theta_{-t} p)\big) = 0.$$

This is a contradiction and finishes the proof of (ii).

The proofs of (iii) and (iv) will be omitted, since they are similar to (i) and (ii). $\qquad \Box$

For simplicity in description, it is our aim to characterize the strength of attractivity or repulsivity not by the above defined fiber-wise sets, but by subsets of the phase space. This reduction is done by the following definition.

**Definition 2.27 (Domains of attraction and repulsion).**

(i) *The* domain of past attraction *of a past attractor $A$ is defined by*

$$\mathcal{A}_A^{\leftarrow} := \text{int} \bigcap_{p \in P^*(A)} \liminf_{t \to \infty} \mathcal{A}_A^{\leftarrow}(\theta_{-t} p).$$

(ii) *The* domain of past repulsion *of a past repeller $R$ is defined by*

$$\mathcal{R}_R^{\leftarrow} := \text{int} \bigcap_{p \in P^*(R)} \liminf_{t \to \infty} \mathcal{R}_R^{\leftarrow}(\theta_{-t} p).$$

(iii) *The* domain of future attraction *of a future attractor $A$ is defined by*

$$\mathcal{A}_A^{\rightarrow} := \text{int} \bigcap_{p \in P^*(A)} \liminf_{t \to \infty} \mathcal{A}_A^{\rightarrow}(\theta_t p).$$

*(iv)* *The* domain of future repulsion *of a future repeller $R$ is defined by*

$$\mathcal{R}_R^{\rightarrow} := \text{int} \bigcap_{p \in P^*(R)} \liminf_{t \to \infty} \mathcal{R}_R^{\rightarrow}(\theta_t p).$$

*(v)* *The* domain of all-time attraction *of an all-time attractor $A$ is defined by*

$$\mathcal{A}_A^{\leftrightarrow} := \Big\{ x \in X : \text{There exists a neighborhood } U \text{ of } x \text{ such that}$$
$$\lim_{t \to \infty} \sup_{p \in P^*(A)} d\big(\varphi(t,p)(A(p) + U)\big| A(\theta_t p)\big) = 0 \Big\}.$$

*(vi)* *The* domain of all-time repulsion *of an all-time repeller $R$ is defined by*

$$\mathcal{R}_R^{\leftrightarrow} := \Big\{ x \in X : \text{There exists a neighborhood } U \text{ of } x \text{ such that}$$
$$\lim_{t \to \infty} \sup_{p \in P^*(R)} d\big(\varphi(-t,p)(R(p) + U)\big| R(\theta_{-t} p)\big) = 0 \Big\}.$$

*Remark 2.28.*

(i) It can be seen immediately from the definitions that all above defined domains of attraction and repulsion are open neighborhoods of zero.

(ii) The relations

$$\mathfrak{A}_A^{\leftarrow} = \hat{d}\big(\mathcal{A}_A^{\leftarrow}\big|\{0\}\big), \quad \mathfrak{A}_A^{\rightarrow} = \hat{d}\big(\mathcal{A}_A^{\rightarrow}\big|\{0\}\big), \quad \mathfrak{R}_R^{\rightarrow} = \hat{d}\big(\mathcal{R}_R^{\rightarrow}\big|\{0\}\big) \quad \text{and}$$
$$\mathfrak{R}_R^{\leftarrow} = \hat{d}\big(\mathcal{R}_R^{\leftarrow}\big|\{0\}\big), \quad \mathfrak{A}_A^{\leftrightarrow} = \hat{d}\big(\mathcal{A}_A^{\leftrightarrow}\big|\{0\}\big), \quad \mathfrak{R}_R^{\leftrightarrow} = \hat{d}\big(\mathcal{R}_R^{\leftrightarrow}\big|\{0\}\big)$$

are fulfilled.

(iii) Given a past attractor $A$, from Proposition 2.26 (i), the relation

$$\mathcal{A}_A^{\leftarrow} = \text{int} \bigcap_{p \in P^*(A)} \mathcal{A}_A^{\leftarrow}(p)$$

follows. Moreover, Proposition 2.26 (iv) implies for a future repeller $R$,

$$\mathcal{R}_R^{\rightarrow} = \text{int} \bigcap_{p \in P^*(R)} \mathcal{R}_R^{\rightarrow}(p).$$

*Example 2.29.* We consider again the linear nonautonomous differential equation (2.2) from Example 2.8. We have already derived criteria for the attractivity and repulsivity of an invariant and compact nonautonomous set in several examples in Subsection 2.3.1. An easy calculation yields that the corresponding domains of attraction and repulsion are maximal for this example, more precisely, an invariant and compact nonautonomous set $M \subset \mathbb{R} \times \mathbb{R}$ is a

- past attractor $(\mathcal{A}_M^{\leftarrow} = \mathbb{R})$ if and only if $\lim_{t \to -\infty} \int_t^0 a(s)\, \mathrm{d}s = -\infty$,

- past repeller ($\mathcal{R}_M^{\leftarrow} = \mathbb{R}$) if and only if $\lim\limits_{t \to -\infty} \int_t^0 a(s)\,\mathrm{d}s = \infty$,

- future attractor ($\mathcal{A}_M^{\rightarrow} = \mathbb{R}$) if and only if $\lim\limits_{t \to \infty} \int_0^t a(s)\,\mathrm{d}s = -\infty$,

- future repeller ($\mathcal{R}_M^{\rightarrow} = \mathbb{R}$) if and only if $\lim\limits_{t \to \infty} \int_0^t a(s)\,\mathrm{d}s = \infty$,

- all-time attractor ($\mathcal{A}_M^{\leftrightarrow} = \mathbb{R}$) if and only if $\lim\limits_{t \to \infty} \sup\limits_{\tau \in \mathbb{R}} \int_\tau^{\tau+t} a(s)\,\mathrm{d}s = -\infty$,

- all-time repeller ($\mathcal{R}_M^{\leftrightarrow} = \mathbb{R}$) if and only if $\lim\limits_{t \to \infty} \sup\limits_{\tau \in \mathbb{R}} \int_\tau^{\tau+t} a(s)\,\mathrm{d}s = \infty$,

- $(p,T)$-attractor ($\mathfrak{A}_M^{(p,T)} = \infty$) if and only if $\int_p^{p+T} a(s)\,\mathrm{d}s < 0$,

- $(p,T)$-repeller ($\mathfrak{R}_M^{(p,T)} = \infty$) if and only if $\int_p^{p+T} a(s)\,\mathrm{d}s > 0$.

*Example 2.30.* In Subsection 2.3.1, we have also analyzed the attractivity and repulsivity of the nonlinear differential equation (2.3). By elementary discussions of the sign of the right hand side of this equation, we are also able to give estimates for the corresponding domains of attraction and repulsion. The trivial solution of (2.3) is

- past attractive with

$$\left( -\liminf_{t \to -\infty} w(t), \; \liminf_{t \to -\infty} w(t) \right) \subset \mathcal{A}_0^{\leftarrow} \subset \left( -\limsup_{t \to -\infty} w(t), \; \limsup_{t \to -\infty} w(t) \right)$$

if $\liminf_{t \to -\infty} -a(t)/b(t) > 0$,

- past repulsive with $\mathcal{R}_0^{\leftarrow} = \mathbb{R}$ if $\limsup_{t \to -\infty} -a(t)/b(t) \le 0$,

- future attractive with

$$\left( -\liminf_{t \to \infty} w(t), \; \liminf_{t \to \infty} w(t) \right) \subset \mathcal{A}_0^{\rightarrow} \subset \left( -\limsup_{t \to \infty} w(t), \; \limsup_{t \to \infty} w(t) \right)$$

if $\liminf_{t \to \infty} -a(t)/b(t) > 0$,

- future repulsive with $\mathcal{R}_0^{\rightarrow} = \mathbb{R}$ if $\limsup_{t \to \infty} -a(t)/b(t) \le 0$,

- all-time attractive with

$$\left( -\inf_{t \in \mathbb{R}} w(t), \; \inf_{t \in \mathbb{R}} w(t) \right) \subset \mathcal{A}_0^{\leftrightarrow} \subset \left( -\sup_{t \in \mathbb{R}} w(t), \; \sup_{t \in \mathbb{R}} w(t) \right)$$

if $\inf_{t \in \mathbb{R}} -a(t)/b(t) > 0$,

- all-time repulsive with $\mathcal{R}_0^{\leftrightarrow} = \mathbb{R}$ if $-a(t)/b(t) \le 0$ for all $t \in \mathbb{R}$,

- $(p,T)$-attractive with

$$\inf_{t \in [p,p+T]} w(t) \le \mathfrak{A}_0^{(p,T)} \le \sup_{t \in [p,p+T]} w(t)$$

if $-a(t)/b(t) > 0$ for all $t \in [p, p+T]$,

- $(p, T)$-repulsive with $\mathfrak{R}_0^{(p,T)} = \infty$ if $-a(t)/b(t) \leq 0$ for all $t \in [p, p+T]$.

These conditions are only sufficient for attractivity or repulsivity of the trivial solution but not necessary.

**Proposition 2.31.** *The following statements are fulfilled:*

(i) *Let* $\mu : \mathcal{O}^-(p) \to X$ *be a past repulsive solution. Then the past non-autonomous set with the* $\theta_{-t}p$*-fibres* $\mathcal{R}_\mu^\leftarrow(\theta_{-t}p) + \mu(\theta_{-t}p)$, $t \geq 0$, *is invariant.*

(ii) *Let* $\mu : \mathcal{O}^+(p) \to X$ *be a future attractive solution. Then the future nonautonomous set with the* $\theta_t p$*-fibres* $A_\mu^\rightarrow(\theta_t p) + \mu(\theta_t p)$, $t \geq 0$, *is invariant.*

*Proof.* (i) We choose $\tau \in \mathbb{T}_0^+$, $\hat{\tau} \in \mathbb{T} \cap (-\infty, \tau]$ and $x \in \mathcal{R}_\mu^\leftarrow(\theta_{-\tau}p) + \mu(\theta_{-\tau}p)$. Let $U$ be a neighborhood of $x - \mu(\theta_{-\tau}p)$ such that

$$\lim_{t\to\infty} d\big(\varphi(-t, \theta_{-\tau}p)(\mu(\theta_{-\tau}p) + U)\big|\{\mu(\theta_{-\tau-t}p)\}\big) = 0. \tag{2.4}$$

Then the set $\varphi(\hat{\tau}, \theta_{-\tau}p)(\mu(\theta_{-\tau}p) + U)$ is a neighborhood of $\varphi(\hat{\tau}, \theta_{-\tau}p)x$, and we have

$$\lim_{t\to\infty} d\big(\varphi(-t, \theta_{\hat{\tau}-\tau}p)\varphi(\hat{\tau}, \theta_{-\tau}p)(\mu(\theta_{-\tau}p) + U))\big|\{\mu(\theta_{\hat{\tau}-\tau-t}p)\}\big)$$

$$= \lim_{t\to\infty} d\big(\varphi(-t, \theta_{-\tau}p)(\mu(\theta_{-\tau}p) + U)\big|\{\mu(\theta_{-\tau-t})\}\big) \overset{(2.4)}{=} 0.$$

This means that $\varphi(\hat{\tau}, \theta_{-\tau}p)x - \mu(\theta_{\hat{\tau}-\tau}) \in \mathcal{R}_\mu^\leftarrow(\theta_{\hat{\tau}-\tau}p)$.
The assertion (ii) can be proved analogously. □

### 2.3.5 Properties of Time Reversal

As the reader may have observed, there are important analogies in the concepts of, say, past repulsivity and future attractivity. The aim of this subsection is to study these relationships.

In addition to the nonautonomous dynamical system $(\theta, \varphi)$, we also consider the system under time reversal, denoted by $(\theta^{-1}, \varphi^{-1})$ and defined by the relations

$$\theta^{-1}(t, p) := \theta(-t, p) \qquad \text{for all } (t, p) \in \mathbb{T} \times P,$$

$$\varphi^{-1}(t, p, x) := \varphi(-t, p, x) \quad \text{for all } (t, p, x) \text{ with } (-t, p, x) \in D.$$

The pair $(\theta, \varphi)^{-1} := (\theta^{-1}, \varphi^{-1})$ is indeed a nonautonomous dynamical system, since we have

$$\varphi^{-1}(t + s, p, x) = \varphi(-t - s, p, x) = \varphi\big(-t, \theta(-s, p), \varphi(-s, p, x)\big)$$

$$= \varphi^{-1}\big(t, \theta^{-1}(s, p), \varphi^{-1}(s, p, x)\big)$$

for all $t, s \in \mathbb{T}$, $(p, x) \in P \times X$ with $-s \in D_{max}(p, x)$ and $-t - s \in D_{max}(p, x)$.

**Proposition 2.32 (Properties of time reversal).** *Let $M$ be a subset of $P \times X$. Then the following statements are fulfilled:*

(i) *$M$ is a past attractor of $(\theta, \varphi)$ if and only if $M$ is a future repeller of $(\theta, \varphi)^{-1}$. We have $\mathfrak{A}_M^{\leftarrow} = \mathfrak{R}_M^{\rightarrow}$. If, in addition, $X$ is a Banach space, then also $\mathcal{A}_M^{\leftarrow} = \mathcal{R}_M^{\rightarrow}$ is fulfilled.*

(ii) *$M$ is a past repeller of $(\theta, \varphi)$ if and only if $M$ is a future attractor of $(\theta, \varphi)^{-1}$. We have $\mathfrak{R}_M^{\leftarrow} = \mathfrak{A}_M^{\rightarrow}$. If, in addition, $X$ is a Banach space, then also $\mathcal{R}_M^{\leftarrow} = \mathcal{A}_M^{\rightarrow}$ is fulfilled.*

(iii) *$M$ is an all-time attractor of $(\theta, \varphi)$ if and only if $M$ is an all-time repeller of $(\theta, \varphi)^{-1}$. We have $\mathfrak{R}_M^{\leftrightarrow} = \mathfrak{A}_M^{\leftrightarrow}$. If, in addition, $X$ is a Banach space, then also $\mathcal{R}_M^{\leftrightarrow} = \mathcal{A}_M^{\leftrightarrow}$ is fulfilled.*

(iv) *$M$ is a $(p, T)$-attractor of $(\theta, \varphi)$ if and only if $M$ is a $(\theta_T p, T)$-repeller of $(\theta, \varphi)^{-1}$. We have $\mathfrak{A}_M^{(p,T)} = \mathfrak{R}_M^{(\theta_T p, T)}$.*

*Proof.* (i) Let $M$ be a past attractor of $(\theta, \varphi)$, i.e., there exists an $\eta > 0$ such that for all $p \in P^*(M)$, there exists a $\hat{p} \in [p] \cap P^*(M)$ with

$$\lim_{t \to \infty} d\big(\varphi(t, \theta_{-\tau-t}\hat{p})U_\eta(M(\theta_{-\tau-t}\hat{p})) \big| M(\theta_{-\tau}\hat{p})\big) = 0 \quad \text{for all } \tau \geq 0.$$

This is equivalent to

$$\lim_{t \to \infty} d\Big(\varphi^{-1}\big(-t, \theta^{-1}(\tau+t, \hat{p})\big)U_\eta\big(M(\theta^{-1}(\tau+t, \hat{p}))\big) \Big| M\big(\theta^{-1}(\tau, \hat{p})\big)\Big) = 0$$

for all $\tau \geq 0$, and this means that $M$ is a future repeller of $(\theta, \varphi)^{-1}$. In case $X$ is a Banach space, the relation $\mathcal{A}_M^{\leftarrow} = \mathcal{R}_M^{\rightarrow}$ follows from

$$\mathcal{A}_M^{\leftarrow}(p) = \Big\{ x \in X : \text{There exists a neighborhood } U \text{ of } x \text{ such that}$$
$$\lim_{t \to \infty} d\big(\varphi(t, \theta_{-t}p)(M(\theta_{-t}p) + U) \big| M(p)\big) = 0\Big\}$$
$$= \Big\{ x \in X : \text{There exists a neighborhood } U \text{ of } x \text{ such that}$$
$$\lim_{t \to \infty} d\Big(\varphi^{-1}\big(-t, \theta^{-1}(t, p)\big)\big(M(\theta^{-1}(t, p)) + U\big) \Big| M(p)\Big) = 0\Big\}$$
$$= \mathcal{R}_M^{\rightarrow}(p)$$

for all $p \in P^*(M)$. The relation $\mathfrak{A}_M^{\leftarrow} = \mathfrak{R}_M^{\rightarrow}$ follows analogously. The proofs of (ii), (iii) and (iv) are similar to that of (i); we therefore omit them. $\qquad\square$

### 2.3.6 Existence and Uniqueness

In this subsection, criteria for the existence of attractors and repellers are formulated, and the question of their uniqueness is discussed.

First, the notions of past absorbing and future rejecting sets are introduced (these definitions are derived from FLANDOLI & SCHMALFUSS [68] and KLOEDEN [94, 95]).

**Definition 2.33 (Past absorbing and future rejecting sets).**

(i) *Let $B$ be a past nonautonomous set and $\mathcal{M}$ be a collection of past nonautonomous sets. Then $B$ is called* past absorbing *with respect to $\mathcal{M}$ if for all $M \in \mathcal{M}$, we have $P^*(M) \subset P^*(B)$, and for all $M \in \mathcal{M}$ and $p \in P^*(M)$, there exists a $t^* > 0$ such that*

$$\varphi(t, \theta_{-t}p)M(\theta_{-t}p) \subset B(p) \quad \text{for all } t \geq t^* . \tag{2.5}$$

(ii) *Let $B$ be a future nonautonomous set and $\mathcal{M}$ be a collection of future nonautonomous sets. Then $B$ is called* future rejecting *with respect to $\mathcal{M}$ if for all $M \in \mathcal{M}$, we have $P^*(M) \subset P^*(B)$, and for all $M \in \mathcal{M}$ and $p \in P^*(M)$, there exists a $t^* > 0$ such that*

$$\varphi(-t, \theta_{t}p)M(\theta_{t}p) \subset B(p) \quad \text{for all } t \geq t^* .$$

*Remark 2.34.* In case the past absorbing set is compact (which is a hypothesis of the next theorem), Definition 2.33 (i) is a nonautonomous generalization of the notion of a dissipative dynamical system (see, e.g., HALE [77]).

The following existence result is adapted from FLANDOLI & SCHMALFUSS [68] (for related results in the context of random attractors, see also CRAUEL & FLANDOLI [58, Theorem 3.11], SCHENK-HOPPÉ [159, Theorem 4.2] and SCHMALFUSS [160, 161]).

**Theorem 2.35 (Existence of $\mathcal{M}$-past attractors and $\mathcal{M}$-future repellers).** *The following statements are fulfilled:*

(i) *Let $\mathcal{M}$ be a collection of past nonautonomous sets and $B$ be a compact past absorbing set with respect to $\mathcal{M}$. Then there exists an $\mathcal{M}$-past attractor $A$ fulfilling the representation*

$$A(p) = \bigcap_{\tau \geq 0} \text{cls} \bigcup_{t \geq \tau} \varphi(t, \theta_{-t}p)B(\theta_{-t}p) \quad \text{for all } p \in P^*(B).$$

*If, in addition, $A \in \mathcal{M}$, then $A$ is uniquely determined. In case $B \in \mathcal{M}$, the relation $A \subset B$ is fulfilled.*

(ii) *Let $\mathcal{M}$ be a collection of future nonautonomous sets and $B$ be a compact future rejecting set with respect to $\mathcal{M}$. Then there exists an $\mathcal{M}$-future repeller $R$ fulfilling the representation*

$$R(p) = \bigcap_{\tau \geq 0} \text{cls} \bigcup_{t \geq \tau} \varphi(-t, \theta_{t}p)B(\theta_{t}p) \quad \text{for all } p \in P^*(B).$$

*If, in addition, $R \in \mathcal{M}$, then $R$ is uniquely determined. In case $B \in \mathcal{M}$, the relation $R \subset B$ is fulfilled.*

*Proof.* For assertion (i), see FLANDOLI & SCHMALFUSS [68, Theorem 3.5], (ii) follows from (i) using Proposition 2.32. □

In case the collection $\mathcal{M}$ of the above theorem contains a neighborhood of the set $B$, existence results for past attractors and future repellers follow directly.

**Corollary 2.36 (Existence of past attractors and future repellers).** *The following statements are fulfilled:*

(i) *Let $\mathcal{M}$ be a collection of past nonautonomous sets, $B$ be a compact past absorbing set with respect to $\mathcal{M}$ and $\eta > 0$ such that the past nonautonomous set $\bar{B}$, defined by*

$$\bar{B}(p) := U_\eta(B(p)) \quad \text{for all } p \in P^*(B)$$

*lies in $\mathcal{M}$. Then the $\mathcal{M}$-past attractor of Theorem 2.35 (i) is also a past attractor.*

(ii) *Let $\mathcal{M}$ be a collection of future nonautonomous sets, $B$ be a compact future rejecting set with respect to $\mathcal{M}$ and $\eta > 0$ such that the future nonautonomous set $\bar{B}$, defined by*

$$\bar{B}(p) := U_\eta(B(p)) \quad \text{for all } p \in P^*(B)$$

*lies in $\mathcal{M}$. Then the $\mathcal{M}$-future repeller of Theorem 2.35 (ii) is also a future repeller.*

*Proof.* The assertions follow directly from the definitions (cf. also Remark 2.7 (iii) and Remark 2.11 (iii)). □

In the following proposition, the question of local uniqueness and nonuniqueness for nonautonomous attractors and repellers is discussed.

**Proposition 2.37 (Local uniqueness and nonuniqueness).** *The following statements are fulfilled:*

(i) *Let $A_1$ and $A_2$ be past attractors such that $A_1(p) \neq A_2(p)$ for all $p \in P$. Then we have*

$$\liminf_{t \to \infty} d_H\big(A_1(\theta_{-t}p), A_2(\theta_{-t}p)\big) \geq \min\big\{\mathfrak{A}^{\leftarrow}_{A_1}, \mathfrak{A}^{\leftarrow}_{A_2}\big\} \quad \text{for all } p \in P.$$

(ii) *Let $R_1$ and $R_2$ be future repellers such that $R_1(p) \neq R_2(p)$ for all $p \in P$. Then we have*

$$\liminf_{t \to \infty} d_H\big(R_1(\theta_t p), R_2(\theta_t p)\big) \geq \min\big\{\mathfrak{R}^{\rightarrow}_{R_1}, \mathfrak{R}^{\rightarrow}_{R_2}\big\} \quad \text{for all } p \in P.$$

(iii) *Let $R_1$ be a past repeller. Then, for all $p \in P^*(R_1)$ and $\delta \in \left(0, \mathfrak{R}^{\leftarrow}_{R_1}\right)$, there exists a $\hat{p} \in [p] \cap P^*(R_1)$ such that for all $\tau \geq 0$ and compact sets $R_2(\theta_{-\tau}\hat{p}) \subset X$ with*

$$R_1(\theta_{-\tau}\hat{p}) \subset R_2(\theta_{-\tau}\hat{p}) \subset \text{cls}\, U_\delta(R_1(\theta_{-\tau}\hat{p})),$$

*the past nonautonomous set $R_2$, defined by*

$$R_2(\bar{p}) := \begin{cases} \varphi(-t, \theta_{-\tau}\hat{p})R_2(\theta_{-\tau}\hat{p}) & : \quad \bar{p} = \theta_{-\tau-t}\hat{p} \ \text{ for some } t \in \mathbb{T}^+_0 \\ R_1(\bar{p}) & : \quad \bar{p} \in P^*(R_1) \setminus [p] \end{cases},$$

*is also a past repeller with $\mathfrak{R}^{\leftarrow}_{R_1} = \mathfrak{R}^{\leftarrow}_{R_2}$.*

(iv) *Let $A_1$ be a future attractor. Then, for all $p \in P^*(A_1)$ and $\delta \in \left(0, \mathfrak{A}^{\rightarrow}_{A_1}\right)$, there exists a $\hat{p} \in [p] \cap P^*(A_1)$ such that for all $\tau \geq 0$ and compact sets $A_2(\theta_\tau\hat{p}) \subset X$ with*

$$A_1(\theta_\tau\hat{p}) \subset A_2(\theta_\tau\hat{p}) \subset \text{cls}\, U_\delta(A_1(\theta_\tau\hat{p})),$$

*the future nonautonomous set $A_2$, defined by*

$$A_2(\bar{p}) := \begin{cases} \varphi(t, \theta_\tau\hat{p})A_2(\theta_\tau\hat{p}) & : \quad \bar{p} = \theta_{\tau+t}\hat{p} \ \text{ for some } t \in \mathbb{T}^+_0 \\ A_1(\bar{p}) & : \quad \bar{p} \in P^*(A_1) \setminus [p] \end{cases},$$

*is also a future attractor with $\mathfrak{A}^{\rightarrow}_{A_1} = \mathfrak{A}^{\rightarrow}_{A_2}$.*

**Remark 2.38.**

(i) The form of (local) nonuniqueness of past repellers and future attractors is weak in the sense that, for instance, the past repellers $R_1$ and $R_2$ from (iii) fulfill

$$\lim_{t\to\infty} d_H\left(R_1(\theta_{-t}p), R_2(\theta_{-t}p)\right) = 0 \quad \text{for all } p \in P.$$

(ii) Since all-time attractors (repellers, respectively) are past attractors (future repellers, respectively) (cf. remark after Definition 2.14), they also fulfill a uniqueness result similar to (i) ((ii), respectively).

(iii) Concerning finite-time attractors and repellers, it is not possible to show uniqueness and nonuniqueness results. More precisely, the situation that in every neighborhood of a $(p, T)$-attractor lies another $(p, T)$-attractor and an invariant $(p, T)$-nonautonomous set which is not a $(p, T)$-attractor can occur.

(iv) In case $X$ is a Banach space, in (iii) ((iv), respectively), not only the relation $\mathfrak{R}^{\leftarrow}_{R_1} = \mathfrak{R}^{\leftarrow}_{R_2}$ ($\mathfrak{A}^{\rightarrow}_{A_1} = \mathfrak{A}^{\rightarrow}_{A_2}$, respectively) but also $\mathcal{R}^{\leftarrow}_{R_1} = \mathcal{R}^{\leftarrow}_{R_2}$ ($\mathcal{A}^{\rightarrow}_{A_1} = \mathcal{A}^{\rightarrow}_{A_2}$, respectively) holds.

*Proof (Proposition 2.37).* We will only prove the statements (i) and (iii), since the proofs of (ii) and (iv) are similar.
(i) Assume to the contrary that there exists a $p \in P$ with

$$\liminf_{t\to\infty} d_H\big(A_1(\theta_{-t}p), A_2(\theta_{-t}p)\big) < \min\big\{\mathfrak{A}_{A_1}^{\leftarrow}, \mathfrak{A}_{A_2}^{\leftarrow}\big\}.$$

Hence, there exist a $\beta < \min\big\{\mathfrak{A}_{A_1}^{\leftarrow}, \mathfrak{A}_{A_2}^{\leftarrow}\big\}$ and a sequence $\{t_n\}_{n\in\mathbb{N}}$ such that $\lim_{n\to\infty} t_n = \infty$ and

$$A_2(\theta_{-t_n}p) \subset U_\beta(A_1(\theta_{-t_n}p)) \quad \text{for all } n \in \mathbb{N}.$$

Due to Definition 2.25, this implies the existence of a $\tau \geq 0$ with

$$\lim_{n\to\infty} d\big(\underbrace{\varphi(t_n - \tau, \theta_{-t_n}p)A_2(\theta_{-t_n}p)}_{= A_2(\theta_{-\tau}p)}\big|A_1(\theta_{-\tau}p)\big) = 0.$$

This means that $d\big(A_2(\theta_{-\tau}p)\big|A_1(\theta_{-\tau}p)\big) = 0$, and hence, $d\big(A_2(p)\big|A_1(p)\big) = 0$. Analogously, one can show $d\big(A_1(p)\big|A_2(p)\big) = 0$. This implies $A_1(p) = A_2(p)$, since $A_1(p)$ and $A_2(p)$ are compact, and this contradiction finishes the proof of (i).

(iii) We choose $p \in P^*(R_1)$ and $\delta, \eta \in (0, \mathfrak{R}_{R_1}^{\leftarrow})$ arbitrarily and define

$$\beta := \frac{1}{2}\big(\max\{\delta, \eta\} + \mathfrak{R}_{R_1}^{\leftarrow}\big) \in \big(\max\{\delta, \eta\}, \mathfrak{R}_{R_1}^{\leftarrow}\big).$$

Due to Definition 2.25, there exists a $\hat{p} \in [p] \cap P^*(R_1)$ such that

$$\lim_{t\to\infty} d\big(\varphi(-t, \theta_{-\tau}\hat{p})U_\beta(R_1(\theta_{-\tau}\hat{p}))\big|R_1(\theta_{-\tau-t}\hat{p})\big) = 0 \quad \text{for all } \tau \geq 0. \quad (2.6)$$

We choose $\tau \geq 0$ and a compact set $R_2(\theta_{-\tau}\hat{p}) \subset X$ with

$$R_1(\theta_{-\tau}\hat{p}) \subset R_2(\theta_{-\tau}\hat{p}) \subset \operatorname{cls} U_\delta(R_1(\theta_{-\tau}\hat{p}))$$

and define the nonautonomous set $R_2$ as stated in the proposition. Because of (2.6), there exists a $t^* \geq 0$ such that

$$d\big(\underbrace{\varphi(-t, \theta_{-\tau}\hat{p})R_2(\theta_{-\tau}\hat{p})}_{= R_2(\theta_{-\tau-t}\hat{p})}\big|R_1(\theta_{-\tau-t}\hat{p})\big) < \frac{\mathfrak{R}_{R_1}^{\leftarrow} - \beta}{2} \quad \text{for all } t \geq t^*.$$

Since $\frac{1}{2}\big(\mathfrak{R}_{R_1}^{\leftarrow} - \beta\big) + \eta < \beta$, this implies using (2.6) the relation

$$\lim_{t\to\infty} d\big(\varphi(-t, \theta_{-\tau-s}\hat{p})U_\eta(R_2(\theta_{-\tau-s}\hat{p}))\big|R_1(\theta_{-\tau-s-t}\hat{p})\big) = 0 \quad \text{for all } s \geq t^*.$$

Because of $R_1 \subset R_2$, we obtain

$$\lim_{t\to\infty} d\big(\varphi(-t, \theta_{-\tau-s}\hat{p})U_\eta(R_2(\theta_{-\tau-s}\hat{p}))\big|R_2(\theta_{-\tau-s-t}\hat{p})\big) = 0 \quad \text{for all } s \geq t^*,$$

and this means that $R_2$ is a past repeller with $\mathfrak{R}_{R_2}^{\leftarrow} \geq \eta$ (please note that $R_1(\bar{p}) = R_2(\bar{p})$ for all $\bar{p} \in P^*(R_1) \setminus [p]$). Hence, $\mathfrak{R}_{R_2}^{\leftarrow} \geq \mathfrak{R}_{R_1}^{\leftarrow}$ ($\eta$ has been chosen arbitrarily). The relation $\mathfrak{R}_{R_2}^{\leftarrow} \leq \mathfrak{R}_{R_1}^{\leftarrow}$ can be obtained from

$$\lim_{t\to\infty} d\big(R_2(\theta_{-t}p)\big|R_1(\theta_{-t}p)\big) = 0.$$

This finishes the proof of this proposition. $\qquad\square$

In the following lemma, domains of attraction (repulsion, respectively) of attractive (repulsive, respectively) solutions lying in repellers (attractors, respectively) are analyzed.

**Lemma 2.39.** *We suppose that $X$ is a Banach space. Then the following statements are fulfilled:*

(i) *Let $R$ be a past repeller and $\mu : \mathcal{O}^-(p) \to X$ be a past attractive solution with*

$$\mu(\theta_{-t}p) \in \operatorname{int} R(\theta_{-t}p) \quad \text{for all } t \geq 0.$$

*Then we have*

$$\liminf_{t \to \infty} \left( R(\theta_{-t}p) - \mu(\theta_{-t}p) \right) \supset \mathcal{A}_\mu^\leftarrow.$$

(ii) *Let $A$ be a past attractor and $\mu : \mathcal{O}^-(p) \to X$ be a past repulsive solution with*

$$\mu(\theta_{-t}p) \in A(\theta_{-t}p) \quad \text{for all } t \geq 0.$$

*Then the relation $A(\theta_{-t}p) - \mu(\theta_{-t}p) \supset \mathcal{R}_\mu^\leftarrow(\theta_{-t}p)$ holds for all $t \geq 0$, and we thus have*

$$\liminf_{t \to \infty} \left( A(\theta_{-t}p) - \mu(\theta_{-t}p) \right) \supset \mathcal{R}_\mu^\leftarrow.$$

(iii) *Let $A$ be a future attractor and $\mu : \mathcal{O}^+(p) \to X$ be a future repulsive solution with*

$$\mu(\theta_t p) \in \operatorname{int} A(\theta_t p) \quad \text{for all } t \geq 0.$$

*Then we have*

$$\liminf_{t \to \infty} \left( A(\theta_t p) - \mu(\theta_t p) \right) \supset \mathcal{R}_\mu^\rightarrow.$$

(iv) *Let $R$ be a future repeller and $\mu : \mathcal{O}^+(p) \to X$ be a future attractive solution with*

$$\mu(\theta_t p) \in R(\theta_t p) \quad \text{for all } t \geq 0.$$

*Then the relation $R(\theta_t p) - \mu(\theta_t p) \supset \mathcal{A}_\mu^\rightarrow(\theta_t p)$ holds for all $t \geq 0$, and we thus have*

$$\liminf_{t \to \infty} \left( R(\theta_t p) - \mu(\theta_t p) \right) \supset \mathcal{A}_\mu^\rightarrow.$$

*Remark 2.40.*

(i) This lemma implies that (past and future) attractors or repellers containing repulsive or attractive solutions, respectively, are nontrivial, i.e., their fibers are not singletons.

(ii) Since the notions of all-time attractivity and repulsivity are stronger than those of past and future attractivity and repulsivity (cf. remark after Definition 2.14), the assertions of the above lemma are also applicable for all-time attractors and repellers.

*Proof (Lemma 2.39).* Because of Proposition 2.32, it is sufficient to show the statements (i) and (ii).
(i) We choose $x \in \mathcal{A}_\mu^\leftarrow$. Due to the hypotheses, there exists an $\eta > 0$ such that

$$\varphi(-t,p)U_\eta(\mu(p)) \subset R(\theta_{-t}p) \quad \text{for all } t \geq 0.$$

Since $x \in \mathcal{A}_\mu^\leftarrow$, there exists a $\tau > 0$ with

$$\varphi(t,\theta_{-t}p)(\mu(\theta_{-t}p)+x) \in U_\eta(\mu(p)) \quad \text{for all } t \geq \tau.$$

Hence, we have

$$\mu(\theta_{-t}p)+x = \varphi(-t,p)\varphi(t,\theta_{-t}p)(\mu(\theta_{-t}p)+x) \in R(\theta_{-t}p) \quad \text{for all } t \geq \tau.$$

Therefore, $x \in \liminf_{t\to\infty}\big(R(\theta_{-t}p) - \mu(\theta_{-t}p)\big)$.
(ii) We choose $x \in \mathcal{R}_\mu^\leftarrow(p)$ and $\delta > 0$ arbitrarily. Since $\mu$ lies in $A$, there exist $\eta > 0$ and $\tau > 0$ such that

$$\varphi(t,\theta_{-t}p)U_\eta(\mu(\theta_{-t}p)) \subset U_\delta(A(p)) \quad \text{for all } t \geq \tau.$$

Since $x \in \mathcal{R}_\mu^\leftarrow(p)$, there exists a $\hat\tau > 0$ with

$$\varphi(-t,p)(\mu(p)+x) \in U_\eta(\mu(\theta_{-t}p)) \quad \text{for all } t \geq \hat\tau.$$

Hence, with $t := \max\{\tau,\hat\tau\}$, the relation

$$\mu(p)+x = \varphi(t,\theta_{-t}p)\varphi(-t,p)(\mu(p)+x) \in U_\delta(A(p))$$

holds. Since $\delta$ has been chosen arbitrarily and $A(p)$ is compact, we have $\mu(p)+x \in A(p)$, and therefore, $\mu(p)+\mathcal{R}_\mu^\leftarrow(p) \subset A(p)$ is fulfilled. The assertion follows directly from Proposition 2.31 (i).  □

In the following theorem, sufficient conditions are derived to guarantee the existence of a nonautonomous attractor (repeller, respectively) which contains a nonautonomous repulsive (attractive, respectively) solution.

**Theorem 2.41 (Existence of nonautonomous attractors and repellers).** *Assume, $X$ is a Banach space. Then the following statements are fulfilled:*

(i) *We suppose that $\mu : \mathcal{O}^-(p) \to X$ is a past attractive solution such that $\mathcal{A}_\mu^\leftarrow$ is bounded and there exist $\varepsilon > 0$ and $s > 0$ such that for all $\tau \geq s$,*

$$\lim_{t\to\infty} d\big(\varphi(-t,\theta_{-\tau}p)U_\varepsilon\big(\mu(\theta_{-\tau}p)+\mathcal{A}_\mu^\leftarrow\big)\big|\mu(\theta_{-\tau-t}p)+\mathcal{A}_\mu^\leftarrow\big) = 0. \quad (2.7)$$

*Then there exists a $\beta > 0$ such that the invariant and compact past nonautonomous set $R$, defined by*

$$R(\theta_{-t-s}p) := \varphi(-t,\theta_{-s}p)\,\mathrm{cls}\,U_\beta(\mu(\theta_{-s}p)) \quad \text{for all } t \geq 0,$$

is a past repeller fulfilling

$$\mathcal{A}_\mu^\leftarrow \subset \liminf_{t\to\infty} \big(R(\theta_{-t}p) - \mu(\theta_{-t}p)\big)$$
$$\subset \limsup_{t\to\infty} \big(R(\theta_{-t}p) - \mu(\theta_{-t}p)\big) \subset \operatorname{cls}\mathcal{A}_\mu^\leftarrow .$$

(ii) *We suppose that* $\mu : \mathcal{O}^-(p) \to X$ *is a past repulsive solution such that* $\mathcal{R}_\mu^\leftarrow$ *is bounded and there exists an* $\eta > 0$ *such that for all* $\varepsilon > 0$, *there exists an* $s > 0$ *such that for all* $\tau \geq s$, *there is a* $T > 0$ *with*

$$\varphi(t, \theta_{-\tau-t}p)U_\eta\big(\mu(\theta_{-\tau-t}p) + \mathcal{R}_\mu^\leftarrow\big) \subset U_\varepsilon\big(\mu(\theta_{-\tau}p) + \mathcal{R}_\mu^\leftarrow\big) \quad \text{for all } t \geq T .$$

*Then there exists a past attractor* $A \subset \mathcal{O}^-(p) \times X$ *fulfilling*

$$\mathcal{R}_\mu^\leftarrow \subset \liminf_{t\to\infty} \big(A(\theta_{-t}p) - \mu(\theta_{-t}p)\big)$$
$$\subset \limsup_{t\to\infty} \big(A(\theta_{-t}p) - \mu(\theta_{-t}p)\big) \subset \operatorname{cls}\mathcal{R}_\mu^\leftarrow .$$

(iii) *We suppose that* $\mu : \mathcal{O}^+(p) \to X$ *is a future repulsive solution such that* $\mathcal{R}_\mu^\rightarrow$ *is bounded and there exist* $\varepsilon > 0$ *and* $s > 0$ *with*

$$\lim_{t\to\infty} d\big(\varphi(t, \theta_\tau p)U_\varepsilon\big(\mu(\theta_\tau p) + \mathcal{R}_\mu^\rightarrow\big)\big|\mu(\theta_{t+\tau}p) + \mathcal{R}_\mu^\rightarrow\big) = 0 \quad \text{for all } \tau \geq s .$$

*Then there exists a* $\beta > 0$ *such that the invariant and compact future nonautonomous set* $A$, *defined by*

$$A(\theta_{t+s}p) := \varphi(t, \theta_s p) \operatorname{cls} U_\beta(\mu(\theta_s p)) \quad \text{for all } t \geq 0 ,$$

*is a future attractor fulfilling*

$$\mathcal{R}_\mu^\rightarrow \subset \liminf_{t\to\infty} \big(A(\theta_t p) - \mu(\theta_t p)\big)$$
$$\subset \limsup_{t\to\infty} \big(A(\theta_t p) - \mu(\theta_t p)\big) \subset \operatorname{cls}\mathcal{R}_\mu^\rightarrow .$$

(iv) *We suppose that* $\mu : \mathcal{O}^+(p) \to X$ *is a future attractive solution such that* $\mathcal{A}_\mu^\rightarrow$ *is bounded and there exists an* $\eta > 0$ *such that for all* $\varepsilon > 0$, *there exists an* $s > 0$ *such that for all* $\tau \geq s$, *there is a* $T > 0$ *with*

$$\varphi(-t, \theta_{\tau+t}p)U_\eta\big(\mu(\theta_{\tau+t}p) + \mathcal{A}_\mu^\rightarrow\big) \subset U_\varepsilon\big(\mu(\theta_\tau p) + \mathcal{A}_\mu^\rightarrow\big) \quad \text{for all } t \geq T .$$

*Then there exists a future repeller* $R \subset \mathcal{O}^+(p) \times X$ *fulfilling*

$$\mathcal{A}_\mu^\rightarrow \subset \liminf_{t\to\infty} \big(R(\theta_t p) - \mu(\theta_t p)\big)$$
$$\subset \limsup_{t\to\infty} \big(R(\theta_t p) - \mu(\theta_t p)\big) \subset \operatorname{cls}\mathcal{A}_\mu^\rightarrow .$$

*Proof.* Due to Proposition 2.32, it is sufficient to show the statements (i) and (ii).

(i) We choose a $\beta > 0$ with $\operatorname{cls} U_\beta(0) \subset U_\varepsilon(A_\mu^\leftarrow)$ and define

$$R(\theta_{-s-t}p) := \varphi(-t, \theta_{-s}p) \operatorname{cls} U_\beta(\mu(\theta_{-s}p)) \quad \text{for all } t \geq 0.$$

This means that

$$R(\theta_{-s-t}p) \subset \varphi(-t, \theta_{-s}p) U_\varepsilon\big(\mu(\theta_{-s}p) + A_\mu^\leftarrow\big) \quad \text{for all } t \geq 0. \tag{2.8}$$

Moreover,

$$\lim_{t\to\infty} d\big(R(\theta_{-t}p)\big|\mu(\theta_{-t}p) + A_\mu^\leftarrow\big)$$

$$\overset{(2.8)}{\leq} \lim_{t\to\infty} d\big(\varphi(-t,\theta_{-s}p)U_\varepsilon\big(\mu(\theta_{-s}p) + A_\mu^\leftarrow\big)\big|\mu(\theta_{-s-t}p) + A_\mu^\leftarrow\big) \overset{(2.7)}{=} 0$$

is fulfilled, and therefore,

$$\limsup_{t\to\infty} \big(R(\theta_{-t}p) - \mu(\theta_{-t}p)\big) \subset \operatorname{cls} A_\mu^\leftarrow$$

holds. Next, we show that $R$ is a past repeller. Suppose that

$$\delta := \liminf_{t\to\infty} d\big(\mu(\theta_{-t}p) + A_\mu^\leftarrow\big|R(\theta_{-t}p)\big) > 0 \tag{2.9}$$

is fulfilled. Since $\operatorname{cls} A_\mu^\leftarrow$ is compact, there exist an $n \in \mathbb{N}$ and elements $x_1, \ldots, x_n \in \operatorname{cls} A_\mu^\leftarrow$ such that

$$\operatorname{cls} A_\mu^\leftarrow \subset \bigcup_{i=1}^n U_{\delta/4}(x_i).$$

For all $i \in \{1, \ldots, n\}$, we choose arbitrary elements

$$y_i \in U_{\delta/4}(x_i) \cap A_\mu^\leftarrow.$$

Then the set $C := \{y_1, \ldots, y_n\}$ is a compact subset of $A_\mu^\leftarrow$ which fulfills

$$d\big(A_\mu^\leftarrow\big|C\big) \leq \frac{\delta}{2}. \tag{2.10}$$

It follows from Proposition 2.26 (i) that there exists a $\tilde{t} \geq 0$ such that

$$\varphi(t, \theta_{-s-t}p)\big(\mu(\theta_{-s-t}p) + C\big) \subset U_\beta(\mu(\theta_{-s}p)) \subset R(\theta_{-s}p) \quad \text{for all } t \geq \tilde{t}.$$

Hence, due to the invariance of $R$, we obtain

$$\lim_{t\to\infty} d\big(\mu(\theta_{-t}p) + C\big|R(\theta_{-t}p)\big) = 0.$$

Using Lemma A.9, this implies

$$\liminf_{t\to\infty} d\big(\mu(\theta_{-t}p) + \mathcal{A}_\mu^\leftarrow \big| R(\theta_{-t}p)\big)$$

$$\leq \liminf_{t\to\infty} \Big( d\big(\mu(\theta_{-t}p) + \mathcal{A}_\mu^\leftarrow \big| \mu(\theta_{-t}p) + C\big) + d\big(\mu(\theta_{-t}p) + C \big| R(\theta_{-t}p)\big)\Big)$$

$$\overset{(2.10)}{\leq} \frac{\delta}{2}.$$

This is a contradiction to (2.9). Therefore,

$$\lim_{t\to\infty} d\big(\mu(\theta_{-t}p) + \mathcal{A}_\mu^\leftarrow \big| R(\theta_{-t}p)\big) = 0 \tag{2.11}$$

is fulfilled. Furthermore, there exists a $\hat{t} \geq 0$ with

$$d\big(R(\theta_{-s-\tau}p) \big| \mu(\theta_{-s-\tau}p) + \mathcal{A}_\mu^\leftarrow\big)$$

$$\overset{(2.8)}{\leq} d\big(\varphi(-\tau,\theta_{-s}p)U_\varepsilon\big(\mu(\theta_{-s}p) + \mathcal{A}_\mu^\leftarrow\big) \big| \mu(\theta_{-s-\tau}p) + \mathcal{A}_\mu^\leftarrow\big) \overset{(2.7)}{<} \frac{\varepsilon}{2}$$

for all $\tau \geq \hat{t}$. Hence, we have

$$U_\varepsilon\big(\mu(\theta_{-s-\tau}p) + \mathcal{A}_\mu^\leftarrow\big) \supset U_{\varepsilon/2}\big(U_{\varepsilon/2}(\mu(\theta_{-s-\tau}p) + \mathcal{A}_\mu^\leftarrow)\big)$$

$$\supset U_{\varepsilon/2}\big(R(\theta_{-s-\tau}p)\big) \qquad \text{for all } \tau \geq \hat{t}. \tag{2.12}$$

For all $\tau \geq \hat{t}$, the inequality

$$\lim_{t\to\infty} d\big(\varphi(-t,\theta_{-s-\tau}p)U_{\varepsilon/2}(R(\theta_{-s-\tau}p)) \big| R(\theta_{-s-\tau-t}p)\big)$$

$$\overset{(2.12)}{\leq} \lim_{t\to\infty} d\big(\varphi(-t,\theta_{-s-\tau}p)U_\varepsilon\big(\mu(\theta_{-s-\tau}p) + \mathcal{A}_\mu^\leftarrow\big) \big| R(\theta_{-s-\tau-t}p)\big)$$

$$\overset{\text{Lemma A.9}}{\leq} \lim_{t\to\infty} d\big(\varphi(-t,\theta_{-s-\tau}p)U_\varepsilon\big(\mu(\theta_{-s-\tau}p) + \mathcal{A}_\mu^\leftarrow\big) \big| \mu(\theta_{-s-\tau-t}p) + \mathcal{A}_\mu^\leftarrow\big)$$

$$+ \lim_{t\to\infty} d\big(\mu(\theta_{-s-\tau-t}p) + \mathcal{A}_\mu^\leftarrow \big| R(\theta_{-s-\tau-t}p)\big)$$

$$\overset{(2.7),\,(2.11)}{=} 0$$

holds, and this means that $R$ is a past repeller. The relation

$$\liminf_{t\to\infty} \big(R(\theta_{-t}p) - \mu(\theta_{-t}p)\big) \supset \mathcal{A}_\mu^\leftarrow$$

follows from Lemma 2.39 (i).

(ii) We define the past nonautonomous set $M$ by its fibers

$$M(\theta_{-t}p) := U_\eta\big(\mu(t) + \mathcal{R}_\mu^\leftarrow\big) \quad \text{for all } t \geq 0.$$

Due to the hypotheses, the fibers of a compact past absorbing set $B$ with respect to $\{M\}$ can be defined with the following property: For all $\varepsilon > 0$, there exists a $\tau \geq 0$ such that

$$\mathcal{R}_\mu^\leftarrow + \mu(\theta_{-t}p) \subset B(\theta_{-t}p) \subset U_\varepsilon\big(\mathcal{R}_\mu^\leftarrow + \mu(\theta_{-t}p)\big) \quad \text{for all } t \geq \tau. \tag{2.13}$$

Therefore, Theorem 2.35 yields the existence of an $\{M\}$-past attractor $A \subset B$ fulfilling

$$\limsup_{t \to \infty} \left( A(\theta_{-t}p) - \mu(\theta_{-t}p) \right) \subset \limsup_{t \to \infty} \left( B(\theta_{-t}p) - \mu(\theta_{-t}p) \right) \overset{(2.13)}{\subset} \text{cls}\, \mathcal{R}_\mu^{\leftarrow}.$$

Due to Corollary 2.36, $A$ is also a past attractor. The relation

$$\liminf_{t \to \infty} \left( A(\theta_{-t}p) - \mu(\theta_{-t}p) \right) \supset \mathcal{R}_\mu^{\leftarrow}$$

follows from Lemma 2.39 (ii).    $\square$

## 2.4 Other Notions of Attractivity and Repulsivity

In this section, other notions of attractivity and repulsivity from the literature are discussed with respect to their relationship to the definitions of the previous section. In the first subsection, the well-known theory of stability in the sense of Lyapunov is treated, and in Subsection 2.4.2, it is indicated that the notions of past (future, all-time, respectively) attractor and repeller are generalizations of the concept of attractor and repeller introduced in CONLEY [53]. Finally, the last subsection is devoted to the theory of nonautonomous attractors.

### 2.4.1 Stability in the Sense of Lyapunov

Several different forms of stability are examined in the literature. Most articles in this area, however, deal with the concept of stability in the sense of Lyapunov, which has been introduced by LYAPUNOV in his thesis [110] (see [112, 113] for translations into French and English). We shortly review the basic definitions of this theory in the context of nonautonomous differential equations (see also the classical books from CESARI [39] and HAHN [74]). An analogous theory exists for nonautonomous difference equations (see, e.g., AGARWAL [2]). Let

$$\dot{x} = f(t, x) \tag{2.14}$$

be a nonautonomous differential equation with a function $f : D \subset \mathbb{R} \times \mathbb{R}^N \to \mathbb{R}^N$ satisfying conditions guaranteeing local existence and uniqueness of solutions (see Appendix A.1). The general solution of (2.14) is denoted by $\lambda$. A solution $\mu : (\tau, \infty) \to \mathbb{R}^N$ is called *Lyapunov-stable* if for all $t_0 > \tau$ and $\varepsilon > 0$, there exists a $\delta = \delta(t_0, \varepsilon) > 0$ with

$$\lambda\big(t, t_0, U_\delta(\mu(t_0))\big) \subset U_\varepsilon(\mu(t)) \quad \text{for all } t \geq t_0.$$

Furthermore, a solution $\mu : (\tau, \infty) \to \mathbb{R}^N$ is called *Lyapunov-attractive* if for all $t_0 > \tau$, there exists an $\eta = \eta(t_0) > 0$ such that

$$\lim_{t \to \infty} \|\lambda(t, t_0, x) - \mu(t)\| = 0 \quad \text{for all } x \in U_\eta(\mu(t_0)).$$

There exist counterexamples in dimensions greater than one which show that not every Lyapunov-attractive solution is Lyapunov-stable (see, e.g., AULBACH [14, Beispiel 7.4.16, p. 325] and BHATIA & SZEGÖ [30, p. 59]). However, if a solution is both Lyapunov-stable and Lyapunov-attractive, we call this solution *Lyapunov-asymptotically stable*. If in the definition of the Lyapunov-stable solution, $\delta$ is independent of $t_0$, we call this solution *uniformly Lyapunov-stable*. In case $\eta$ is independent of $t_0$ in the definition of the Lyapunov-attractive solution, we call this solution *uniformly Lyapunov-attractive*. A solution which is both uniformly Lyapunov-stable and uniformly Lyapunov-attractive is called *uniformly Lyapunov-asymptotically stable* (see, e.g., SELL [167, p. 130]).

The concept of uniform asymptotically stability is a very strong form of stability in the sense of Lyapunov. It is easy to prove that any future attractive solution of (2.14) is uniformly Lyapunov-asymptotically stable.

### 2.4.2 Autonomous Attractors and Repellers

There are many different notions of attractor and repeller for (autonomous) dynamical systems (see SIENZ [174] for a summary). Many authors use various properties such as irreducibility, topological transitivity or connectivity in their definitions. As stated below, the concept of nonautonomous attractor and repeller in this book is closely related to the autonomous definitions used in CONLEY [53]. There, the main building blocks of attractor and repeller are *invariance*, *compactness* and *local attractivity and repulsivity*. We shortly review the definitions. Let $\phi : \mathbb{T} \times X \to X$ be a discrete (i.e., $\mathbb{T} = \mathbb{Z}$) or continuous (i.e., $\mathbb{T} = \mathbb{R}$) dynamical system on a metric space $(X, d)$. A compact set $A \subset X$ is called *attractor* of $\phi$ if $A$ is *invariant*, i.e.,

$$\phi(t, A) = A \quad \text{for all } t \in \mathbb{T},$$

and if $A$ is the $\omega$-*limit set* of some neighborhood $V$ of $A$, i.e.,

$$A = \omega(V) := \bigcap_{t \geq 0} \overline{\phi([t, \infty), V)}.$$

An invariant and compact set $R \subset X$ is called *repeller* if it is the $\alpha$-limit set of some neighborhood $W$ of $R$, i.e.,

$$R = \alpha(W) := \bigcap_{t \leq 0} \overline{\phi((-\infty, t], W)}.$$

One easily verifies that the definitions of past (future, all-time, respectively) attractor and repeller are indeed proper generalizations of this concept of attractor and repeller. In case of finite-time attractors and repellers, the situation is more subtle, since, given $T > 0$, not every attractor is a $(0, T)$-attractor.

However, given an attractor $A$, there exists a $\tau > 0$ such that for all $T \geq \tau$, $A$ is a $(0, T)$-attractor.

### 2.4.3 Nonautonomous Attractors

Since the 1990s, the attractivity of nonautonomous sets is intensively discussed. In particular, the notions of *pullback attractor* and *forward attractor* have been introduced (see, e.g., CHEBAN & KLOEDEN & SCHMALFUSS [40, 41] or KLOEDEN & KELLER & SCHMALFUSS [98]). Pullback and forward attractors whose attraction rate is uniform with respect to the time are called *uniform attractors* (such attractors are discussed in the monograph CHEPYZHOV & VISHIK [44]). Closely related to pullback attractors are the so-called random attractors (see, e.g., ARNOLD [5], CRAUEL & DEBUSSCHE & FLANDOLI [56], CRAUEL & FLANDOLI [58] and SCHENK-HOPPÉ [159]). The most general form of a pullback attractor (see, e.g., ARNOLD [5, Definition 9.3.1, p. 483]) coincides basically with the notion of the $\mathcal{M}$-past attractor as introduced in Definition 2.6 (iii). In the literature, $\mathcal{M}$ is called the *attraction universe*. Global pullback attractors are considered often, e.g., in CHEBAN & KLOEDEN & SCHMALFUSS [41, Definition 2.4]. In this case, the universe $\mathcal{M}$ is supposed to contain all fiber-wise constant and compact nonautonomous sets. The past attractor as introduced in Definition 2.6 (i), however, is a local form of a pullback attractor. Here, the universe contains a neighborhood of the attractor itself. Another form of a local pullback attractor is introduced in LANGA & ROBINSON & SUÁREZ [103, 105].

In contrast to pullback attractors, forward attractors play a minor role in the literature. Usually, only global forward attractors are considered (for an exception, see AULBACH & RASMUSSEN & SIEGMUND [16, Definition 3.4]). The $\mathcal{M}$-future attractor of Definition 2.10 (iii) provides a very general form of a forward attractor. By choosing $\mathcal{M}$ as the set of all fiber-wise constant and compact nonautonomous sets, one obtains the usual definition of a global forward attractor. A local form of a forward attractor, however, is provided by the future attractor as introduced in Definition 2.10 (i).

Apart from these classes of attractors, pullback and forward attractors which are allowed to be noncompact are introduced in AULBACH & RASMUSSEN & SIEGMUND [16, Definition 3.4]. Instead to be compact, attractors of this type are supposed to be "compactly generated". This notion includes some classes of noncompact nonautonomous invariant manifolds (see AULBACH & RASMUSSEN & SIEGMUND [17, 18]), but is no proper generalization of a compact attractor, since a compact attractor is not compactly generated in general.

## 2.5 Bifurcation and Transition

This section is devoted to the introduction of various nonautonomous concepts of bifurcation and transition based on the notions of attractivity and repulsivity from Section 2.3.

Throughout this section, let $\big(\theta : \mathbb{T} \times P \to P,\ \varphi_\alpha : D_\alpha \subset \mathbb{T} \times P \times X \to X\big)$, $\alpha \in (\alpha^-, \alpha^+)$, be a family of nonautonomous dynamical systems with a base set $P$ and a metric space $(X, d)$.

### 2.5.1 Definitions

In addition to the four different time domains (past, future, all-time and finite-time), we also distinguish between bifurcations of radii of attraction and repulsion and transitions of attractors and repellers. Attractor transitions are studied in, e.g., MA & WANG [114], or see KLOEDEN & SIEGMUND [99], where also nonautonomous attractors are considered.

We begin with the definitions concerning the past of the system.

**Definition 2.42 (Past bifurcation and transition).** *Let $\alpha_0 \in (\alpha^-, \alpha^+)$. We say, $(\theta, \varphi_\alpha)$ admits a supercritical past bifurcation at $\alpha_0$ if there exist an $\hat{\alpha} > \alpha_0$ and a continuous function $\mu : D \subset \mathcal{O}(p) \times (\alpha_0, \hat{\alpha}) \to X$ such that one of the following two statements is fulfilled:*

*(i) $\mu(\cdot, \alpha)$ is a past attractive solution of $(\theta, \varphi_\alpha)$ for all $\alpha \in (\alpha_0, \hat{\alpha})$, and*

$$\lim_{\alpha \searrow \alpha_0} \mathfrak{A}^{\leftarrow}_{\mu(\cdot,\alpha)} = 0$$

*is fulfilled. In case $X$ is a Banach space, we call this bifurcation total if*

$$\lim_{\alpha \searrow \alpha_0} d\big(\mathcal{A}^{\leftarrow}_{\mu(\cdot,\alpha)}\big|\{0\}\big) = 0$$

*holds, otherwise, we call this bifurcation partial.*

*(ii) $\mu(\cdot, \alpha)$ is a past repulsive solution of $(\theta, \varphi_\alpha)$ for all $\alpha \in (\alpha_0, \hat{\alpha})$, and*

$$\lim_{\alpha \searrow \alpha_0} \mathfrak{R}^{\leftarrow}_{\mu(\cdot,\alpha)} = 0$$

*is fulfilled. In case $X$ is a Banach space, we call this bifurcation total if*

$$\lim_{\alpha \searrow \alpha_0} d\big(\mathcal{R}^{\leftarrow}_{\mu(\cdot,\alpha)}\big|\{0\}\big) = 0$$

*holds, otherwise, we call this bifurcation partial.*

*We say, $(\theta, \varphi_\alpha)$ admits a supercritical past attractor (past repeller, respectively) transition at $\alpha_0$ if there exist an $\hat{\alpha} > \alpha_0$ and past attractors (past repellers, respectively) $M_\alpha$ of $(\theta, \varphi_\alpha)$ for $\alpha \in (\alpha_0, \hat{\alpha})$ with*

$$\lim_{\alpha \searrow \alpha_0} \limsup_{t \to \infty} \operatorname{diam} M_\alpha(\theta_{-t}p) = 0 \quad \text{for all } p \in P.$$

*Accordingly, subcritical past bifurcations and past attractor (past repeller, respectively) transitions are defined by considering the limit $\alpha \nearrow \alpha_0$.*

The following definition is devoted to the introduction of future bifurcations and transitions.

**Definition 2.43 (Future bifurcation and transition).** *Let $\alpha_0 \in (\alpha^-, \alpha^+)$. We say, $(\theta, \varphi_\alpha)$ admits a supercritical future bifurcation at $\alpha_0$ if there exist an $\hat{\alpha} > \alpha_0$ and a continuous function $\mu : D \subset \mathcal{O}(p) \times (\alpha_0, \hat{\alpha}) \to X$ such that one of the following two statements is fulfilled:*

(i) *$\mu(\cdot, \alpha)$ is a future attractive solution of $(\theta, \varphi_\alpha)$ for all $\alpha \in (\alpha_0, \hat{\alpha})$, and*

$$\lim_{\alpha \searrow \alpha_0} \mathfrak{A}^{\to}_{\mu(\cdot,\alpha)} = 0$$

*is fulfilled. In case $X$ is a Banach space, we call this bifurcation total if*

$$\lim_{\alpha \searrow \alpha_0} d\big(\mathcal{A}^{\to}_{\mu(\cdot,\alpha)} \big| \{0\}\big) = 0$$

*holds, otherwise, we call this bifurcation partial.*

(ii) *$\mu(\cdot, \alpha)$ is a future repulsive solution of $(\theta, \varphi_\alpha)$ for all $\alpha \in (\alpha_0, \hat{\alpha})$, and*

$$\lim_{\alpha \searrow \alpha_0} \mathfrak{R}^{\to}_{\mu(\cdot,\alpha)} = 0$$

*is fulfilled. In case $X$ is a Banach space, we call this bifurcation total if*

$$\lim_{\alpha \searrow \alpha_0} d\big(\mathcal{R}^{\to}_{\mu(\cdot,\alpha)} \big| \{0\}\big) = 0$$

*holds, otherwise, we call this bifurcation partial.*

*We say, $(\theta, \varphi_\alpha)$ admits a supercritical future attractor (future repeller, respectively) transition at $\alpha_0$ if there exist an $\hat{\alpha} > \alpha_0$ and future attractors (future repellers, respectively) $M_\alpha$ of $(\theta, \varphi_\alpha)$ for $\alpha \in (\alpha_0, \hat{\alpha})$ with*

$$\lim_{\alpha \searrow \alpha_0} \limsup_{t \to \infty} \operatorname{diam} M_\alpha(\theta_t p) = 0 \quad \text{for all } p \in P.$$

*Accordingly, subcritical future bifurcations and future attractor (future repeller, respectively) transitions are defined by considering the limit $\alpha \nearrow \alpha_0$.*

In the next definition, the notions of all-time bifurcation and transition are explained.

**Definition 2.44 (All-time bifurcation and transition).** *For a given $\alpha_0 \in (\alpha^-, \alpha^+)$, we say, $(\theta, \varphi_\alpha)$ admits a supercritical all-time bifurcation at $\alpha_0$ if there exist an $\hat{\alpha} > \alpha_0$ and a continuous function $\mu : \mathcal{O}(p) \times (\alpha_0, \hat{\alpha}) \to X$ such that one of the following two statements is fulfilled:*

(i) $\mu(\cdot, \alpha)$ is an all-time attractive solution of $(\theta, \varphi_\alpha)$ for all $\alpha \in (\alpha_0, \hat{\alpha})$, and

$$\lim_{\alpha \searrow \alpha_0} \mathfrak{A}^{\leftrightarrow}_{\mu(\cdot, \alpha)} = 0$$

is fulfilled. In case $X$ is a Banach space, we call this bifurcation total if

$$\lim_{\alpha \searrow \alpha_0} d\big(\mathcal{A}^{\leftrightarrow}_{\mu(\cdot, \alpha)} \big| \{0\}\big) = 0$$

holds, otherwise, we call this bifurcation partial.

(ii) $\mu(\cdot, \alpha)$ is an all-time repulsive solution of $(\theta, \varphi_\alpha)$ for all $\alpha \in (\alpha_0, \hat{\alpha})$, and

$$\lim_{\alpha \searrow \alpha_0} \mathfrak{R}^{\leftrightarrow}_{\mu(\cdot, \alpha)} = 0$$

is fulfilled. In case $X$ is a Banach space, we call this bifurcation total if

$$\lim_{\alpha \searrow \alpha_0} d\big(\mathcal{R}^{\leftrightarrow}_{\mu(\cdot, \alpha)} \big| \{0\}\big) = 0$$

holds, otherwise, we call this bifurcation partial.

We say, $(\theta, \varphi_\alpha)$ admits a supercritical all-time attractor (all-time repeller, respectively) transition at $\alpha_0$ if there exist an $\hat{\alpha} > \alpha_0$ and all-time attractors (all-time repellers, respectively) $M_\alpha$ of $(\theta, \varphi_\alpha)$ for $\alpha \in (\alpha_0, \hat{\alpha})$ with

$$\lim_{\alpha \searrow \alpha_0} \sup_{p \in P} \operatorname{diam} M_\alpha(p) = 0 \,.$$

Accordingly, subcritical all-time bifurcations and all-time attractor (all-time repeller, respectively) transitions are defined by considering the limit $\alpha \nearrow \alpha_0$.

Finally, the following definition treats the concept of finite-time bifurcation and transition.

**Definition 2.45 (Finite-time bifurcation and transition).** *For a given* $\alpha_0 \in (\alpha^-, \alpha^+)$, *we say,* $(\theta, \varphi_\alpha)$ *admits a supercritical* $(p, T)$-*bifurcation at* $\alpha_0$ *if there exist an* $\hat{\alpha} > \alpha_0$ *and a continuous function* $\mu : \mathcal{O}^T(p) \times (\alpha_0, \hat{\alpha}) \to X$ *such that one of the following two statements is fulfilled:*

(i) $\mu(\cdot, \alpha)$ is a $(p, T)$-attractive solution of $(\theta, \varphi_\alpha)$ for all $\alpha \in (\alpha_0, \hat{\alpha})$, and

$$\lim_{\alpha \searrow \alpha_0} \mathfrak{A}^{(p,T)}_{\mu(\cdot, \alpha)} = 0$$

is fulfilled.

(ii) $\mu(\cdot, \alpha)$ is a $(p, T)$-repulsive solution of $(\theta, \varphi_\alpha)$ for all $\alpha \in (\alpha_0, \hat{\alpha})$, and

$$\lim_{\alpha \searrow \alpha_0} \mathfrak{R}^{(p,T)}_{\mu(\cdot, \alpha)} = 0$$

holds.

*We say, $(\theta, \varphi_\alpha)$ admits a supercritical $(p, T)$-attractor $((p, T)$-repeller, respectively) transition at $\alpha_0 \in (\alpha^-, \alpha^+)$ if there exist an $\hat{\alpha} > \alpha_0$ and $(p, T)$-attractors $((p, T)$-repellers, respectively) $M_\alpha$ of $(\theta, \varphi_\alpha)$ for $\alpha \in (\alpha_0, \hat{\alpha})$ with*

$$\lim_{\alpha \searrow \alpha_0} \operatorname{diam} M_\alpha(p) = 0.$$

*Accordingly, subcritical $(p, T)$-bifurcations and $(p, T)$-attractor $((p, T)$-repeller, respectively) transitions are defined by considering the limit $\alpha \nearrow \alpha_0$.*

## 2.5.2 Examples

In this subsection, two nonautonomous differential equations are discussed which are closely related to standard examples of equations admitting an autonomous bifurcation. The first example is of pitchfork type and leads to a total nonautonomous bifurcation; the second one is of transcritical type and gives rise to a partial nonautonomous bifurcation.

*Example 2.46 (Nonautonomous pitchfork bifurcation).* We consider the nonautonomous differential equation

$$\dot{x} = \alpha a(t)x + b(t)x^3 = x\big(\alpha a(t) + b(t)x^2\big) \tag{2.15}$$

depending on a real parameter $\alpha$ with continuous functions $a : \mathbb{R} \to \mathbb{R}$ and $b : \mathbb{R} \to \mathbb{R}_\kappa^+$ for some $\kappa > 0$. The equation (2.15) is a nonautonomous version of the well-known autonomous differential equation

$$\dot{x} = \alpha x + x^3 = x\big(\alpha + x^2\big),$$

which admits a pitchfork bifurcation (see, e.g., GUCKENHEIMER & HOLMES [72, p. 150]). For fixed $\alpha \in \mathbb{R}$, (2.15) has already been discussed in Example 2.30, where we have derived sufficient conditions concerning the attractivity and repulsivity of the trivial solution. The following statements are direct consequences of these observations. The above nonautonomous differential equation admits a

- total supercritical past bifurcation at $\alpha = 0$ if

$$\liminf_{t \to -\infty} -\frac{a(t)}{b(t)} > 0 \quad \text{and} \quad \limsup_{t \to -\infty} -\frac{a(t)}{b(t)} < \infty,$$

- total subcritical past bifurcation at $\alpha = 0$ if

$$\liminf_{t \to -\infty} \frac{a(t)}{b(t)} > 0 \quad \text{and} \quad \limsup_{t \to -\infty} \frac{a(t)}{b(t)} < \infty,$$

- total supercritical future bifurcation at $\alpha = 0$ if

$$\liminf_{t \to \infty} -\frac{a(t)}{b(t)} > 0 \quad \text{and} \quad \limsup_{t \to \infty} -\frac{a(t)}{b(t)} < \infty,$$

- total subcritical future bifurcation at $\alpha = 0$ if

$$\liminf_{t\to\infty} \frac{a(t)}{b(t)} > 0 \quad \text{and} \quad \limsup_{t\to\infty} \frac{a(t)}{b(t)} < \infty \,,$$

- total supercritical all-time bifurcation at $\alpha = 0$ if

$$\inf_{t\in\mathbb{R}} -\frac{a(t)}{b(t)} > 0 \quad \text{and} \quad \sup_{t\in\mathbb{R}} -\frac{a(t)}{b(t)} < \infty \,,$$

- total subcritical all-time bifurcation at $\alpha = 0$ if

$$\inf_{t\in\mathbb{R}} \frac{a(t)}{b(t)} > 0 \quad \text{and} \quad \sup_{t\in\mathbb{R}} \frac{a(t)}{b(t)} < \infty \,,$$

- total supercritical $(\tau, T)$-bifurcation at $\alpha = 0$ if

$$-\frac{a(t)}{b(t)} > 0 \quad \text{for all } t \in [\tau, \tau + T] \,,$$

- total subcritical $(\tau, T)$-bifurcation at $\alpha = 0$ if

$$\frac{a(t)}{b(t)} > 0 \quad \text{for all } t \in [\tau, \tau + T] \,.$$

A generalization of this equation is discussed in Section 6.2. It is also shown there that this example admits attractor and repeller transitions.

*Remark 2.47.* A special form of the above example (for constant functions $a$) is discussed in LANGA & ROBINSON & SUÁREZ [103, Proposition 3.1] and CARABALLO & LANGA [37, Subsection 4.1].

*Example 2.48 (Nonautonomous transcritical bifurcation).* We consider the nonautonomous differential equation

$$\dot{x} = \alpha a(t)x + b(t)x^2 = x\big(\alpha a(t) + b(t)x\big)$$

depending on a real parameter $\alpha$ with continuous functions $a : \mathbb{R} \to \mathbb{R}$ and $b : \mathbb{R} \to \mathbb{R}_\kappa^+$ for some $\kappa > 0$. This equation is a nonautonomous version of the well-known autonomous differential equation

$$\dot{x} = \alpha x + x^2 = x(\alpha + x) \,,$$

which admits a transcritical bifurcation (see HALE & KOÇAK [78, Example 2.3, p. 28]). Analogously to Example 2.46, we see that the above non-autonomous differential equation admits a

- partial supercritical and subcritical past bifurcation at $\alpha = 0$ if

$$\liminf_{t \to -\infty} \left| \frac{a(t)}{b(t)} \right| > 0 \quad \text{and} \quad \limsup_{t \to -\infty} \left| \frac{a(t)}{b(t)} \right| < \infty,$$

- partial supercritical and subcritical future bifurcation at $\alpha = 0$ if

$$\liminf_{t \to \infty} \left| \frac{a(t)}{b(t)} \right| > 0 \quad \text{and} \quad \limsup_{t \to \infty} \left| \frac{a(t)}{b(t)} \right| < \infty,$$

- partial supercritical and subcritical all-time bifurcation at $\alpha = 0$ if

$$\inf_{t \in \mathbb{R}} \left| \frac{a(t)}{b(t)} \right| > 0 \quad \text{and} \quad \sup_{t \in \mathbb{R}} \left| \frac{a(t)}{b(t)} \right| < \infty,$$

- partial supercritical and subcritical $(\tau, T)$-bifurcation at $\alpha = 0$ if

$$\left| \frac{a(t)}{b(t)} \right| > 0 \quad \text{for all } t \in [\tau, \tau + T].$$

A generalization of this nonautonomous differential equation is discussed in Section 6.1.

## 2.6 Other Notions of Bifurcation and Transition

In this section, several notions of bifurcation for (nonautonomous) dynamical systems are discussed with respect to their relationship to the concept of bifurcation and transition introduced in the previous section.

In the first subsection of this section, the autonomous bifurcation theory is treated. As mentioned in Section 2.2, the notion of the nonautonomous dynamical system is an abstraction of both topological skew product flows and random dynamical systems. In the recent studies of bifurcations of nonautonomous dynamical systems, one should also distinguish between topological skew product flows (cf. Subsection 2.6.2) and random dynamical systems (cf. Subsection 2.6.3). So far, there are only few approaches to the nonautonomous bifurcation theory without imposing special hypotheses on the base set $P$ such as compactness or existence of an invariant measure (cf. Subsection 2.6.4).

Please note that in Section 5.4, a relationship between the concept of finite-time bifurcation and the bifurcation theory of adiabatic systems is pointed out.

### 2.6.1 The Autonomous Case

As mentioned in CHOW & HALE [46] and MARSDEN & HUGHES [116], there are two distinct aspects of autonomous bifurcation theory: static and dynamic. The static point of view is concentrated on the qualitative changes in the structure of the set of zeros of a function as parameters are varied. The dynamic bifurcation theory, however, is concerned with dynamical changes that occur in invariant sets (such as equilibria, periodic orbits, heteroclinic orbits and invariant tori).

Since the concept of bifurcation and transition used in this book is based on notions of attractivity and repulsivity, the static approach is too narrow in our situation, and we hope that the center manifold theory, which had been so fruitful in dynamic bifurcation theory, will be a method for future research in a higher dimensional nonautonomous bifurcation theory (cf. Example 7.14). It is not clear a priori to what extent the method of Lyapunov-Schmidt (see, e.g., HALE [76, Section 1]) is able to give a contribution in the nonautonomous context.

In the Introduction (Chapter 1), an easy example already indicated that autonomous bifurcation phenomena can be described in terms of the concepts of nonautonomous bifurcation and transition. Please note also that in Chapter 6, one-dimensional nonautonomous bifurcations are studied, and Example 6.3 shows that the nonautonomous patterns are applicable also in the autonomous context. Moreover, it is shown in Chapter 7 that the classical bifurcation scenarios of saddle node, pitchfork, transcritical or Hopf type can be transferred to asymptotically autonomous equations. By regarding an autonomous system which admits a bifurcation of this type as an asymptotically autonomous system, one sees that the autonomous situation fits well into our context.

Interesting books on the topic of autonomous bifurcation theory are CHOW & HALE [46], GUCKENHEIMER & HOLMES [72], HALE & KOÇAK [78], KUZNETSOW [101] and LUO & WANG & ZHU & HAN [109]. For a brief introduction, see also CRAWFORD [60].

### 2.6.2 Topological Skew Product Flows

In the bifurcation theory of nonautonomous dynamical systems where the base set is supposed to have a certain topological structure, one distinguishes between attractor-repeller bifurcations and bifurcations of solutions.

An attractor-repeller bifurcation either occurs if a nontrivial attractor or repeller, respectively, shrinks down to a trivial object by variation of the parameter (this corresponds to the notion of transition), or if an attractor bifurcates from a repeller in the sense of Hausdorff distance. Please note that the attractors and repellers under consideration are autonomous objects of the skew product flow.

In JOHNSON & MANTELLINI [86], FABBRI & JOHNSON & MANTELLINI [67] and FABBRI & JOHNSON [66], for one-dimensional nonautonomous differential equations with strictly ergodic time dependence (e.g., quasi-periodic equations are of this type), attractor-repeller bifurcations are considered. Bifurcations of attractors and repellers are also studied in JOHNSON & KLOEDEN & PAVANI [85] and JOHNSON [84] for deterministic counterparts of the *Two-Step-Bifurcation-Pattern*. These considerations are based on the studies of Ludwig Arnold and his coworkers in the context of stochastic differential equations (see ARNOLD [5]). In GLENDINNING [70], a bifurcation of nonchaotic strange attractors of a quasi-periodic differential equation is verified, both numerically and analytically.

A bifurcation (of pitchfork and transcritical type) of almost periodic solutions of an almost periodic ordinary differential equation is examined in KLOEDEN [97].

We also mention autonomous bifurcations of invariant sets on which the dynamics of the system is nonperiodic, because the analysis requires—by means of the equation of perturbed motion—nonautonomous techniques. In JOHNSON [84] and JOHNSON & YI [87], former studies concerning the bifurcation of invariant tori (see, e.g., SELL [169] and CHENCINER & IOOSS [42, 43]) are continued. The authors consider for an autonomous differential equation the loss of stability of an invariant set (which is, for instance, the closure of a nonperiodic and bounded trajectory). The bifurcation theory of tori with quasi-periodic flows whose frequencies satisfy the Diophantine condition is well-developed (see BROER & HUITEMA & TAKENS & BRAAKSMA [36], BROER [35] and BRAAKSMA & BROER [33]).

### 2.6.3 Random Dynamical Systems

To study bifurcation phenomena of random dynamical systems, two different concepts have been pursued so far: the so-called *phenomenological approach* (P-bifurcation) and the *dynamical approach* (D-bifurcation). For fundamental explanations and comparisons, we refer to ARNOLD & NAMACHCHIVAYA & SCHENK-HOPPÉ [8], ARNOLD [5, 6] and SCHENK-HOPPÉ [156, 157] (see also SCHENK-HOPPÉ [158, 159]).

P-bifurcations describe changes in stationary probability densities in special families of random dynamical systems. For instance, these densities exhibit transitions from one-peak to two-peak or crater-like structures. The concept of P-bifurcation can be formalized using the notion of equivalent probability densities, introduced by ZEEMAN [185, 186], which gives rise to a notion of structural stability. There are many drawbacks to this phenomenological approach, which are mentioned, e.g., in ARNOLD [5, Subsection 9.2.2]. Since P-bifurcations are static in the sense that there is no connection to stability properties obtained by Lyapunov exponents, we cannot expect a relationship to the concept of bifurcation introduced in this chapter.

Recently, the study of random bifurcation phenomena concentrated on D-bifurcations. A D-bifurcation occurs if from an invariant reference measure, another invariant measure bifurcates in the sense of weak convergence. It has been shown that this concept links the local bifurcation of invariant measures with the stability determined by the Lyapunov exponents.

### 2.6.4 General Nonautonomous Dynamical Systems

So far, there have been two approaches in the study of bifurcation phenomena of nonautonomous dynamical systems where no special hypotheses concerning the base set are made.

In KLOEDEN & SIEGMUND [99], a nonautonomous bifurcation is understood as a (continuous or discontinuous) transition from a nontrivial (global) pullback attractor to a trivial pullback attractor.

In LANGA & ROBINSON & SUÁREZ [103], for nonautonomous differential equations, notions of Lyapunov pullback-stable and Lyapunov pullback-unstable solutions are introduced, and bifurcations in form of merging processes of two distinct solutions with different stability behavior are studied by means of relatively simple examples. In their recent paper [105], the three authors found sufficient conditions for the Taylor coefficients of the right hand side of one-dimensional differential equations which guarantee the existence of such bifurcations. These conditions, however, are of a quite different form than the results obtained in Chapter 6 (cf. also the introduction of Chapter 6). Note that in CRAUEL & IMKELLER & STEINKAMP [59], also one-dimensional bifurcation patterns are studied, but in the context of random dynamical systems. Necessary and sufficient conditions are obtained for stochastic pitchfork and transcritical bifurcations.

# 3

# Nonautonomous Morse Decompositions

The global asymptotic behavior of dynamical systems on compact metric spaces can be described via *Morse decompositions*. Their components, the so-called *Morse sets*, are obtained as intersections of attractors and repellers. In this chapter, nonautonomous generalizations of the Morse decomposition are established with respect to the notions of past and future attractivity and repulsivity. The dynamical properties of these decompositions are discussed, and nonautonomous Lyapunov functions which are constant on the Morse sets are constructed explicitly. Furthermore, Morse decompositions of one-dimensional and linear systems are analyzed.

For a discussion of elementary properties of Morse decompositions, we refer to the original work of CONLEY [53] and to RYBAKOWSKI [149, Chapter 3] (see also COLONIUS & KLIEMANN [50, Appendix B2], ROBINSON [146], AKIN [3] and SCHMIDT [162]). Recently, OCHS [123] used the notion of weak attractor to construct Morse decompositions for random dynamical systems (see also CRAUEL & DUC & SIEGMUND [57]).

In this chapter, we suppose that $(\theta : \mathbb{T} \times P \to P, \varphi : \mathbb{T} \times P \times X \to X)$ is an invertible nonautonomous dynamical system with an arbitrary base set $P$ and a compact metric space $(X, d)$. Note that invertibility implies that $\mathbb{T} = \mathbb{R}$ or $\mathbb{T} = \mathbb{Z}$. Moreover, we assume that all (past or future) attractors and repellers $M$ under consideration fulfill either $P^*(M) = P$ or $M = \emptyset$.

Since Morse decompositions for the future are obtained via time reversal from Morse decompositions for the past, only the results concerning past Morse decompositions are proved in this chapter.

## 3.1 Attractor-Repeller Pairs

In this section, it is analyzed if for a given nonautonomous attractor, there exists a corresponding nonautonomous repeller and vice versa.

Due to the Axiom of Choice, there exists a set $P^* \subset P$ such that $[p] \cap P^*$ is a singleton for all $p \in P$. We write $P^* = P_p^* \cup P_n^*$ with $P_p^*$ containing all periodic points in $P^*$, i.e., $p^* \in P_p^*$ if and only if there exists a $\tau \in \mathbb{T}^+$ with $p^* = \theta_\tau p^*$, and $P_n^* := P^* \setminus P_p^*$.

Let $R$ be a past repeller, i.e., there exists an $\eta > 0$ such that for all $p^* \in P^*$, there exists a $t^*(p^*) > 0$ with

$$\lim_{t \to \infty} d\big(\varphi(-t, \theta_{-\tau}p^*)U_\eta(R(\theta_{-\tau}p^*))\big|R(\theta_{-\tau-t}p^*)\big) = 0 \quad \text{for all } \tau \ge t^*(p^*).$$
(3.1)

For $\zeta \in (0, \eta]$, we define the compact nonautonomous set $B_\zeta$ by

$$B_\zeta(\theta_{-t}p^*) := \begin{cases} X \setminus U_\zeta(R(\theta_{-t}p^*)) & : t \ge t^*(p^*) \\ X & : t < t^*(p^*) \end{cases} \quad \text{for all } p^* \in P_n^* \text{ and } t \in \mathbb{T}$$

and

$$B_\zeta(\theta_{-t}p^*) := X \setminus U_\zeta(R(\theta_{-t}p^*)) \quad \text{for all } p^* \in P_p^* \text{ and } t \in \mathbb{T}.$$

**Theorem 3.1 (Existence of a past attractor-repeller pair).** *Let $R$ be a past repeller, and set $\mathcal{M} := \{B_\zeta : \zeta \in (0, \eta]\}$ with $B_\zeta$ defined as above. Then there exists a uniquely determined $\mathcal{M}$-past attractor $R^* \subset B_\eta$, which is also a past attractor. Furthermore, $R^*$ is the maximal past attractor outside of $R$ in the following sense: Any other past attractor $A \supsetneq R^*$ has nonempty intersection with $R$. We call $(R^*, R)$ a past attractor-repeller pair.*

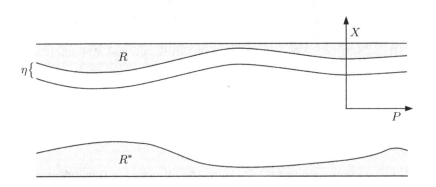

**Fig. 3.1.** Past attractor-repeller pair

*Proof.* We show that the hypotheses of Theorem 2.35 (i) are fulfilled by setting $B := B_\eta$. Thereto, let $\zeta \in (0, \eta]$ and $p \in P$. In case $B(p) = X$, the condition (2.5) certainly holds, otherwise, there exist $p^* \in P^*$ and $\tau \ge t^*(p^*)$ with $\theta_{-\tau}p^* = p$. Due to (3.1), there exists a $\hat{t} \ge 0$ with

$$d\big(\varphi(-t,p)U_\eta(R(p))\big|R(\theta_{-t}p)\big) < \frac{\zeta}{2} \quad \text{for all } t \geq \hat{t}.$$

This means that $\varphi(-t,p)U_\eta(R(p)) \subset U_{\zeta/2}(R(\theta_{-t}p))$ for all $t \geq \hat{t}$. Thus, we have

$$\varphi(-t,p)B_\eta(p) = X \setminus \varphi(-t,p)U_\eta(R(p)) \supset B_\zeta(\theta_{-t}p) \quad \text{for all } t \geq \hat{t}.$$

This implies the desired relation $\varphi(t,\theta_{-t}p)B_\zeta(\theta_{-t}p) \subset B(p)$ for all $t \geq \hat{t}$. Therefore, Theorem 2.35 (i) guarantees the existence of an $\mathcal{M}$-past attractor $R^* \subset B_\eta$. Due to Corollary 2.36, $R^*$ is also a past attractor. Let $A \supsetneq R^*$ be another past attractor. Then there exists a $p \in P$ with $A(p) \supsetneq R^*(p)$. We choose an $x \in A(p) \setminus R^*(p)$. Since $A \ni (p,x)$ is a past attractor, there exists an $\tilde{\eta} > 0$ with

$$\lim_{t\to\infty} d\big(\varphi(t,\theta_{-t}p)U_{\tilde{\eta}}(\varphi(-t,p)x)\big|A(p)\big) = 0.$$

Due to $\lim_{t\to\infty} d\big(\varphi(-t,p)x, R(\theta_{-t}p)\big) = 0$ (we will see this in Theorem 3.5 (ii)), there exists a sequence $\{y_n\}_{n\in\mathbb{N}}$ in $R(p)$ with

$$\lim_{n\to\infty} d\big(y_n, A(p)\big) = 0.$$

Since $R(p)$ and $A(p)$ are compact, this implies that their intersection is non-empty.                                                                 $\square$

Based on Proposition 2.32, the construction of a future attractor-repeller pair is not difficult.

**Corollary 3.2 (Existence of a future attractor-repeller pair).** *Let $A$ be a future attractor. Then there exists a uniquely determined future repeller $A^* \subset (P \times X) \setminus A$, which is the maximal future repeller outside of $A$ in the following sense: Any future repeller $R \supsetneq A^*$ has nonempty intersection with $A$. We call $(A, A^*)$ a future attractor-repeller pair.*

*Proof.* Because of Proposition 2.32, $A$ is a past repeller of $(\theta, \varphi)^{-1}$. Due to Theorem 3.1, there exists a corresponding past attractor $A^*$ of $(\theta, \varphi)^{-1}$. Then $A^*$ is a future repeller of $(\theta, \varphi)$. The property that $A^*$ is the maximal future repeller outside of $A$ follows easily from Theorem 3.1.                                              $\square$

It is natural to ask if a past attractor implies the existence of a past repeller and, equivalently, if a future repeller implies the existence of a future attractor. The following example shows that this does not follow.

*Example 3.3.* The nonautonomous differential equation

$$\dot{x} = f(t,x) \tag{3.2}$$

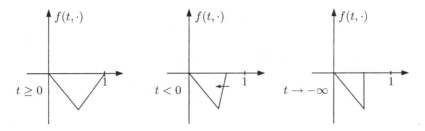

**Fig. 3.2.** The right hand side of (3.2)

with the function $f : \mathbb{R} \times [0, 1] \to \mathbb{R}$, defined by

$$
f(t, x) := \begin{cases}
|2x - 1| - 1 & : \quad t \geq 0 \text{ and } x \in [0, 1] \\
|2x - 1| - 1 & : \quad t < 0 \text{ and } x \in \left[0, \frac{1}{2}\right] \\
(2 - 2t)\left(x - \frac{1}{2}\right) - 1 & : \quad t < 0 \text{ and } x \in \left[\frac{1}{2}, \frac{2-t}{2-2t}\right] \\
0 & : \quad t < 0 \text{ and } x \in \left(\frac{2-t}{2-2t}, 1\right]
\end{cases},
$$

generates a nonautonomous dynamical system with $P = \mathbb{R}$ and $X = [0, 1]$. The invariant nonautonomous set $A := \mathbb{R} \times \{0\}$ is a past (as well as a future and an all-time) attractor. Assume, there exists a past repeller $A^* \subset \mathbb{R} \times (0, 1]$. Due to the invariance of $A^*$, the form of the right hand side implies that there exist $\gamma > \frac{1}{2}$ and $\tau_1 < 0$ with

$$
A^*(t) \subset (\gamma, 1] \quad \text{for all } t \leq \tau_1 .
$$

Thus, there exists a $\tau_2 < \tau_1$ with

$$
A^*(s) = A^*(t) \quad \text{and} \quad f(t, [\gamma, 1]) = \{0\} \quad \text{for all } t, s \leq \tau_2 .
$$

This contradicts the fact that $A^*$ is a past repeller.

*Remark 3.4.* This example shows that there is no possibility to construct an *all-time attractor-repeller pair*: The past attractor $A$ is also an all-time attractor, and no corresponding all-time repeller exists, since this would be also a past repeller. Furthermore, $A$ is an all-time repeller for the system under time reversal (see Proposition 2.32), and there is no corresponding all-time attractor, since this would be an all-time repeller for the original system.

Now, some properties of nonautonomous attractor-repeller pairs are derived.

**Theorem 3.5 (Properties of nonautonomous attractor-repeller pairs).** *Let $(R^*, R)$ be a past attractor-repeller pair. Then the following statements are fulfilled:*

(i)  Past isolation. *There exists a $\beta > 0$ such that for all $p \in P$, there exists a $\tau > 0$ with*

$$U_\beta(R^*(\theta_{-t}p)) \cap U_\beta(R(\theta_{-t}p)) = \emptyset \quad \text{for all } t \geq \tau.$$

(ii)  Backward convergence. *Let $p \in P$ and $C \subset X \setminus R^*(p)$ be a compact set. Then we have*

$$\lim_{t \to \infty} d\big(\varphi(-t,p)C \big| R(\theta_{-t}p)\big) = 0.$$

(iii)  Pullback convergence. *For all $p \in P$ and all functions $\gamma : \mathbb{T}^+ \to X$ with*

$$\liminf_{t \to \infty} d\big(\gamma(t), R(\theta_{-t}p)\big) > 0,$$

*we have*

$$\lim_{t \to \infty} d\big(\varphi(t, \theta_{-t}p)\gamma(t), R^*(p)\big) = 0.$$

*Let $(A, A^*)$ be a future attractor-repeller pair. Then the following statements are fulfilled:*

(i)  Future isolation. *There exists a $\beta > 0$ such that for all $p \in P$, there exists a $\tau > 0$ with*

$$U_\beta(A(\theta_t p)) \cap U_\beta(A^*(\theta_t p)) = \emptyset \quad \text{for all } t \geq \tau.$$

(ii)  Forward convergence. *Let $p \in P$ and $C \subset X \setminus A^*(p)$ be a compact set. Then we have*

$$\lim_{t \to \infty} d\big(\varphi(t,p)C \big| A(\theta_t p)\big) = 0.$$

(iii)  Pushforward convergence. *For all $p \in P$ and all functions $\gamma : \mathbb{T}^+ \to X$ with*

$$\liminf_{t \to \infty} d\big(\gamma(t), A(\theta_t p)\big) > 0,$$

*we have*

$$\lim_{t \to \infty} d\big(\varphi(-t, \theta_t p)\gamma(t), A^*(p)\big) = 0.$$

*Proof.* Let $(R^*, R)$ be a past attractor-repeller pair with $\eta$ and $\mathcal{M}$ defined as in the introduction of this section.

(i) Theorem 3.1 implies $R^* \subset B_\eta$. The assertion follows by choosing $\beta := \eta/2$.

(ii) Let $p \in P$ and $C \subset X \setminus R^*(p)$ be a compact set. Since $R^*$ is an $\mathcal{M}$-past attractor and thus a $\{B_\eta\}$-past attractor, there exists a $\tau > 0$ such that

$$C \cap \varphi(t, \theta_{-t}p)B_\eta(\theta_{-t}p) = \emptyset \quad \text{for all } t \geq \tau.$$

Hence, for all $t \geq \tau$, the relation $\varphi(-t,p)C \cap B_\eta(\theta_{-t}p) = \emptyset$ is fulfilled, and therefore, we have $\lim_{t \to \infty} d\big(\varphi(-t,p)C, R(\theta_{-t}p)\big) = 0$.

(iii) We set $\zeta := \frac{1}{2} \min \big\{\eta, \liminf_{t \to \infty} d(\gamma(t), R(\theta_{-t}p))\big\}$ and see that there exists a $\tau > 0$ with

$$\gamma(t) \in B_\zeta(\theta_{-t}p) \quad \text{for all } t \geq \tau.$$

This finishes the proof, since $B_\zeta \in \mathcal{M}$ and $R^*$ is a $\mathcal{M}$-past attractor. □

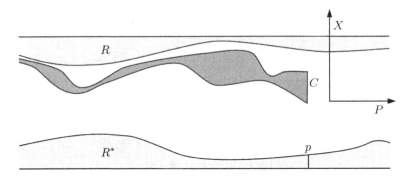

**Fig. 3.3.** Backward convergence

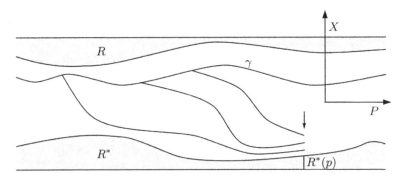

**Fig. 3.4.** Pullback convergence

Theorem 3.1 implies that, given a past repeller $R$, the set $R^*$ is the uniquely determined past attractor outside of $R$ with the property of pullback convergence as described in Theorem 3.5 (iii). It is easy to see that such a uniqueness result is not valid for past repellers, i.e., it is possible that $(A, R_1)$ and $(A, R_2)$ are past attractor-repeller pairs with $R_1 \neq R_2$. The following proposition says that in this case, $R_1$ and $R_2$ are converging to each other when time tends to the past.

**Proposition 3.6 (Form of nonuniqueness of nonautonomous attractor-repeller pairs).** *Let $R_1$ and $R_2$ be past repellers with $R_1^* = R_2^*$. Then we have*

$$\lim_{t \to \infty} d_H\big(R_1(\theta_{-t}p), R_2(\theta_{-t}p)\big) = 0 \quad \text{for all } p \in P.$$

*Let $A_1$ and $A_2$ be future attractors with $A_1^* = A_2^*$. Then we have*

$$\lim_{t \to \infty} d_H\big(A_1(\theta_t p), A_2(\theta_t p)\big) = 0 \quad \text{for all } p \in P.$$

*Proof.* Suppose, there exist a $p \in P$ and sequences $\{t_n\}_{n \in \mathbb{N}}$ in $\mathbb{T}$ and $\{\gamma_n\}_{n \in \mathbb{N}}$ in $X$ with $\lim_{n \to \infty} t_n = \infty$ and $\gamma_n \in R_1(\theta_{-t_n} p)$ such that

$$\liminf_{n \to \infty} d(\gamma_n, R_2(\theta_{-t_n} p)) > 0.$$

Hence, Theorem 3.5 (iii), applied to the attractor-repeller pair $(R_2^*, R_2)$, implies that

$$\lim_{n \to \infty} d(\varphi(t_n, \theta_{-t_n} p)\gamma_n, R_2^*(p)) = 0.$$

Since $\varphi(t_n, \theta_{-t_n} p)\gamma_n \in R_1(p)$ and $R_1$ and $R_1^* = R_2^*$ are compact nonautonomous sets, we obtain $R_1(p) \cap R_1^*(p) \neq \emptyset$. This is a contradiction. $\qquad \square$

## 3.2 Morse Decompositions

In this section, the notion of the attractor-repeller pair is generalized by considering Morse decompositions.

**Definition 3.7 (Nonautonomous Morse decompositions).** *A family* $\{M_1, M_2, \ldots, M_n\}$ *of nonautonomous sets, the so-called* Morse sets, *is called* past Morse decomposition *if the representation*

$$M_i = R_i^* \cap R_{i-1} \quad \text{for all } i \in \{1, \ldots, n\}$$

*holds with past repellers*

$$P \times X = R_0 \supsetneq R_1 \supsetneq \cdots \supsetneq R_n = \emptyset$$

*fulfilling* $\emptyset = R_0^* \subsetneq R_1^* \subsetneq \cdots \subsetneq R_n^* = P \times X$.
*A family* $\{M_1, M_2, \ldots, M_n\}$ *of nonautonomous sets, the so-called* Morse sets, *is called* future Morse decomposition *if the representation*

$$M_i = A_i \cap A_{i-1}^* \quad \text{for all } i \in \{1, \ldots, n\}$$

*holds with future attractors*

$$\emptyset = A_0 \subsetneq A_1 \subsetneq \cdots \subsetneq A_n = P \times X$$

*fulfilling* $P \times X = A_0^* \supsetneq A_1^* \supsetneq \cdots \supsetneq A_n^* = \emptyset$.

*Remark 3.8.* Let $(A, R)$ be a past (future, respectively) attractor-repeller pair such that the relation $\emptyset \subsetneq A \subsetneq P \times X$ is fulfilled. Then $\{A, R\}$ is a past (future, respectively) Morse decomposition.

**Proposition 3.9 (Basic    properties    of    nonautonomous    Morse decompositions).**    *The Morse sets of a past Morse decomposition* $\{M_1, \ldots, M_n\}$ *are nonempty, invariant, pairwise disjoint and past isolated, i.e., there exists a* $\beta > 0$ *such that for all* $1 \leq i < j \leq n$ *and* $p \in P$, *there exists a* $\tau > 0$ *with*

$$U_\beta(M_i(\theta_{-t}p)) \cap U_\beta(M_j(\theta_{-t}p)) = \emptyset \quad \text{for all } t \geq \tau.$$

*The Morse sets of a future Morse decomposition* $\{M_1, \ldots, M_n\}$ *are nonempty, invariant, pairwise disjoint and future isolated, i.e., there exists a* $\beta > 0$ *such that for all* $1 \leq i < j \leq n$ *and* $p \in P$, *there exists a* $\tau > 0$ *with*

$$U_\beta(M_i(\theta_t p)) \cap U_\beta(M_j(\theta_t p)) = \emptyset \quad \text{for all } t \geq \tau.$$

*Proof.* Let $M_i = R_i^* \cap R_{i-1}$ be a Morse set. Since $R_{i-1}^* \subsetneqq R_i^*$, we can choose a $p \in P$ and an $x \in R_i^*(p) \setminus R_{i-1}^*(p)$. Since $R_i^* \ni (p, x)$ is a past attractor, there exists an $\eta > 0$ with

$$\lim_{t \to \infty} d\big(\varphi(t, \theta_{-t}p)U_\eta(\varphi(-t, p)x)\big| R_i^*(p)\big) = 0 \,.$$

Due to $\lim_{t \to \infty} d\big(\varphi(-t, p)x, R_{i-1}(\theta_{-t}p)\big) = 0$ (cf. Theorem 3.5 (ii)), this means that there exists a sequence $\{y_n\}_{n \in \mathbb{N}}$ in $R_{i-1}(p)$ with

$$\lim_{n \to \infty} d\big(y_n, R_i^*(p)\big) = 0 \,.$$

Since $R_{i-1}(p)$ and $R_i^*(p)$ are compact, this implies $M_i = R_i^* \cap R_{i-1} \neq \emptyset$. Furthermore, $M_i$ is the intersection of two invariant nonautonomous sets and thus invariant. Choose another Morse set $M_j = R_j^* \cap R_{j-1}$. W.l.o.g, we assume $j > i$. Then we get

$$M_i \cap M_j = R_i^* \cap R_{i-1} \cap R_j^* \cap R_{j-1} = R_{i-1}^* \cap R_{j-1} \subset R_{j-1}^* \cap R_{j-1} = \emptyset \,.$$

The fact that the Morse sets are past isolated is an easy consequence of Theorem 3.5 (i).    □

As in the autonomous case, nonautonomous Morse decompositions are not uniquely determined.

**Definition 3.10.** *We say, the past Morse decomposition* $\{M_1, \ldots, M_n\}$ *is finer than the past Morse decomposition* $\{\tilde{M}_1, \ldots, \tilde{M}_m\}$ *if*

$$\lim_{t \to \infty} d\left( \bigcup_{i=1}^n M_i(\theta_{-t}p) \,\middle|\, \bigcup_{i=1}^m \tilde{M}_i(\theta_{-t}p) \right) = 0 \quad \text{for all } p \in P$$

*is fulfilled.*
*We say, the future Morse decomposition* $\{M_1, \ldots, M_n\}$ *is finer than the future Morse decomposition* $\{\tilde{M}_1, \ldots, \tilde{M}_m\}$ *if*

$$\lim_{t \to \infty} d \left( \bigcup_{i=1}^{n} M_i(\theta_t p) \,\middle|\, \bigcup_{i=1}^{m} \tilde{M}_i(\theta_t p) \right) = 0 \quad \text{for all } p \in P$$

*is fulfilled.*

**Remark 3.11.**

(i) The above definition is a generalization of the notion of a finer (autonomous) Morse decomposition. In the autonomous case, a Morse decomposition $\{M_1, \ldots, M_n\}$ is called *finer* than the Morse decomposition $\{\tilde{M}_1, \ldots, \tilde{M}_m\}$ if for all $j \in \{1, \ldots, m\}$, there exists an $i \in \{1, \ldots, n\}$ such that $M_i \subset \tilde{M}_j$ (see, e.g., COLONIUS & KLIEMANN [50, p. 542]). It is easy to see that this is equivalent to

$$d \left( \bigcup_{i=1}^{n} M_i \,\middle|\, \bigcup_{i=1}^{m} \tilde{M}_i \right) = 0.$$

The additional limit in our nonautonomous context is motivated by Proposition 3.6.

(ii) There are different forms of nonuniqueness for the Morse sets. As seen in Proposition 3.6, two past attractor-repeller pairs are converging to each other in case the past attractors are equal. One can find examples to show that such a (weak) form of nonuniqueness is not valid for arbitrary Morse decompositions (i.e., those consisting of more than two sets). However, in the special cases of one-dimensional and linear systems (cf. Section 3.4 and 3.5), one obtains similar results as in Proposition 3.6 (cf. Proposition 3.18 and 3.23).

The following theorem shows that Morse sets are important for the asymptotic behavior of nonautonomous dynamical systems.

**Theorem 3.12 (Dynamical properties of nonautonomous Morse decompositions).** *Pullback convergence. Let $\{M_1, \ldots, M_n\}$ be a past Morse decomposition obtained by the finite sequence of past repellers $R_0 \supset \cdots \supset R_n$. Then, for all $p \in P$ and all functions $\gamma : \mathbb{T}^+ \to X$ with*

$$\liminf_{t \to \infty} d \left( \gamma(t), \bigcup_{j=1}^{n} \partial R_j(\theta_{-t} p) \right) > 0,$$

*we have*

$$\lim_{t \to \infty} d \left( \varphi(t, \theta_{-t} p) \gamma(t), \bigcup_{j=1}^{n} M_j(p) \right) = 0.$$

*Pushforward convergence. Let $\{M_1, \ldots, M_n\}$ be a future Morse decomposition obtained by the finite sequence of future attractors $A_0 \subset \cdots \subset A_n$. Then, for all $p \in P$ and all functions $\gamma : \mathbb{T}^+ \to X$ with*

$$\liminf_{t\to\infty} d\left(\gamma(t), \bigcup_{j=1}^{n} \partial A_j(\theta_t p)\right) > 0,$$

*we have*

$$\lim_{t\to\infty} d\left(\varphi(-t, \theta_t p)\gamma(t), \bigcup_{j=1}^{n} M_j(p)\right) = 0.$$

*Proof.* We assume w.l.o.g. that there exists an $i \in \{1, \ldots, n\}$ with

$$\gamma(t) \in R_{i-1}(\theta_{-t}p) \quad \text{and} \quad \gamma(t) \notin R_i(\theta_{-t}p) \quad \text{for all } t > 0.$$

Then $\liminf_{t\to\infty} d(\gamma(t), \partial R_i(\theta_{-t}p)) > 0$ yields

$$\liminf_{t\to\infty} d(\gamma(t), R_i(\theta_{-t}p)) > 0.$$

Therefore, Theorem 3.5 implies that

$$\lim_{t\to\infty} d(\varphi(t, \theta_{-t}p)\gamma(t), R_i^*(p)) = 0. \tag{3.3}$$

Assume, there exist $\varepsilon > 0$ and a sequence $\{t_n\}_{n\in\mathbb{N}}$ in $\mathbb{T}^+$ with $\lim_{n\to\infty} t_n = \infty$ and

$$d(\varphi(t_n, \theta_{-t_n}p)\gamma(t_n), M_i(p)) \geq \varepsilon \quad \text{for all } n \in \mathbb{N}. \tag{3.4}$$

W.l.o.g., the sequence $\{\varphi(t_n, \theta_{-t_n}p)\gamma(t_n)\}_{n\in\mathbb{N}}$ in $R_{i-1}(p)$ is convergent with limit $x_0 \in R_{i-1}(p)$ ($R_{i-1}(p)$ is compact). Moreover, $x_0 \in R_i^*(p)$, since (3.3) holds and $R_i^*(p)$ is compact. Thus, $x_0 \in M_i(p) = R_i^*(p) \cap R_{i-1}(p)$. This contradicts (3.4) and finishes the proof of this theorem. □

*Remark 3.13.* In contrast to attractor-repeller pairs, backward and forward convergence conditions as described in Theorem 3.5 do not hold for arbitrary Morse decompositions. However, in the special cases of one-dimensional and linear systems (cf. Section 3.4 and 3.5), one obtains similar results as in Theorem 3.5 (cf. Theorem 3.17 and 3.22).

If the backward (in case of a past Morse decomposition) or forward (in case of a future Morse decomposition) convergence holds, the following uniqueness result concerning the past attractors or future repellers, respectively, is fulfilled.

**Proposition 3.14.** *Let* $\{M_1, \ldots, M_n\}$ *be a past Morse decomposition obtained by the finite sequence of past repellers* $R_0 \supset \cdots \supset R_n$. *We assume that the backward convergence holds, i.e., for all* $(p, x) \in P \times X$, *there exists an* $i \in \{1, \ldots, n\}$ *with*

$$\lim_{t\to\infty} d(\varphi(-t, p)x, M_i(\theta_{-t}p)) = 0.$$

*Then the representation*

$$R_i^* = \left\{ (p,x) \in P \times X : \lim_{t \to \infty} d\left( \varphi(-t,p)x, \bigcup_{j=1}^{i} M_j(\theta_{-t}p) \right) = 0 \right\}$$

*holds for all* $i \in \{1, \ldots, n\}$, *i.e., the past attractors of the past Morse decomposition are uniquely determined.*
*Let* $\{M_1, \ldots, M_n\}$ *be a future Morse decomposition obtained by the finite sequence of future attractors* $A_0 \subset \cdots \subset A_n$. *We assume that the forward convergence holds, i.e., for all* $(p,x) \in P \times X$, *there exists an* $i \in \{1, \ldots, n\}$ *with*

$$\lim_{t \to \infty} d\big(\varphi(t,p)x, M_i(\theta_t p)\big) = 0 \,.$$

*Then the representation*

$$A_i^* = \left\{ (p,x) \in P \times X : \lim_{t \to \infty} d\left( \varphi(t,p)x, \bigcup_{j=i+1}^{n} M_j(\theta_t p) \right) = 0 \right\}$$

*holds for all* $i \in \{1, \ldots, n\}$, *i.e., the future repellers of the future Morse decomposition are uniquely determined.*

*Proof.* ($\subseteq$) Let $(p,x) \in R_i^*$. We choose $j \in \{1, \ldots, n\}$ such that

$$0 = \lim_{t \to \infty} d\big(\varphi(-t,p)x, M_j(\theta_{-t}p)\big) = \lim_{t \to \infty} d\big(\varphi(-t,p)x, R_{j-1}(\theta_{-t}p)\big) \,.$$

The assumption $j > i$ leads to

$$\lim_{t \to \infty} d\big(\varphi(-t,p)x, R_i(\theta_{-t}p)\big) = 0 \,.$$

This contradicts Theorem 3.5 (i), since $\varphi(-t,p)x \in R_i^*(\theta_{-t}p)$ for all $t \in \mathbb{T}$.
($\supseteq$) Let $(p,x) \in (P \times X) \setminus R_i^*$. Then Theorem 3.5 (ii) implies

$$\lim_{t \to \infty} d\big(\varphi(-t,p)x, R_i(\theta_{-t}p)\big) = 0 \,. \tag{3.5}$$

The assumption

$$\lim_{t \to \infty} d\left( \varphi(-t,p)x, \bigcup_{j=1}^{i} M_j(\theta_{-t}p) \right) = 0$$

leads to

$$\lim_{t \to \infty} d\big(\varphi(-t,p)x, R_i^*(\theta_{-t}p)\big) = 0 \,,$$

since $M_j \subset R_i^*$ for $j \in \{1, \ldots, i\}$. Because of Theorem 3.5 (i), this is a contradiction to (3.5). $\qquad\square$

## 3.3 Lyapunov Functions

In this section, nonautonomous Lyapunov functions which are constant on the Morse sets and which strictly decrease outside them are obtained explicitly. A similar construction is used in CONLEY [53, §5 and §6 of Chapter II] (see also FRANKS [69, §1], ROBINSON [146, Chapter X] and NORTON [122]), and this technique has also been adapted in KLOEDEN [94, 95] and ARNOLD & SCHMALFUSS [7] in the nonautonomous setting.

First, the case that the nonautonomous Morse decomposition is given by a nonautonomous attractor-repeller pair is treated.

**Lemma 3.15.** *Let* $(A, R)$ *be a past (future, respectively) attractor-repeller pair. Then there exists a function* $L : P \times X \to [0, 1]$ *which is continuous with respect to* $x \in X$ *such that* $L|_A \equiv 0$, $L|_R \equiv 1$ *and*

$$L\big(\theta_t p, \varphi(t, p)x\big) < L(p, x) \quad \text{for all } t > 0 \text{ and } (p, x) \in (P \times X) \setminus (A \cup R)$$

*is satisfied.*

*Proof.* In case of a *past* attractor-repeller pair, $R^* = A$ is fulfilled, and we define the function $V : P \times X \to [0, 1]$ by

$$V(p, x) := \frac{d(x, R^*(p))}{d(x, R^*(p)) + d(x, R(p))} \quad \text{for all } (p, x) \in P \times X .$$

This function is continuous with respect to $x \in X$ and fulfills $V|_{R^*} \equiv 0$, $V|_R \equiv 1$, but is not necessarily decreasing along solutions. Therefore, we define by

$$V^*(p, x) := \inf_{s \geq 0} V\big(\theta_{-s} p, \varphi(-s, p)x\big) \quad \text{for all } (p, x) \in P \times X$$

a function $V^* : P \times X \to [0, 1]$, which obviously satisfies $V^*|_{R^*} \equiv 0$, $V^*|_R \equiv 1$ and $V^*\big(\theta_t p, \varphi(t, p)x\big) \leq V^*(p, x)$ for all $t \geq 0$ and $(p, x) \in P \times X$. To prove that $V^*(p, \cdot)$ is continuous for all $p \in P$, we first choose $\xi \in X \setminus R^*(p)$ and $\varepsilon > 0$. Then there exists a $\hat{\delta} > 0$ such that $C := \operatorname{cls} U_{\hat{\delta}}(\xi) \subset X \setminus R^*(p)$. It follows that

$$\lim_{s \to \infty} \inf_{x \in C} V\big(\theta_{-s} p, \varphi(-s, p)x\big) = 1 \,,$$

since $\lim_{s \to \infty} d\big(\varphi(-s, p)C, R(\theta_{-s} p)\big) = 0$ (cf. Theorem 3.5 (ii)) and there exist a $\beta > 0$ and an $\hat{s} > 0$ such that $d\big(\varphi(-s, p)C, R^*(\theta_{-s} p)\big) \geq \beta/2$ for all $s \geq \hat{s}$ (cf. Theorem 3.5 (i), (ii)). Thus, there exists an $s_0 > 0$ such that

$$\inf_{x \in C} V\big(\theta_{-s} p, \varphi(-s, p)x\big) > 1 - \varepsilon \quad \text{for all } s \geq s_0 \,.$$

Due to the continuity of $V\big(\theta(\cdot, p), \varphi(\cdot, p, \cdot)\big) : \mathbb{T} \times X \to \mathbb{R}$, there exists a $\delta \in (0, \hat{\delta})$ such that

$$\left| V\big(\theta_{-s}p, \varphi(-s,p)\xi\big) - V\big(\theta_{-s}p, \varphi(-s,p)x\big) \right| < \varepsilon$$

for all $x \in U_\delta(\xi)$ and $0 \le s \le s_0$. This implies that $V^*(p, \cdot)$ is continuous in $\xi \notin R^*(p)$. The continuity of $V^*(p, \cdot)$ in $\xi \in R^*(p)$ follows directly from the continuity of $V$. Please note that $V$ is not strictly decreasing along solutions in $(P \times X) \setminus (R^* \cup R)$. Therefore, we define $L$ to be a weighted average of $V^*$ over the backward solution:

$$L(p,x) := \int_0^\infty e^{-s} V^*\big(\theta_{-s}p, \varphi(-s,p)x\big) \, ds \quad \text{for all } (p,x) \in P \times X.$$

This function is obviously continuous with respect to $x \in X$, and we have

$$\begin{aligned} L\big(\theta_t p, \varphi(t,p,x)\big) &= \int_0^\infty e^{-s} V^*\big(\theta_{-s}\theta_t p, \varphi(-s,\theta_t p)\varphi(t,p)x\big) \, ds \\ &= \int_0^\infty e^{-s} V^*\big(\theta_t \theta_{-s}p, \varphi(t,\theta_{-s}p)\varphi(-s,p)x\big) \, ds \\ &\le \int_0^\infty e^{-s} V^*\big(\theta_{-s}p, \varphi(-s,p)x\big) \, ds = L(p,x). \end{aligned}$$

To prove that this function is also strictly decreasing along solutions in the set $(P \times X) \setminus (R^* \cup R)$, we assume that $L\big(\theta_t p, \varphi(t,p)x\big) = L(p,x)$ for some $t > 0$ and $x \in (P \times X) \setminus (R^* \cup R)$. Then we have

$$V^*\big(\theta_{-s}p, \varphi(-s,p)x\big) = V^*\big(\theta_{t-s}p, \varphi(t-s,p)x\big) \quad \text{for all } s \ge 0.$$

This is impossible, since we have both $\lim_{s\to\infty} V^*\big(\theta_{-s}p, \varphi(-s,p,x)\big) = 0$ and $V^*(p,x) \in (0,1)$. $\qquad\square$

In the following theorem, the above Lyapunov function for attractor-repeller pairs is extended to Morse decompositions.

**Theorem 3.16 (Lyapunov functions for nonautonomous Morse decompositions).** *Let $\{M_1, \ldots, M_n\}$ be a past (future, respectively) Morse decomposition. Then there exists a function $L : P \times X \to [0,1]$ which is continuous with respect to $x \in X$ such that $L|_{M_i} \equiv \frac{i-1}{n-1}$ for $i \in \{1, \ldots, n\}$ and*

$$L\big(\theta_t p, \varphi(t,p)x\big) < L(p,x) \quad \text{for all } t > 0 \text{ and } (p,x) \in (P \times X) \setminus \cup_{i=1}^n M_i$$

*is satisfied.*

*Proof.* Let $P \times X = R_0 \supsetneq R_1 \supsetneq \cdots \supsetneq R_n = \emptyset$ be the sequence of past repellers leading to the given past Morse decomposition, i.e.,

$$M_i = R_i^* \cap R_{i-1} \quad \text{for all } i \in \{1, \ldots, n\}.$$

Furthermore, let $L_i$, $i \in \{1, \ldots, n-1\}$, be the Lyapunov function corresponding to the past attractor-repeller pair $(R_i^*, R_i)$ as introduced in Lemma 3.15. We define

$$L(p, x) := \frac{1}{n-1} \sum_{i=1}^{n-1} L_i(p, x) \quad \text{for all } (p, x) \in P \times X.$$

Choose $(p, x) \in M_i$ arbitrarily, and let $j \in \{1, \ldots, n-1\}$. Then $(p, x) \in R_j$ if and only if $j \in \{1, \ldots, i-1\}$, and $(p, x) \in R_j^*$ if and only if $j \in \{i, \ldots, n-1\}$. This implies $L|_{M_i} \equiv \frac{i-1}{n-1}$ for all $i \in \{1, \ldots, n\}$. Now choose $(p, x) \in (P \times X) \setminus (M_1 \cup \cdots \cup M_n)$. Then there exists a $j \in \{1, \ldots, n\}$ with $(p, x) \notin R_j^* \cup R_j$. This means that $L_j(\theta_t p, \varphi(t, p)x) < L_j(p, x)$ for all $t > 0$ and finishes the proof of this theorem. $\qquad \square$

## 3.4 Morse Decompositions in Dimension One

In this section, Morse decompositions of nonautonomous dynamical systems whose phase space is a compact interval are studied. In this special case, stronger results concerning the convergence behavior of the system and the nonuniqueness of the Morse sets are obtained.

Let $\mathbb{I} \subset \mathbb{R}$ be a compact interval and $(\theta : \mathbb{T} \times P \to P, \varphi : \mathbb{T} \times P \times \mathbb{I} \to \mathbb{I})$ be a nonautonomous dynamical system.

**Theorem 3.17 (Dynamical properties of nonautonomous Morse decompositions in dimension one).** *Let $\{M_1, \ldots, M_n\}$ be a past Morse decomposition obtained by the finite sequence of past repellers $R_0 \supset \cdots \supset R_n$. Then the following statements are fulfilled:*

(i) *Pullback convergence. For all $p \in P$ and all functions $\gamma : \mathbb{T}^+ \to \mathbb{I}$ with*

$$\liminf_{t \to \infty} d\left(\gamma(t), \bigcup_{j=1}^{n} \partial R_j(\theta_{-t} p)\right) > 0,$$

*we have*

$$\lim_{t \to \infty} d\left(\varphi(t, \theta_{-t} p)\gamma(t), \bigcup_{j=1}^{n} M_j(p)\right) = 0.$$

(ii) *Backward convergence. For all $(p, x) \in P \times \mathbb{I}$, there exists an $i \in \{1, \ldots, n\}$ with*

$$\lim_{t \to \infty} d\left(\varphi(-t, p)x, M_i(\theta_{-t} p)\right) = 0.$$

*Let $\{M_1, \ldots, M_n\}$ be a future Morse decomposition obtained by the finite sequence of future attractors $A_0 \subset \cdots \subset A_n$. Then the following statements are fulfilled:*

(i) *Forward convergence. For all* $(p, x) \in P \times \mathbb{I}$, *there exists an* $i \in \{1, \ldots, n\}$ *with*

$$\lim_{t \to \infty} d\big(\varphi(t, p)x, M_i(\theta_t p)\big) = 0.$$

(ii) *Pushforward convergence. For all* $p \in P$ *and all functions* $\gamma : \mathbb{T}^+ \to \mathbb{I}$ *with*

$$\liminf_{t \to \infty} d\left(\gamma(t), \bigcup_{j=1}^{n} \partial A_j(\theta_t p)\right) > 0,$$

*we have*

$$\lim_{t \to \infty} d\left(\varphi(-t, \theta_t p)\gamma(t), \bigcup_{j=1}^{n} M_j(p)\right) = 0.$$

*Proof.* (i) This assertion is also valid for general Morse decompositions and was proved in Theorem 3.12.

(ii) Choose $(p, x) \in P \times \mathbb{I}$ arbitrarily. Then there exists an $i \in \{1, \ldots, n\}$ such that

$$x \in R_i^*(p) \quad \text{and} \quad x \notin R_{i-1}^*(p).$$

In case $x \in R_{i-1}(p)$, the asserted limit relation follows, since then $x \in M_i(p)$ and $M_i$ is invariant. We therefore assume $x \notin R_{i-1}(p)$ from now on. Due to the topology of $\mathbb{I}$, $\varphi$ is *order preserving* in the following sense: For fixed $t \in \mathbb{T}$, exactly one of the following two statements is fulfilled:

- $y_1 < y_2$ implies $\varphi(t, p)y_1 < \varphi(t, p)y_2$,

- $y_1 < y_2$ implies $\varphi(t, p)y_1 > \varphi(t, p)y_2$.

Since $\lim_{t \to \infty} d\big(\varphi(-t, p)x, R_{i-1}(\theta_{-t}p)\big) = 0$ (cf. Theorem 3.5 (ii)), this implies that there exists a $y \in R_{i-1}(p)$ such that

$$\lim_{t \to \infty} \big|\varphi(-t, p)x - \varphi(-t, p)y\big| = 0. \tag{3.6}$$

Because $R_i^*$ is a past attractor, there exists an $\eta > 0$ such that

$$R_i^*(p) = \limsup_{t \to \infty} \varphi(t, \theta_{-t}p)U_\eta(R_i^*(\theta_{-t}p))$$

(cf. Remark 2.7 (ii)). This implies $\limsup_{t \to \infty} \varphi(t, \theta_{-t}p)U_\eta(\varphi(-t, p)x) \subset R_i^*(p)$. Due to (3.6), this leads to $y \in R_i^*(p)$. Hence, $y \in M_i^*(p)$, and this finishes the proof of this theorem. $\square$

In our special situation, Proposition 3.6 can be generalized.

**Proposition 3.18 (Form of nonuniqueness of the Morse sets).** *Let* $\{M_1, \ldots, M_n\}$ *and* $\{\hat{M}_1, \ldots, \hat{M}_n\}$ *be past Morse decompositions obtained by the finite sequences of past repellers* $R_0 \supset \cdots \supset R_n$ *and* $\hat{R}_0 \supset \cdots \supset \hat{R}_n$. *We assume that*

$$R_i^* = \hat{R}_i^* \quad \text{for all } i \in \{1, \ldots, n-1\}.$$

*Then the relation*

$$\lim_{t \to \infty} d_H\big(M_i(\theta_{-t}p), \hat{M}_i(\theta_{-t}p)\big) = 0 \quad \text{for all } i \in \{1, \ldots, n\} \text{ and } p \in P$$

*is fulfilled.*
Let $\{M_1, \ldots, M_n\}$ *and* $\{\hat{M}_1, \ldots, \hat{M}_n\}$ *be future Morse decompositions obtained by the finite sequences of future attractors* $A_0 \subset \cdots \subset A_n$ *and* $\hat{A}_0 \subset \cdots \subset \hat{A}_n$. *We assume that*

$$A_i^* = \hat{A}_i^* \quad \text{for all } i \in \{1, \ldots, n-1\}.$$

*Then the relation*

$$\lim_{t \to \infty} d_H\big(M_i(\theta_t p), \hat{M}_i(\theta_t p)\big) = 0 \quad \text{for all } i \in \{1, \ldots, n\} \text{ and } p \in P$$

*is fulfilled.*

*Proof.* Choose $i \in \{1, \ldots, n\}$ and $p \in P$ arbitrarily. W.l.o.g., we only show the relation

$$\lim_{t \to \infty} d\big(M_i(\theta_{-t}p) \big| \hat{M}_i(\theta_{-t}p)\big) = 0.$$

The proof is divided into three steps.
*Step 1.* There exists a past repeller $\bar{R}_{i-1} \supset R_{i-1}$ with $\bar{R}_{i-1}^* = R_{i-1}^*$ such that

$$\bar{M}_i(\bar{p}) := R_i^*(\bar{p}) \cap \bar{R}_{i-1}(\bar{p}) \quad \text{for all } \bar{p} \in [p]$$

*has only finitely many connected components.*
Since $R_i^*$ is a past attractor, there exists an $\eta > 0$ such that

$$R_i^*(p) = \limsup_{t \to \infty} \varphi(t, \theta_{-t}p) U_\eta(R_i^*(\theta_{-t}p))$$

(cf. Remark 2.7 (ii)). Since $\varphi$ is continuous and $U_\eta(R_i^*(\theta_{-t}p))$ has only finitely many connected components for $t \in \mathbb{T}$, this implies that $R_i^*(p)$ has only finitely many connected components. Because $R_{i-1}$ is a past repeller, there exists a $\beta > 0$ such that

$$\lim_{t \to \infty} d\big(\varphi(-t, p) U_\beta(R_{i-1}(p)) \big| R_{i-1}(\theta_{-t}p)\big) = 0.$$

Hence, the nonautonomous set $\bar{R}_{i-1}$, defined by

$$\bar{R}_{i-1}(\bar{p}) := \begin{cases} \varphi(t, p) \operatorname{cls} U_{\beta/2}(R_{i-1}(p)) & : \quad \bar{p} = \theta_t p \quad \text{for some } t \in \mathbb{T} \\ R_{i-1}(\bar{p}) & : \quad \bar{p} \notin [p] \end{cases}$$

is also a past repeller fulfilling $\bar{R}_{i-1}^* = R_{i-1}^*$ (cf. Proposition 2.37 (iii)). Moreover, for all $\bar{p} \in [p]$, the set $\bar{R}_{i-1}(\bar{p})$ has only finitely many connected components, since $\varphi$ is continuous and $\operatorname{cls} U_{\beta/2}(R_{i-1}(p))$ has only finitely many connected components. This implies the assertion.

*Step 2. For all connected components $C$ of $\bar{M}_i(p)$, we have*

$$\lim_{t\to\infty} d\big(\varphi(-t,p)C\big|\hat{M}_i(\theta_{-t}p)\big) = 0.$$

Let $C = [c_1, c_2]$ be a connected component of $\bar{M}_i(p)$, and choose $\varepsilon > 0$ arbitrarily. Due to Theorem 3.17 (ii), there exists a $\tau_1 \geq 0$ such that we have

$$d\big(\varphi(-t,p)c_j, \hat{M}_i(\theta_{-t}p)\big) \leq \frac{\varepsilon}{2} \quad \text{for all } t \geq \tau_1 \text{ and } j \in \{1,2\}. \tag{3.7}$$

Furthermore, because of Proposition 3.6, there exists a $\tau_2 \geq \tau_1$ with

$$d_H\big(\bar{R}_{i-1}(\theta_{-t}p), \hat{R}_{i-1}(\theta_{-t}p)\big) \leq \frac{\varepsilon}{2} \quad \text{for all } t \geq \tau_2. \tag{3.8}$$

Let $t \geq \tau_2$ and $x \in \varphi(-t,p)C$. In case

$$\min\big\{|x - \varphi(-t,p)c_1|, |x - \varphi(-t,p)c_2|\big\} \leq \frac{\varepsilon}{2},$$

the inequality (3.7) implies that $d\big(x, \hat{M}_i(\theta_{-t}p)\big) \leq \varepsilon$. Otherwise, since we have $x \in \bar{R}_{i-1}(\theta_{-t}p)$ and due to (3.8), there exists a $y \in \hat{R}_{i-1}(\theta_{-t}p)$ with $|x - y| \leq \varepsilon/2$. Obviously,

$$y \in \varphi(-t,p)C \subset \varphi(-t,p)\bar{M}_i(p) \subset R_i^*(\theta_{-t}p)$$

is fulfilled. Hence, $y \in \hat{M}_i(\theta_{-t}p)$, and thus, $d\big(x, \hat{M}_i(\theta_{-t}p)\big) \leq \varepsilon$. This finishes the proof of this step.

*Step 3. The relation*

$$\lim_{t\to\infty} d\big(M_i(\theta_{-t}p)\big|\hat{M}_i(\theta_{-t}p)\big) = 0$$

*is fulfilled.*

Since $\bar{M}_i(p)$ has only finitely many connected components, this assertion follows from Step 2 and the fact that $\bar{M}_i \supset M_i$.    □


## 3.5 Morse Decompositions of Linear Systems

In this section, Morse decompositions of linear nonautonomous dynamical systems are analyzed. Under the assumption that the base space is chain recurrent, such (autonomous) Morse decompositions of the corresponding skew product flow have been studied in SELGRADE [164], SALAMON & ZEHNDER [155] and COLONIUS & KLIEMANN [50, Chapter 5] (see also COLONIUS & KLIEMANN [49, 51] and BRAGA BARROS & SAN MARTIN [34]).

Given $N \in \mathbb{N}$, and let $\left(\theta : \mathbb{T} \times P \to P, \; \varphi : \mathbb{T} \times P \times \mathbb{R}^N \to \mathbb{R}^N\right)$ be a *linear nonautonomous dynamical system*, i.e., for all $\alpha, \beta \in \mathbb{R}$, $t \in \mathbb{T}, p \in P$ and $x, y \in \mathbb{R}^N$, we have

$$\varphi(t, p, \alpha x + \beta y) = \alpha \varphi(t, p, x) + \beta \varphi(t, p, y).$$

Thus, there exists a function $\Phi : \mathbb{T} \times P \to \mathbb{R}^{N \times N}$ with $\Phi(t, p)x = \varphi(t, p, x)$ for all $t \in \mathbb{T}$, $p \in P$ and $x \in \mathbb{R}^N$. We suppose that $(\theta, \varphi)$ is invertible, which implies $\mathbb{T} = \mathbb{R}$ or $\mathbb{T} = \mathbb{Z}$.

For our purpose, $\mathbb{R}^N$ is equipped with the Euclidean norm $\| \cdot \|$, induced by the Euclidean scalar product (see Section 2.1). The NDS $(\theta, \varphi)$ canonically induces a nonautonomous dynamical system $(\theta, \mathbb{P}\Phi)$ on the real projective space $\mathbb{P}^{N-1}$ of the vector space $\mathbb{R}^N$ by defining

$$\mathbb{P}\Phi(t, p)\mathbb{P}x := \mathbb{P}(\Phi(t, p)x) \quad \text{for all } t \in \mathbb{T}, p \in P \text{ and } x \in \mathbb{R}^N$$

(see COLONIUS & KLIEMANN [50, Lemma 5.2.1, p. 149]). For basic properties of the projective space and notation, we refer to Appendix A.3.

The main observation of the following lemma is that past attractors and future repellers in $\mathbb{P}^{N-1}$ are linear nonautonomous invariant manifolds in $\mathbb{R}^N$ (cf. Definition 4.1). For a similar result, see SALAMON & ZEHNDER [155, Proposition 2.9] and COLONIUS & KLIEMANN [50, Lemma 5.2.2., p. 149].

**Proposition 3.19 (Past attractors and future repellers in $\mathbb{P}^{N-1}$).** *Let $A$ be a past attractor of $(\theta, \mathbb{P}\Phi)$. Then, for all $p \in P$ and all compact sets $C \subset \mathbb{S}^{N-1} \setminus \mathbb{P}^{-1}A(p)$, we have*

$$\lim_{t \to \infty} \frac{\sup_{v \in \mathbb{S}^{N-1} \cap \mathbb{P}^{-1}A(p)} \|\Phi(-t, p)v\|}{\inf_{w \in C} \|\Phi(-t, p)w\|} = 0 \, .$$

*Moreover, $\mathbb{P}^{-1}A$ is a linear nonautonomous invariant manifold in $\mathbb{R}^N$, i.e., $\mathbb{P}^{-1}A$ is an invariant nonautonomous set and for all $p \in P$, the set $\mathbb{P}^{-1}A(p)$ is a linear subspace of the $\mathbb{R}^N$.*
*Let $R$ be a future repeller of $(\theta, \mathbb{P}\Phi)$. Then, for all $p \in P$ and all compact sets $C \subset \mathbb{S}^{N-1} \setminus \mathbb{P}^{-1}R(p)$, we have*

$$\lim_{t \to \infty} \frac{\sup_{v \in \mathbb{S}^{N-1} \cap \mathbb{P}^{-1}R(p)} \|\Phi(t, p)v\|}{\inf_{w \in C} \|\Phi(t, p)w\|} = 0 \, .$$

*Moreover, $\mathbb{P}^{-1}R$ is a linear nonautonomous invariant manifold in $\mathbb{R}^N$.*

*Proof.* Let $A$ be a past attractor of $(\theta, \mathbb{P}\Phi)$, and choose a $p \in P$ and a compact set $C \subset \mathbb{S}^{N-1} \setminus \mathbb{P}^{-1}A(p)$ arbitrarily. First, we define for $0 \neq v \in \mathbb{P}^{-1}A(p)$ and $w \in C$ the two-dimensional linear subspace $L_{v,w} \subset \mathbb{R}^N$ by

$$L_{v,w} := \left\{rv + sw : r, s \in \mathbb{R}\right\}.$$

The proof of this proposition is divided into five steps.

*Step 1. For all $0 \neq v \in \mathbb{P}^{-1} A(p)$ and $w \in C$ such that $\mathbb{P}v$ is a boundary point of $A(p) \cap \mathbb{P}L_{v,w}$ relative to $\mathbb{P}L_{v,w}$, we have*

$$\lim_{t \to \infty} \frac{\|\Phi(-t, p)v\|}{\|\Phi(-t, p)w\|} = 0.$$

Since $A$ is a past attractor, there exists an $\eta > 0$ such that

$$\lim_{t \to \infty} d\big(\mathbb{P}\Phi(t, \theta_{-t}p)U_{2\eta}(A(\theta_{-t}p))\big|A(p)\big) = 0. \qquad (3.9)$$

Due to Lemma A.11, there exists a $\delta \in (0, 1)$ such that for all $0 \neq u_1, u_2 \in \mathbb{R}^N$ with

$$\frac{\langle u_1, u_2 \rangle^2}{\|u_1\|^2 \|u_2\|^2} \geq 1 - \delta,$$

we have

$$d_{\mathbb{P}}(\mathbb{P}u_1, \mathbb{P}u_2) \leq \eta.$$

We argue negatively and suppose that there exist a $\gamma > 0$ and a sequence $\{t_n\}_{n \in \mathbb{N}}$ with $\lim_{n \to \infty} t_n = -\infty$ such that

$$\frac{\|\Phi(t_n, p)w\|}{\|\Phi(t_n, p)v\|} \leq \gamma \quad \text{for all } n \in \mathbb{N}.$$

For nonzero $c \in \mathbb{R}$ with $|c|$ sufficiently small, this implies that for all $n \in \mathbb{N}$,

$$\frac{\langle \Phi(t_n,p)(cw+v), \Phi(t_n,p)v \rangle^2}{\|\Phi(t_n,p)(cw+v)\|^2 \|\Phi(t_n,p)v\|^2}$$
$$= \frac{c^2 \langle \Phi(t_n,p)w, \Phi(t_n,p)v \rangle^2 + 2c\|\Phi(t_n,p)v\|^2 \langle \Phi(t_n,p)w, \Phi(t_n,p)v \rangle + \|\Phi(t_n,p)v\|^4}{c^2\|\Phi(t_n,p)w\|^2 \|\Phi(t_n,p)v\|^2 + 2c\|\Phi(t_n,p)v\|^2 \langle \Phi(t_n,p)w, \Phi(t_n,p)v \rangle + \|\Phi(t_n,p)v\|^4}$$
$$\geq 1 - \delta$$

holds. Hence, for $|c| > 0$ sufficiently small, we have

$$d_{\mathbb{P}}\big(\mathbb{P}\Phi(t_n, p)\mathbb{P}(cw + v), A(\theta_{t_n}p)\big) \leq \eta \quad \text{for all } n \in \mathbb{N}.$$

This implies

$$d_{\mathbb{P}}\big(\mathbb{P}(cw + v), A(p)\big) = \lim_{n \to \infty} d_{\mathbb{P}}\big(\mathbb{P}(cw + v), A(p)\big)$$
$$= \lim_{n \to \infty} d_{\mathbb{P}}\big(\mathbb{P}\Phi(-t_n, \theta_{t_n}p) \underbrace{\mathbb{P}\Phi(t_n, p)\mathbb{P}(cw + v)}_{\in U_{2\eta}(A(\theta_{t_n}p))}, A(p)\big)$$
$$\overset{(3.9)}{=} 0.$$

This is a contradiction, since $\mathbb{P}v$ is a boundary point of $A(p) \cap \mathbb{P}L_{v,w}$ in $\mathbb{P}L_{v,w}$, and thus, the first step of this proof is finished.

*Step 2. For all nonzero $v \in \mathbb{P}^{-1} A(p)$ and $w \in C$, the set $A(p) \cap \mathbb{P}L_{v,w}$ is a singleton.*

Please note that any point in $\mathbb{P}L_{v,w} \setminus \{\mathbb{P}v\}$ is given by $\mathbb{P}(w + cv)$ for some $c \in \mathbb{R}$. It follows from Step 1 that

$$\lim_{t\to-\infty} \frac{\langle \Phi(t,p)(w+cv),\Phi(t,p)w\rangle^2}{\|\Phi(t,p)(w+cv)\|^2\|\Phi(t,p)w\|^2}$$

$$= \lim_{t\to-\infty} \frac{\|\Phi(t,p)w\|^4+2c\|\Phi(t,p)w\|^2\langle\Phi(t,p)v,\Phi(t,p)w\rangle+c^2\langle\Phi(t,p)v,\Phi(t,p)w\rangle^2}{\|\Phi(t,p)w\|^4+2c\|\Phi(t,p)w\|^2\langle\Phi(t,p)v,\Phi(t,p)w\rangle+c^2\|\Phi(t,p)v\|^2\|\Phi(t,p)w\|^2}$$

$$= 1$$

in case $\mathbb{P}v$ is a boundary point of $A(p) \cap \mathbb{P}L_{v,w}$ relative to $\mathbb{P}L_{v,w}$. This implies with Lemma A.11 that

$$\lim_{t\to\infty} d_\mathbb{P}\big(\mathbb{P}\Phi(-t,p)\mathbb{P}(w+cv), \mathbb{P}\Phi(-t,p)\mathbb{P}w\big) = 0\,,$$

and hence, $\mathbb{P}(w + cv) \notin A(p)$. Therefore, $A(p) \cap \mathbb{P}L_{v,w}$ consists of a single point.

*Step 3.* For all nonzero $v \in \mathbb{P}^{-1}A(p)$ and $w \in C$, we have

$$\lim_{t\to\infty} \frac{\|\Phi(-t,p)v\|}{\|\Phi(-t,p)w\|} = 0\,.$$

This follows directly from Step 1 and Step 2.

*Step 4.* $\mathbb{P}^{-1}A(p)$ is a linear subspace of $\mathbb{R}^N$.

We have shown that for any two-dimensional linear subspace $L_{v,w}$, the set $A(p) \cap \mathbb{P}L_{v,w}$ is either empty, equals $\mathbb{P}L_{v,w}$ or consists of a single point. This implies that $\mathbb{P}^{-1}A$ intersects each fiber in a linear subspace.

*Step 5.* We have

$$\lim_{t\to\infty} \frac{\sup_{v\in\mathbb{S}^{N-1}\cap\mathbb{P}^{-1}A(p)} \|\Phi(-t,p)v\|}{\inf_{w\in C} \|\Phi(-t,p)w\|} = 0\,.$$

We assume to the contrary that there exist sequences $\{t_n\}_{n\in\mathbb{N}}$ in $\mathbb{R}$, $\{v_n\}_{n\in\mathbb{N}}$ in $\mathbb{S}^{N-1}\cap\mathbb{P}^{-1}A(p)$ and $\{w_n\}_{n\in\mathbb{N}}$ in $C$ such that $\lim_{n\to\infty} t_n = -\infty$ and, w.l.o.g., $\lim_{n\to\infty} v_n = v$ and $\lim_{n\to\infty} w_n = w$ for some $v \in \mathbb{P}^{-1}A(p)\cap\mathbb{S}^{N-1}$ and $w \in C$, and the following property is fulfilled: There exists a $\gamma > 0$ such that

$$\frac{\|\Phi(t_n,p)w_n\|}{\|\Phi(t_n,p)v_n\|} \leq \gamma \quad \text{for all } n \in \mathbb{N}\,.$$

We write $\Phi_n := \Phi(t_n,p)$. Similarly to Step 1, for nonzero $c \in \mathbb{R}$ with $|c|$ sufficiently small, this implies that for all $n \in \mathbb{N}$,

$$\frac{\langle\Phi_n(cw_n+v_n),\Phi_nv_n\rangle^2}{\|\Phi_n(cw_n+v_n)\|^2\|\Phi_nv_n\|^2}$$

$$= \frac{c^2\langle\Phi_nw_n,\Phi_nv_n\rangle^2 + 2c\|\Phi_nv_n\|^2\langle\Phi_nw_n,\Phi_nv_n\rangle + \|\Phi_nv_n\|^4}{c^2\|\Phi_nw_n\|^2\|\Phi_nv_n\|^2 + 2c\|\Phi_nv_n\|^2\langle\Phi_nw_n,\Phi_nv_n\rangle + \|\Phi_nv_n\|^4}$$

$$\geq 1 - \delta$$

holds, with $\delta \in (0,1)$ chosen as in Step 1. Hence, for $|c| > 0$ sufficiently small, we have

$$d_{\mathbb{P}}\big(\mathbb{P}\Phi(t_n, p)\mathbb{P}(cw_n + v_n), A(\theta_{t_n}p)\big) \leq \eta \quad \text{for all } n \in \mathbb{N}.$$

Since $\mathbb{P}(cw + v) \notin A(p)$ (due to Step 2, $A(p) \cap \mathbb{P}L_{v,w}$ is a singleton), there exist an $n_0 \in \mathbb{N}$ and a $\beta > 0$ such that $cw_n + v_n \notin \mathbb{P}^{-1}U_\beta(A(p))$ for all $n \geq n_0$. Similarly to Step 1, using (3.9), this implies a contradiction.    □

Concerning past repellers and future attractors, we can not expect that their fibers give rise to linear subspaces, since they are intrinsically nonunique (cf. Proposition 2.37). The following lemma, however, says that for any past attractor or future repeller, a linear counterpart in form of a past repeller or future attractor, respectively, can be found easily.

**Proposition 3.20 (Past repellers and future attractors in $\mathbb{P}^{N-1}$).** *Let $A$ be a past attractor of $(\theta, \mathbb{P}\Phi)$ and $R \subset P \times \mathbb{P}^{N-1}$ be an invariant nonautonomous set such that $\mathbb{P}^{-1}R(p)$ is a linear subspace of the $\mathbb{R}^N$ and*

$$\mathbb{P}^{-1}A(p) \oplus \mathbb{P}^{-1}R(p) = \mathbb{R}^N \quad \text{for all } p \in P.$$

*Then $R$ is a past repeller, and the relation $A = R^*$ is fulfilled.*
*Let $R$ be a future repeller of $(\theta, \mathbb{P}\Phi)$ and $A \subset P \times \mathbb{P}^{N-1}$ be an invariant nonautonomous set such that $\mathbb{P}^{-1}A(p)$ is linear subspace of the $\mathbb{R}^N$ and*

$$\mathbb{P}^{-1}A(p) \oplus \mathbb{P}^{-1}R(p) = \mathbb{R}^N \quad \text{for all } p \in P.$$

*Then $A$ is a future attractor, and the relation $R = A^*$ is fulfilled.*

*Proof.* The proof of this proposition is divided into five steps.
*Step 1.* For all $p \in P$ and compact sets $C \subset \mathbb{P}^{N-1}$ with $C \cap A(p) = \emptyset$, we have

$$\lim_{t \to \infty} \inf_{0 \neq v \in \mathbb{P}^{-1}C} \frac{\|\Phi(-t,p)v_r\|}{\|\Phi(-t,p)v\|} = \lim_{t \to \infty} \sup_{0 \neq v \in \mathbb{P}^{-1}C} \frac{\|\Phi(-t,p)v_r\|}{\|\Phi(-t,p)v\|} = 1,$$

where $v = v_a + v_r$ with $v_a \in \mathbb{P}^{-1}A(p)$ and $v_r \in \mathbb{P}^{-1}R(p)$.
The first assertion follows from

$$\lim_{t \to \infty} \inf_{0 \neq v \in \mathbb{P}^{-1}C} \frac{\|\Phi(-t,p)v_r\|}{\|\Phi(-t,p)v\|}$$

$$\geq \left( \lim_{t \to \infty} \sup_{0 \neq v \in \mathbb{P}^{-1}C} \frac{\|\Phi(-t,p)v_a\|}{\|\Phi(-t,p)v_r\|} + 1 \right)^{-1}$$

$$= \left( \lim_{t \to \infty} \sup_{v \in \mathbb{P}^{-1}C,\, v_a \neq 0} \frac{\|v_a\| \left\|\Phi(-t,p)\frac{v_a}{\|v_a\|}\right\|}{\|v_r\| \left\|\Phi(-t,p)\frac{v_r}{\|v_r\|}\right\|} + 1 \right)^{-1}$$

$$\overset{\text{Prop. 3.19}}{=} 1$$

and

$$\lim_{t\to\infty} \inf_{0\neq v\in\mathbb{P}^{-1}C} \frac{\|\Phi(-t,p)v_r\|}{\|\Phi(-t,p)v\|}$$

$$\leq \left(\lim_{t\to\infty} \sup_{0\neq v\in\mathbb{P}^{-1}C} \left|1 - \frac{\|\Phi(-t,p)v_a\|}{\|\Phi(-t,p)v_r\|}\right|\right)^{-1}$$

$$= \left(\lim_{t\to\infty} \sup_{v\in\mathbb{P}^{-1}C,\, v_a\neq 0} \left|1 - \frac{\|v_a\|\left\|\Phi(-t,p)\frac{v_a}{\|v_a\|}\right\|}{\|v_r\|\left\|\Phi(-t,p)\frac{v_r}{\|v_r\|}\right\|}\right|\right)^{-1}$$

$$\overset{\text{Prop. 3.19}}{=} 1.$$

Proposition 3.19 is applicable, because the set $\{v_a : v \in \mathbb{P}^{-1}C \cap \mathbb{S}^{N-1}\}$ is compact and the set $\{v_r : v \in \mathbb{P}^{-1}C \cap \mathbb{S}^{N-1}\}$ is bounded away from zero. This is due to the fact that the projector $Q \in \mathbb{R}^{N\times N}$ with range $\mathbb{P}^{-1}A(p)$ and null space $\mathbb{P}^{-1}R(p)$ satisfies

$$\{v_a : v \in \mathbb{P}^{-1}C \cap \mathbb{S}^{N-1}\} = Q(\mathbb{P}^{-1}C \cap \mathbb{S}^{N-1}) \qquad \text{and}$$
$$\{v_r : v \in \mathbb{P}^{-1}C \cap \mathbb{S}^{N-1}\} = (1-Q)(\mathbb{P}^{-1}C \cap \mathbb{S}^{N-1})$$

(cf. also Step 3 of the proof of Lemma 4.14). The assertion

$$\lim_{t\to\infty} \sup_{0\neq v\in\mathbb{P}^{-1}C} \frac{\|\Phi(-t,p)v_r\|}{\|\Phi(-t,p)v\|} = 1$$

follows analogously.

*Step 2.* For all $p \in P$ and compact sets $C \subset \mathbb{P}^{N-1}$ with $C \cap A(p) = \emptyset$, we have

$$\lim_{t\to\infty} d_\mathbb{P}\big(\mathbb{P}\Phi(-t,p)C\big|R(\theta_{-t}p)\big) = 0.$$

With $v_a$ and $v_r$ defined as in Step 1, for all $t \geq 0$ and $v \in \mathbb{S}^{N-1} \cap \mathbb{P}^{-1}C$, we consider the expression

$$\frac{\langle\Phi(-t,p)v,\Phi(-t,p)v_r\rangle^2}{\|\Phi(-t,p)v\|^2\|\Phi(-t,p)v_r\|^2} = \frac{(\langle\Phi(-t,p)v_a,\Phi(-t,p)v_r\rangle+\langle\Phi(-t,p)v_r,\Phi(-t,p)v_r\rangle)^2}{\|\Phi(-t,p)v\|^2\|\Phi(-t,p)v_r\|^2}$$

$$= \frac{\langle\Phi(-t,p)v_a,\Phi(-t,p)v_r\rangle^2+\|\Phi(-t,p)v_r\|^4+2\langle\Phi(-t,p)v_a,\Phi(-t,p)v_r\rangle\|\Phi(-t,p)v_r\|^2}{\|\Phi(-t,p)v\|^2\|\Phi(-t,p)v_r\|^2}$$

$$= \frac{\langle\Phi(-t,p)v_a,\Phi(-t,p)v_r\rangle^2}{\|\Phi(-t,p)v\|^2\|\Phi(-t,p)v_r\|^2} + \frac{\|\Phi(-t,p)v_r\|^2}{\|\Phi(-t,p)v\|^2} + \frac{2\langle\Phi(-t,p)v_a,\Phi(-t,p)v_r\rangle}{\|\Phi(-t,p)v\|^2}.$$

Using the Cauchy-Schwartz inequality, we obtain the following relations:

$$0 \leq \lim_{t\to\infty} \sup_{v\in\mathbb{S}^{N-1}\cap\mathbb{P}^{-1}C} \frac{\langle\Phi(-t,p)v_a,\Phi(-t,p)v_r\rangle^2}{\|\Phi(-t,p)v\|^2\|\Phi(-t,p)v_r\|^2} \leq$$

$$\lim_{t\to\infty} \sup_{v\in\mathbb{S}^{N-1}\cap\mathbb{P}^{-1}C} \frac{\|\Phi(-t,p)v_a\|^2}{\|\Phi(-t,p)v\|^2} \overset{\text{Proposition 3.19}}{=} 0$$

and

$$
\begin{aligned}
0 \quad &\le \quad \lim_{t\to\infty} \sup_{v\in\mathbb{S}^{N-1}\cap\mathbb{P}^{-1}C} \frac{2\big|\langle\Phi(-t,p)v_a,\Phi(-t,p)v_r\rangle\big|}{\|\Phi(-t,p)v\|^2} \\
&\le \quad \lim_{t\to\infty} \sup_{v\in\mathbb{S}^{N-1}\cap\mathbb{P}^{-1}C} 2\,\frac{\|\Phi(-t,p)v_a\|\,\|\Phi(-t,p)v_r\|}{\|\Phi(-t,p)v\|\,\|\Phi(-t,p)v\|} \\
&\overset{\text{Step 1}}{=} \quad \lim_{t\to\infty} \sup_{v\in\mathbb{S}^{N-1}\cap\mathbb{P}^{-1}C} \frac{2\|\Phi(-t,p)v_a\|}{\|\Phi(-t,p)v\|} \\
&\overset{\text{Proposition 3.19}}{=} \quad 0 \,.
\end{aligned}
$$

Hence, we obtain

$$
\begin{aligned}
&\lim_{t\to\infty} \inf_{v\in\mathbb{S}^{N-1}\cap\mathbb{P}^{-1}C} \frac{\big\langle\Phi(-t,p)v,\Phi(-t,p)v_r\big\rangle^2}{\|\Phi(-t,p)v\|^2\|\Phi(-t,p)v_r\|^2} \\
&= \lim_{t\to\infty} \inf_{v\in\mathbb{S}^{N-1}\cap\mathbb{P}^{-1}C} \left( \frac{\big\langle\Phi(-t,p)v_a,\Phi(-t,p)v_r\big\rangle^2}{\|\Phi(-t,p)v\|^2\|\Phi(-t,p)v_r\|^2} + \frac{\|\Phi(-t,p)v_r\|^2}{\|\Phi(-t,p)v\|^2} \right. \\
&\qquad\qquad\qquad\qquad \left. + \frac{2\big\langle\Phi(-t,p)v_a,\Phi(-t,p)v_r\big\rangle}{\|\Phi(-t,p)v\|^2} \right) \\
&\overset{\text{Step 1}}{=} 1\,.
\end{aligned}
$$

Using Lemma A.11, this implies the assertion.

*Step 3. A and R are past isolated, i.e., there exists a $\beta > 0$ such that for all $p \in P$, there exists a $\tau > 0$ with*

$$
U_\beta(A(\theta_{-t}p)) \cap U_\beta(R(\theta_{-t}p)) = \emptyset \quad \text{for all } t \ge \tau\,.
$$

Since $A$ is a past attractor, there exists an $\eta > 0$ such that for all $p \in P$, we have

$$
\lim_{t\to\infty} d_\mathbb{P}\big(\mathbb{P}\Phi(t,\theta_{-t}p)U_\eta(A(\theta_{-t}p))\big|A(p)\big) = 0
$$

(cf. Remark 2.7 (iv)). Defining $\beta := \eta/2$ and using the invariance of $R$, this implies the assertion.

*Step 4. R is a past repeller.*
This is a direct consequence of Step 2 and Step 3.

*Step 5. The relation $A = R^*$ is fulfilled.*
We define $\eta > 0$, $P^*$ and $B_\zeta$ for $\zeta \in (0,\eta]$ as in the introduction of Section 3.1. We also consider the collection $\mathcal{M} := \{B_\zeta : \zeta \in (0,\eta]\}$. Due to Theorem 3.1, it is sufficient to show that $A$ is an $\mathcal{M}$-past attractor. Thereto, we fix an element $\zeta \in (0,\eta]$ and $p \in P^*$. Furthermore, we choose $\varepsilon > 0$ arbitrarily and consider the compact set $C := \mathbb{P}^{N-1} \setminus U_\varepsilon(A(p))$. Due to Step 2, we have

$$
\lim_{t\to\infty} d\big(\mathbb{P}\Phi(-t,p)C\big|R(\theta_{-t}p)\big) = 0\,.
$$

This implies that there exists a $\tau > 0$ such that $\mathbb{P}\Phi(-t,p)C \cap B_\zeta(\theta_{-t}p) = \emptyset$ for all $t \geq \tau$. Thus,

$$d_\mathbb{P}\big(\mathbb{P}\Phi(t,\theta_{-t}p)B_\zeta(\theta_{-t}p)\big|A(p)\big) \leq \varepsilon \quad \text{for all } t \geq \tau,$$

and hence, $A$ is an $\mathcal{M}$-past attractor. This finishes the proof of this proposition. □

**Lemma 3.21.** *For all $n \in \mathbb{N}$, let $\{0\} \subsetneq W_n \subsetneq V_n \subsetneq \mathbb{R}^N$ be nontrivial linear subspaces. Furthermore, let $\{x_n\}_{n\in\mathbb{N}}$ be a sequence in $\mathbb{R}^N$ such that the following hypotheses are fulfilled:*

*(i) $x_n \notin V_n$ for all $n \in \mathbb{N}$,*

*(ii) $\lim_{n\to\infty} d_\mathbb{P}(\mathbb{P}x_n, \mathbb{P}V_n) = 0$,*

*(iii) there exists an $\varepsilon > 0$ such that $d_\mathbb{P}(\mathbb{P}x_n, \mathbb{P}W_n) \geq \varepsilon$ for all $n \in \mathbb{N}$.*

*For all $n \in \mathbb{N}$, we define $C_n := W_n \oplus \{\lambda x_n : \lambda \in \mathbb{R}\}$. Then the limit relation*

$$\lim_{n\to\infty} d_\mathbb{P}(\mathbb{P}C_n|\mathbb{P}V_n) = 0$$

*is fulfilled.*

*Proof.* W.l.o.g., we assume that $\|x_n\| = 1$ for all $n \in \mathbb{N}$. Due to Hypothesis (ii), there exists a sequence $\{v_n\}_{n\in\mathbb{N}}$ with $v_n \in V_n$ and $\|v_n\| = 1$ for all $n \in \mathbb{N}$ such that $\lim_{n\to\infty} \|x_n - v_n\| = 0$. Since $\mathbb{P}C_n$ is a compact subset of $\mathbb{P}^{N-1}$, there exists a sequence $\{c_n\}_{n\in\mathbb{N}}$ with $c_n \in C_n$ for all $n \in \mathbb{N}$ such that $d_\mathbb{P}(\mathbb{P}C_n|\mathbb{P}V_n) = d_\mathbb{P}(\mathbb{P}c_n, \mathbb{P}V_n)$. W.l.o.g., we assume that $c_n$ is of the form

$$c_n = x_n + w_n \quad \text{for all } n \in \mathbb{N},$$

where $\{w_n\}_{n\in\mathbb{N}}$ is a sequence with $w_n \in W_n$ for all $n \in \mathbb{N}$, and we define

$$r_n := v_n + w_n \quad \text{for all } n \in \mathbb{N}$$

and $\beta_n := \langle x_n, w_n \rangle$, $\delta_n := \langle v_n, w_n \rangle$ and $\gamma_n := \langle x_n, v_n \rangle$ for all $n \in \mathbb{N}$. Then, for all $n \in \mathbb{N}$, we have

$$\frac{\langle c_n, r_n \rangle^2}{\|c_n\|^2 \|r_n\|^2}$$

$$= \frac{\overbrace{\gamma_n^2 + \beta_n^2 + \delta_n^2 + \|w_n\|^4 + 2\gamma_n\|x_n\|^2}^{=:\,\xi_n} + \overbrace{2\beta_n\delta_n + 2(\beta_n + \delta_n)(\|w_n\|^2 + \gamma_n)}^{=:\,\eta_n}}{\underbrace{1 + 2\|w_n\|^2 + \|w_n\|^4}_{=:\,\bar{\xi}_n} + \underbrace{4\beta_n\delta_n + 2(\beta_n + \delta_n)(\|w_n\|^2 + 1)}_{=:\,\bar{\eta}_n}},$$

and it is easy to see that we have $\lim_{n\to\infty} \gamma_n = 1$, $\lim_{n\to\infty} \xi_n/\bar{\xi}_n = 1$ and $\lim_{n\to\infty} \eta_n/\bar{\eta}_n = 1$. This implies that

$$\lim_{n \to \infty} \frac{\langle c_n, r_n \rangle^2}{\|c_n\|^2 \|r_n\|^2} = 1$$

under the condition that $\|c_n\| \|r_n\|$ is bounded away from 0 in the limit $n \to \infty$. To see that this is fulfilled, we need Hypothesis (iii), which says that there exists a $\delta \in (0,1)$ with

$$\frac{\langle x_n, w \rangle}{\|w\|} \leq \delta \quad \text{for all } n \in \mathbb{N} \text{ and } w \in W_n$$

(cf. Lemma A.11). This means that for all $n \in \mathbb{N}$ and $w \in W_n$, we have

$$\|x_n - w\|^2 = 1 - 2\langle x_n, w \rangle + \|w\|^2 \geq 1 - 2\delta \|w\| + \|w\|^2 \geq \gamma$$

for some $\gamma > 0$, and using $\lim_{n \to \infty} \|x_n - v_n\| = 0$, this finishes the proof of this lemma.  $\square$

In our special situation, convergence in both directions to the Morse sets is satisfied.

**Theorem 3.22 (Dynamical properties of nonautonomous Morse decompositions of linear systems).** *Let $\{M_1, \ldots, M_n\}$ be a past Morse decomposition obtained by the finite sequence of past repellers $R_0 \supset \cdots \supset R_n$ such that $\mathbb{P}^{-1} R_i(p)$ is a linear subspace of $\mathbb{R}^N$ for $i \in \{1, \ldots, n-1\}$ and $p \in P$. Then the following statements are fulfilled:*

(i) *Pullback convergence. For all $p \in P$ and all functions $\gamma : \mathbb{T}^+ \to \mathbb{P}^{N-1}$ with*

$$\liminf_{t \to \infty} d_{\mathbb{P}} \left( \gamma(t), \bigcup_{j=1}^{n} \partial R_j(\theta_{-t} p) \right) > 0,$$

*we have*

$$\lim_{t \to \infty} d_{\mathbb{P}} \left( \mathbb{P}\Phi(t, \theta_{-t} p) \gamma(t), \bigcup_{j=1}^{n} M_j(p) \right) = 0.$$

(ii) *Backward convergence. For all $(p, x) \in P \times \mathbb{P}^{N-1}$, there exists an $i \in \{1, \ldots, n\}$ with*

$$\lim_{t \to \infty} d_{\mathbb{P}} \big( \mathbb{P}\Phi(-t, p) x, M_i(\theta_{-t} p) \big) = 0.$$

*Let $\{M_1, \ldots, M_n\}$ be a future Morse decomposition obtained by the finite sequence of future attractors $A_0 \subset \cdots \subset A_n$ such that $\mathbb{P}^{-1} A_i(p)$ is a linear subspace of $\mathbb{R}^N$ for $i \in \{1, \ldots, n-1\}$ and $p \in P$. Then the following statements are fulfilled:*

(i) *Forward convergence. For all $(p, x) \in P \times \mathbb{P}^{N-1}$, there exists an $i \in \{1, \ldots, n\}$ with*

$$\lim_{t \to \infty} d_{\mathbb{P}} \big( \mathbb{P}\Phi(t, p) x, M_i(\theta_t p) \big) = 0.$$

(ii) Pushforward convergence. *For all $p \in P$ and all functions $\gamma : \mathbb{T}^+ \to \mathbb{P}^{N-1}$ with*

$$\liminf_{t \to \infty} d_{\mathbb{P}} \left( \gamma(t), \bigcup_{j=1}^{n} \partial A_j(\theta_t p) \right) > 0,$$

*we have*

$$\lim_{t \to \infty} d_{\mathbb{P}} \left( \mathbb{P}\Phi(-t, \theta_t p)\gamma(t), \bigcup_{j=1}^{n} M_j(p) \right) = 0.$$

*Proof.* (i) This assertion is also valid for general Morse decompositions and was proved in Theorem 3.12.

(ii) Choose $(p, x) \in P \times \mathbb{P}^{N-1}$ arbitrarily. Then there exists an $i \in \{1, \ldots, n\}$ such that

$$x \in R_i^*(p) \quad \text{and} \quad x \notin R_{i-1}^*(p).$$

In case $x \in R_{i-1}(p)$, the above limit relation follows, since then $x \in M_i(p)$ and $M_i$ is invariant. We therefore assume $x \notin R_{i-1}(p)$ from now on. To obtain a contradiction, we also assume that there exist an $\varepsilon > 0$ and a sequence $\{t_n\}_{n \in \mathbb{N}}$ in $\mathbb{R}$ with $\lim_{n \to \infty} t_n = \infty$ such that

$$d_{\mathbb{P}}\big(\mathbb{P}\Phi(-t_n, p)x, M_i(\theta_{-t_n} p)\big) \geq \varepsilon \quad \text{for all } n \in \mathbb{N}.$$

We define $C := \mathbb{P}^{-1}M_i(p) \oplus \mathbb{P}^{-1}\{x\}$. Since

$$\lim_{t \to \infty} d_{\mathbb{P}}\big(\mathbb{P}\Phi(-t, p)x, R_{i-1}(\theta_{-t} p)\big) = 0$$

(cf. Theorem 3.5), Lemma 3.21 implies

$$\lim_{n \to \infty} d_{\mathbb{P}}\big(\mathbb{P}\Phi(-t_n, p)\mathbb{P}C \,\big|\, R_{i-1}(\theta_{-t_n} p)\big) = 0. \tag{3.10}$$

We define $\bar{C} := \mathbb{P}^{-1}R_{i-1}(p) \oplus \mathbb{P}^{-1}\{x\}$. Then

$$\dim\big(\bar{C} \cap \mathbb{P}^{-1}R_{i-1}^*(p)\big) = \dim \bar{C} + \dim \mathbb{P}^{-1}R_{i-1}^*(p) - \dim\big(\bar{C} + \mathbb{P}^{-1}R_{i-1}^*(p)\big)$$
$$= N + 1 - N = 1.$$

Let $y = v + w$ be a nonzero element of $\bar{C} \cap \mathbb{P}^{-1}R_{i-1}^*(p)$ with $v \in \mathbb{P}^{-1}\{x\}$ and $w \in \mathbb{P}^{-1}R_{i-1}(p)$. Since $y$ and $v$ are in $\mathbb{P}^{-1}R_i^*(p)$, $w$ is also an element of $\mathbb{P}^{-1}R_i^*(p)$. Hence, $w \in \mathbb{P}^{-1}M_i(p)$. This implies $y \in C$, and hence, we get from (3.10) the relation

$$\lim_{n \to \infty} d_{\mathbb{P}}\big(\mathbb{P}\Phi(-t_n, p)\mathbb{P}y, R_{i-1}(\theta_{-t_n} p)\big) = 0.$$

This is a contradiction, since $\mathbb{P}y \in R_{i-1}^*(p)$ and $R_{i-1}$ and $R_{i-1}^*$ are past isolated (cf. Theorem 3.5 (i)). □

In our special situation, Proposition 3.6 can be generalized.

**Proposition 3.23 (Form of nonuniqueness of the Morse sets).** *Let* $\{M_1, \ldots, M_n\}$ *and* $\{\hat{M}_1, \ldots, \hat{M}_n\}$ *be past Morse decompositions obtained by the finite sequences of past repellers* $R_0 \supset \cdots \supset R_n$ *and* $\hat{R}_0 \supset \cdots \supset \hat{R}_n$ *such that* $\mathbb{P}^{-1}R_i(p)$ *and* $\mathbb{P}^{-1}\hat{R}_i(p)$ *are linear subspaces of* $\mathbb{R}^N$ *for* $i \in \{1, \ldots, n-1\}$ *and* $p \in P$. *We assume that*

$$R_i^* = \hat{R}_i^* \quad \text{for all } i \in \{1, \ldots, n-1\}.$$

*Then the relation*

$$\lim_{t \to \infty} d_{\mathbb{P}H}\big(M_i(\theta_{-t}p), \hat{M}_i(\theta_{-t}p)\big) = 0 \quad \text{for all } i \in \{1, \ldots, n\} \text{ and } p \in P$$

*is fulfilled.*

*Let* $\{M_1, \ldots, M_n\}$ *and* $\{\hat{M}_1, \ldots, \hat{M}_n\}$ *be future Morse decompositions obtained by the finite sequences of future attractors* $A_0 \subset \cdots \subset A_n$ *and* $\hat{A}_0 \subset \cdots \subset \hat{A}_n$ *such that* $\mathbb{P}^{-1}A_i(p)$ *and* $\mathbb{P}^{-1}\hat{A}_i(p)$ *are linear subspaces of* $\mathbb{R}^N$ *for* $i \in \{1, \ldots, n-1\}$ *and* $p \in P$. *We assume that*

$$A_i^* = \hat{A}_i^* \quad \text{for all } i \in \{1, \ldots, n-1\}.$$

*Then the relation*

$$\lim_{t \to \infty} d_{\mathbb{P}H}\big(M_i(\theta_t p), \hat{M}_i(\theta_t p)\big) = 0 \quad \text{for all } i \in \{1, \ldots, n\} \text{ and } p \in P$$

*is fulfilled.*

*Proof.* For $i \in \{1, n\}$, the above limit relation follows from $M_1 = R_1^* = \hat{R}_1^* = \hat{M}_1$ and from Proposition 3.6, since $M_n = R_{n-1}$ and $\hat{M}_n = \hat{R}_{n-1}$. We argue negatively and assume w.l.o.g. that there exist an $i \in \{2, \ldots, n-1\}$ and a $p \in P$ such that

$$\limsup_{t \to \infty} d_{\mathbb{P}}\big(\hat{M}_i(\theta_{-t}p)\big| M_i(\theta_{-t}p)\big) > 0.$$

Since Proposition 3.6 implies that

$$\lim_{t \to \infty} d_{\mathbb{P}H}\big(R_{i-1}(\theta_{-t}p), \hat{R}_{i-1}(\theta_{-t}p)\big) = 0,$$

and $\hat{R}_{i-1} \supset \hat{M}_i$, this means that there exist a $\gamma > 0$ and sequences $\{t_n\}_{n \in \mathbb{N}}$ (with $\lim_{n \to \infty} t_n = \infty$) and $\{x_n\}_{n \in \mathbb{N}}$ (with $x_n \in R_{i-1}(\theta_{-t_n}p) \setminus M_i(\theta_{-t_n}p)$) such that

$$d_{\mathbb{P}}\big(x_n, M_i(\theta_{-t_n}p)\big) \geq \gamma \quad \text{for all } n \in \mathbb{N}$$

and

$$\lim_{n \to \infty} d_{\mathbb{P}}\big(x_n, \hat{M}_i(\theta_{-t_n}p)\big) = 0.$$

The last formula implies $\lim_{n \to \infty} d_{\mathbb{P}}\big(x_n, R_i^*(\theta_{-t_n}p)\big) = 0$. We define

$$C_n := \mathbb{P}^{-1}M_i(\theta_{-t_n}p) \oplus \mathbb{P}^{-1}\{x_n\} \quad \text{for all } n \in \mathbb{N}.$$

Due to Lemma 3.21, the relation

$$\lim_{n\to\infty} d_\mathbb{P}\big(\mathbb{P}C_n \big| R_i^*(\theta_{-t_n}p)\big) = 0$$

holds. Since $R_i^*$ is a past attractor, we thus get

$$\lim_{n\to\infty} d_\mathbb{P}\big(\mathbb{P}\Phi(t_n, \theta_{-t_n}p)\mathbb{P}C_n \big| R_i^*(p)\big) = 0. \tag{3.11}$$

Due to Lemma A.12, we have $d_\mathbb{P}\big(\mathbb{P}\Phi(t_n, \theta_{-t_n}p)\mathbb{P}C_n \big| M_i(p)\big) = \sqrt{2}$ for all $n \in \mathbb{N}$, since $\mathbb{P}^{-1}\mathbb{P}\Phi(t_n, \theta_{-t_n}p)\mathbb{P}C_n$ has a higher dimension than $\mathbb{P}^{-1}M_i(p)$. This means that there exists a sequence $\{y_n\}_{n\in\mathbb{N}}$ with

$$y_n \in \mathbb{P}\Phi(t_n, \theta_{-t_n}p)\mathbb{P}C_n \quad \text{and} \quad d_\mathbb{P}\big(y_n, M_i(p)\big) \geq 1 \quad \text{for all } n \in \mathbb{N}. \tag{3.12}$$

Since $y_n \in R_{i-1}(p)$ for all $n \in \mathbb{N}$, we assume w.l.o.g. that this sequence is convergent with limit $y \in R_{i-1}(p)$. Due to (3.11), we also have $y \in R_i^*(p)$. Hence, $y \in M_i(p)$, and this contradicts (3.12). $\qquad\square$

For the rest of this chapter, attention is restricted to the situation $P = \mathbb{T}$ and $\theta(t, s) = t + s$ for all $t, s \in \mathbb{T}$. As described in Section 2.2, this setting includes arbitrary nonautonomous differential and difference equations. Under this assumption, an analogon to the Theorem of Selgrade (see SELGRADE [164, Theorem 9.7] and COLONIUS & KLIEMANN [50, Theorem 5.2.5]) can be proved.

**Theorem 3.24 (Finest nonautonomous Morse decomposition).** *We suppose that $P = \mathbb{T}$ and $\theta(t, s) = t + s$ for all $t, s \in \mathbb{T}$. Then the following statements are fulfilled:*

(i) *There exists a* finest past Morse decomposition $\{M_1, \ldots, M_n\}$, *i.e., any other past Morse decomposition $\{\tilde{M}_1, \ldots, \tilde{M}_m\}$ fulfills*

$$\lim_{t\to\infty} d_\mathbb{P}\left( \bigcup_{i=1}^{n} M_i(-t) \,\middle|\, \bigcup_{i=1}^{m} \tilde{M}_i(-t) \right) = 0.$$

*Moreover, we have $n \leq N$, and the following decomposition in a Whitney sum holds (cf. the definition on p. 82):*

$$\mathbb{P}^{-1}M_1 \oplus \cdots \oplus \mathbb{P}^{-1}M_n = \mathbb{T} \times \mathbb{R}^N.$$

(ii) *There exists a* finest future Morse decomposition $\{M_1, \ldots, M_n\}$, *i.e., any other future Morse decomposition $\{\tilde{M}_1, \ldots, \tilde{M}_m\}$ fulfills*

$$\lim_{t\to\infty} d_\mathbb{P}\left( \bigcup_{i=1}^{n} M_i(t) \,\middle|\, \bigcup_{i=1}^{m} \tilde{M}_i(t) \right) = 0.$$

*Moreover, we have $n \leq N$, and the following decomposition in a Whitney sum holds:*

$$\mathbb{P}^{-1}M_1 \oplus \cdots \oplus \mathbb{P}^{-1}M_n = \mathbb{T} \times \mathbb{R}^N .$$

*Proof.* First, we prove that any past attractors $A$ and $\hat{A}$ either fulfill

$$A \subset \hat{A} \quad \text{or} \quad A \supset \hat{A}.$$

Supposing the contrary, due to $P = \mathbb{T}$, there exist a $\tau \in \mathbb{T}$ and elements

$$x \in \mathbb{S}^{N-1} \cap \left( \mathbb{P}^{-1}A(\tau) \setminus \mathbb{P}^{-1}\hat{A}(\tau) \right) \quad \text{and} \quad \hat{x} \in \mathbb{S}^{N-1} \cap \left( \mathbb{P}^{-1}\hat{A}(\tau) \setminus \mathbb{P}^{-1}A(\tau) \right).$$

Because of Proposition 3.19, we obtain

$$\lim_{t \to \infty} \frac{\|\Phi(-t,\tau)x\|}{\|\Phi(-t,\tau)\hat{x}\|} = 0 \quad \text{and} \quad \lim_{t \to \infty} \frac{\|\Phi(-t,\tau)\hat{x}\|}{\|\Phi(-t,\tau)x\|} = 0 .$$

This is a contradiction. Proposition 3.19 also implies that the fibers of past attractors correspond to linear subspaces. Thus, there are at most $N+1$ past attractors of $(\theta, \mathbb{P}\Phi)$, namely

$$\emptyset = A_0 \subsetneq A_1 \subsetneq \cdots \subsetneq A_n = \mathbb{T} \times \mathbb{P}^{N-1}$$

with $n \leq N$. Due to Proposition 3.20, it is possible to choose a sequence of past repellers $\mathbb{T} \times \mathbb{P}^{N-1} = R_0 \supsetneq R_1 \supsetneq \cdots \supsetneq R_n = \emptyset$ such that $R_i^* = A_i$ for $i \in \{0, \ldots, n\}$. We denote by $\{M_1, \ldots, M_n\}$ the corresponding past Morse decomposition. Let $\{\tilde{M}_1, \ldots, \tilde{M}_m\}$ be another past Morse decomposition, obtained by the sequence $\mathbb{T} \times \mathbb{P}^{N-1} = \tilde{R}_0 \supsetneq \tilde{R}_1 \supsetneq \cdots \supsetneq \tilde{R}_m = \emptyset$ of past repellers. Then, for each $i \in \{0, \ldots, m\}$, there exists an $n_i \in \{0, \ldots, n\}$ such that $\tilde{R}_i^* = A_{n_i}$. We consider now the past Morse decomposition $\{\hat{M}_1, \ldots, \hat{M}_n\}$ which is obtained by the past repellers $R_{n_0}, \ldots, R_{n_m}$. Due to Proposition 3.23, we have

$$\lim_{t \to \infty} d_{\mathbb{P}H}\left( \bigcup_{i=1}^{m} \hat{M}_i(-t), \bigcup_{i=1}^{m} \tilde{M}_i(-t) \right) = 0 .$$

Moreover, it is easy to see that $\cup_{i=1}^n M_i \subset \cup_{i=1}^m \hat{M}_i$ holds, and this finishes the proof of the first assertion of this theorem. To show

$$\mathbb{P}^{-1}M_1 \oplus \cdots \oplus \mathbb{P}^{-1}M_n = \mathbb{T} \times \mathbb{R}^N ,$$

we first note that for $1 \leq i < j \leq n$, we have $\mathbb{P}^{-1}M_i \cap \mathbb{P}^{-1}M_j = \mathbb{T} \times \{0\}$ (cf. Proposition 3.9). Furthermore, Proposition 3.20 and Lemma A.10 implies

$$\begin{aligned}
\mathbb{T} \times \mathbb{R}^N &= \mathbb{P}^{-1}R_1^* + \mathbb{P}^{-1}R_1 \\
&= \mathbb{P}^{-1}M_1 + \left( \mathbb{P}^{-1}R_1 \cap (\mathbb{P}^{-1}R_2^* + \mathbb{P}^{-1}R_2) \right) \\
&= \mathbb{P}^{-1}M_1 + \left( \mathbb{P}^{-1}R_1 \cap \mathbb{P}^{-1}R_2^* \right) + \mathbb{P}^{-1}R_2 \\
&= \mathbb{P}^{-1}M_1 + \mathbb{P}^{-1}M_2 + \mathbb{P}^{-1}R_2 .
\end{aligned}$$

It follows inductively that

$$\mathbb{T} \times \mathbb{R}^N = \mathbb{P}^{-1} M_1 + \cdots + \mathbb{P}^{-1} M_n + \mathbb{P}^{-1} R_n = \mathbb{P}^{-1} M_1 + \cdots + \mathbb{P}^{-1} M_n \,.$$

This finishes the proof of this theorem. □

*Remark 3.25.* A finest past Morse decomposition $\{M_1, \ldots, M_n\}$ is not uniquely determined, but it follows directly from the above theorem that any other finest Morse decomposition $\{\tilde{M}_1, \ldots, \tilde{M}_m\}$ satisfies

$$\lim_{t \to \infty} d_{\mathbb{P}H} \left( \bigcup_{i=1}^{n} M_i(-t), \bigcup_{i=1}^{m} \tilde{M}_i(-t) \right) = 0 \,.$$

Moreover, the relation $n = m$ is fulfilled. A similar statement holds for finest future Morse decompositions.

# 4

# Linear Systems

In the qualitative theory, the study of linear systems is very important, since a comprehensive analysis of nonlinear systems via perturbation techniques requires linear theory. This is due to the fact that in many cases, stability properties of solutions can be derived from the linearization along the solution, the so-called variational equation. In this chapter, methods are provided for the analysis of linear systems with respect to the notions of attractivity and repulsivity which have been introduced in Chapter 2.

Throughout this chapter, let $\left( \theta : \mathbb{T} \times P \to P, \varphi : \mathbb{T} \times P \times \mathbb{R}^N \to \mathbb{R}^N \right)$ be a linear nonautonomous dynamical system, i.e., for all $\alpha, \beta \in \mathbb{R}$, $t \in \mathbb{T}, p \in P$ and $x, y \in \mathbb{R}^N$, we have

$$\varphi(t, p)(\alpha x + \beta y) = \alpha \varphi(t, p)x + \beta \varphi(t, p)y \,.$$

We suppose that $(\theta, \varphi)$ is invertible, which implies $\mathbb{T} = \mathbb{R}$ or $\mathbb{T} = \mathbb{Z}$. Moreover, let $\Phi : \mathbb{T} \times P \to \mathbb{R}^{N \times N}$ be the matrix function with $\Phi(t, p)x = \varphi(t, p, x)$ for all $t \in \mathbb{T}$, $p \in P$ and $x \in \mathbb{R}^N$.

## 4.1 Notions of Dichotomy

In this section, several notions of dichotomy are introduced for the different time domains. The classical concept of exponential dichotomy for nonautonomous linear differential equations has been established by PERRON [130, 131] in the late 1920s. In the sequel, many authors developed the theory; for fundamental work on this topic, we refer to COPPEL [55], DALECKIĬ & KREIN [61], MASSERA & SCHÄFFER [118], PALMER [125, 126, 127] and SACKER & SELL [151, 152, 153, 150] (see also PAPASCHINOPOULOS [129] for difference equations). The noninvertible case is treated in HENRY [79, Section 7.6], KALKBRENNER [89], AULBACH & KALKBRENNER [15] and AULBACH & SIEGMUND [20].

**Definition 4.1 (Linear nonautonomous invariant manifold).** *An invariant nonautonomous set* $M \subset P \times \mathbb{R}^N$ *is called* linear nonautonomous invariant manifold *of* $(\theta, \varphi)$ *if* $M(p)$ *is a linear subspace of* $\mathbb{R}^N$ *for all* $p \in P$.

Given linear nonautonomous invariant manifolds $M_1, M_2$ of $(\theta, \varphi)$, the sets

$$M_1 \cap M_2 := \left\{ (p, \xi) \in P \times \mathbb{R}^N : \xi \in M_1(p) \cap M_2(p) \right\} \quad \text{and}$$
$$M_1 + M_2 := \left\{ (p, \xi) \in P \times \mathbb{R}^N : \xi \in M_1(p) + M_2(p) \right\}$$

are also linear nonautonomous invariant manifolds of $(\theta, \varphi)$. A finite sum $M_1 + \cdots + M_n$ of linear nonautonomous invariant manifolds is called *Whitney sum* $M_1 \oplus \cdots \oplus M_n$ if the relation $M_i \cap M_j = P \times \{0\}$ is satisfied for $i \neq j$.

Linear nonautonomous invariant manifolds can be described via invariant projectors.

**Definition 4.2 (Invariant projector).** *An* invariant projector *of* $(\theta, \varphi)$ *is a function* $Q : P \to \mathbb{R}^{N \times N}$ *with*

$$Q(p) = Q(p)^2 \qquad \text{for all } p \in P,$$
$$Q(\theta_t p)\Phi(t, p) = \Phi(t, p)Q(p) \quad \text{for all } p \in P \text{ and } t \in \mathbb{T}.$$

*Remark 4.3.* In case the nonautonomous dynamical system $(\theta, \varphi)$ is a topological skew product flow, i.e., $P$ is a topological space, one usually supposes additionally that an invariant projector is continuous (see, e.g., SACKER & SELL [151]).

The *range*

$$\mathcal{R}(Q) := \left\{ (p, \xi) \in P \times \mathbb{R}^N : \xi \in \mathcal{R}(Q(p)) \right\}$$

and the *null space*

$$\mathcal{N}(Q) := \left\{ (p, \xi) \in P \times \mathbb{R}^N : \xi \in \mathcal{N}(Q(p)) \right\}$$

of an invariant projector $Q$ are linear nonautonomous invariant manifolds of $(\theta, \varphi)$ such that $\mathcal{R}(Q) \oplus \mathcal{N}(Q) = P \times \mathbb{R}^N$.

Next, several notions of dichotomy are introduced for the linear system $(\theta, \varphi)$.

**Definition 4.4 (Notions of dichotomy).** *Let* $Q : P \to \mathbb{R}^{N \times N}$ *be an invariant projector of* $(\theta, \varphi)$.

(i) *We say that* $(\theta, \varphi)$ *admits a* past exponential dichotomy *with constants* $\alpha > 0, K \geq 1$ *and projector* $Q$ *if for all* $p \in P$, *there exists a* $\hat{p} \in [p]$ *with*

$$\left\| \Phi(t, \theta_{-\tau}\hat{p})Q(\theta_{-\tau}\hat{p}) \right\| \leq Ke^{-\alpha t} \quad \text{for all } \tau \geq t \geq 0,$$
$$\left\| \Phi(-t, \theta_{-\tau}\hat{p})(\mathbb{1} - Q(\theta_{-\tau}\hat{p})) \right\| \leq Ke^{-\alpha t} \quad \text{for all } \tau, t \geq 0.$$

(ii) *We say that $(\theta, \varphi)$ admits a future exponential dichotomy with constants $\alpha > 0, K \geq 1$ and projector $Q$ if for all $p \in P$, there exists a $\hat{p} \in [p]$ with*

$$\|\Phi(t, \theta_\tau \hat{p}) Q(\theta_\tau \hat{p})\| \leq K e^{-\alpha t} \quad \text{for all } \tau, t \geq 0,$$
$$\|\Phi(-t, \theta_\tau \hat{p})(\mathbb{1} - Q(\theta_\tau \hat{p}))\| \leq K e^{-\alpha t} \quad \text{for all } \tau \geq t \geq 0.$$

(iii) *We say that $(\theta, \varphi)$ admits an all-time exponential dichotomy with constants $\alpha > 0$, $K \geq 1$ and projector $Q$ if for all $p \in P$, we have*

$$\|\Phi(t, p) Q(p)\| \leq K e^{-\alpha t} \quad \text{for all } t \geq 0,$$
$$\|\Phi(-t, p)(\mathbb{1} - Q(p))\| \leq K e^{-\alpha t} \quad \text{for all } t \geq 0.$$

(iv) *Given $p \in P$ and $T \in \mathbb{T}^+$, we say that $(\theta, \varphi)$ admits a $(p, T)$-dichotomy with projector $Q$ if we have*

$$\|\Phi(T, p)\xi\| < \|\xi\| \quad \text{for all } 0 \neq \xi \in \mathcal{R}(Q(p)),$$
$$\|\Phi(-T, \theta_T p)\xi\| < \|\xi\| \quad \text{for all } 0 \neq \xi \in \mathcal{N}(Q(\theta_T p)).$$

Having these definitions at hand, some remarks are in order.

*Remark 4.5.*

(i) In the literature (see the references cited in the introduction of this section), an all-time exponential dichotomy is simply called *exponential dichotomy*.

(ii) In case the nonautonomous dynamical system $(\theta, \varphi)$ is generated by a nonautonomous differential or difference equation, i.e., $P = \mathbb{T}$, a past or future exponential dichotomy is called exponential dichotomy on half line $\mathbb{R}_0^-$, $\mathbb{Z}_0^-$ or $\mathbb{R}_0^+$, $\mathbb{Z}_0^+$, respectively (see, e.g., COPPEL [55] and Proposition 4.16).

(iii) In contrast to past, future or all-time exponential dichotomies, the notion of $(p, T)$-dichotomy is not invariant with respect to a change of the norm to an equivalent norm (cf. also Remark 2.18).

(iv) In the scalar case ($N = 1$), $(\theta, \varphi)$ admits a $(p, T)$-dichotomy if and only if $|\Phi(T, p)| \neq 1$.

In the following proposition, the relationship between the above introduced notions of dichotomies is examined.

**Proposition 4.6.** *The following statements are fulfilled:*

(i) *If $(\theta, \varphi)$ admits an all-time exponential dichotomy, then it also admits a past exponential dichotomy and a future exponential dichotomy.*

(ii) *Suppose, $(\theta, \varphi)$ is generated by a nonautonomous differential or difference equation, i.e., $P = \mathbb{T}$, and $(\theta, \varphi)$ admits a past exponential dichotomy and a future exponential dichotomy with the same invariant* projector $Q : \mathbb{T} \to \mathbb{R}^{N \times N}$. Then $(\theta, \varphi)$ also admits an all-time exponential dichotomy.

*Proof.* Statement (i) is obvious; for (ii), see COPPEL [55, p. 19]. □

**Definition 4.7 (Nonhyperbolic dichotomies).** *Let* $\gamma \in \mathbb{R}$, *and consider the linear nonautonomous dynamical system* $(\theta, \varphi_\gamma)$, *defined by*

$$\varphi_\gamma(t, p, x) := e^{-\gamma t} \varphi(t, p, x) \quad \text{for all } t \in \mathbb{T}, p \in P \text{ and } x \in \mathbb{R}^N.$$

*We say that* $(\theta, \varphi)$ *admits a nonhyperbolic past exponential (future exponential, all-time exponential, $(p, T)$-, respectively) dichotomy with growth rate* $\gamma$, *constants* $\alpha > 0, K \geq 1$ *and projector* $Q$ *if* $(\theta, \varphi_\gamma)$ *admits a past exponential (future exponential, all-time exponential, $(p, T)$-, respectively) dichotomy with constants* $\alpha > 0, K \geq 1$ *and projector* $Q$.

*Remark 4.8.* The nonautonomous dynamical system $(\theta, \varphi)$ admits a nonhyperbolic past exponential (future exponential, all-time exponential, $(p, T)$-, respectively) dichotomy with growth rate $\gamma = 0$ if and only if it admits a past exponential (future exponential, all-time exponential, $(p, T)$-, respectively) dichotomy.

**Lemma 4.9 (Criteria for nonhyperbolic dichotomies).** *Suppose,* $(\theta, \varphi)$ *admits a nonhyperbolic past exponential (future exponential, all-time exponential, $(p, T)$-, respectively) dichotomy with growth rate* $\gamma$ *and projector* $Q_\gamma$. *Then the following statements are fulfilled:*

(i) *If* $Q_\gamma \equiv \mathbb{1}$, *then* $(\theta, \varphi)$ *admits a nonhyperbolic past exponential (future exponential, all-time exponential, $(p, T)$-, respectively) dichotomy with growth rate* $\zeta$ *and projector* $Q_\zeta \equiv \mathbb{1}$ *for all* $\zeta > \gamma$.

(ii) *If* $Q_\gamma \equiv 0$, *then* $(\theta, \varphi)$ *admits a nonhyperbolic past exponential (future exponential, all-time exponential, $(p, T)$-, respectively) dichotomy with growth rate* $\zeta$ *and projector* $Q_\zeta \equiv 0$ *for all* $\zeta < \gamma$.

*Proof.* The assertions follow directly from the monotonicity of the exponential function. □

We make use of the following equivalent characterizations of nonhyperbolic dichotomies.

**Proposition 4.10 (Equivalent characterizations of nonhyperbolic dichotomies).** *Let* $Q : P \to \mathbb{R}^{N \times N}$ *be an invariant projector of* $(\theta, \varphi)$. *Then the following statements are fulfilled:*

(i) $(\theta, \varphi)$ admits a nonhyperbolic past exponential dichotomy with growth rate $\gamma \in \mathbb{R}$, constants $\alpha > 0, K \geq 1$ and projector $Q$ if and only if for all $p \in P$, there exists a $\hat{p} \in [p]$ with

$$\|\Phi(t, \theta_{-\tau}\hat{p})Q(\theta_{-\tau}\hat{p})\| \leq Ke^{(\gamma-\alpha)t} \quad \text{for all } \tau \geq t \geq 0,$$
$$\|\Phi(-t, \theta_{-\tau}\hat{p})(\mathbb{1} - Q(\theta_{-\tau}\hat{p}))\| \leq Ke^{-(\gamma+\alpha)t} \quad \text{for all } \tau, t \geq 0.$$

(ii) $(\theta, \varphi)$ admits a nonhyperbolic future exponential dichotomy with growth rate $\gamma \in \mathbb{R}$, constants $\alpha > 0, K \geq 1$ and projector $Q$ if and only if for all $p \in P$, there exists a $\hat{p} \in [p]$ with

$$\|\Phi(t, \theta_{\tau}\hat{p})Q(\theta_{\tau}\hat{p})\| \leq Ke^{(\gamma-\alpha)t} \quad \text{for all } \tau, t \geq 0,$$
$$\|\Phi(-t, \theta_{\tau}\hat{p})(\mathbb{1} - Q(\theta_{\tau}\hat{p}))\| \leq Ke^{-(\gamma+\alpha)t} \quad \text{for all } \tau \geq t \geq 0.$$

(iii) $(\theta, \varphi)$ admits a nonhyperbolic all-time exponential dichotomy with growth rate $\gamma \in \mathbb{R}$, constants $\alpha > 0$, $K \geq 1$ and projector $Q$ if and only if for all $p \in P$, we have

$$\|\Phi(t, p)Q(p)\| \leq Ke^{(\gamma-\alpha)t} \quad \text{for all } t \geq 0,$$
$$\|\Phi(-t, p)(\mathbb{1} - Q(p))\| \leq Ke^{-(\gamma+\alpha)t} \quad \text{for all } t \geq 0.$$

(iv) Given $p \in P$ and $T \in \mathbb{T}^+$, $(\theta, \varphi)$ admits a nonhyperbolic $(p, T)$-dichotomy with growth rate $\gamma \in \mathbb{R}$ and projector $Q$ if and only if we have

$$\|\varphi(T, p)\xi\| < e^{\gamma T}\|\xi\| \quad \text{for all } 0 \neq \xi \in \mathcal{R}(Q(p)),$$
$$\|\varphi(-T, \theta_T p)\xi\| < e^{-\gamma T}\|\xi\| \quad \text{for all } 0 \neq \xi \in \mathcal{N}(Q(\theta_T p)).$$

For $\gamma \in \mathbb{R}$, we define

$$\mathcal{S}_{\gamma} := \left\{(p, \xi) \in P \times \mathbb{R}^N : \Phi(\cdot, p)\xi \text{ is } \gamma^+\text{-quasibounded}\right\}$$

and

$$\mathcal{U}_{\gamma} := \left\{(p, \xi) \in P \times \mathbb{R}^N : \Phi(\cdot, p)\xi \text{ is } \gamma^-\text{-quasibounded}\right\}.$$

It is obvious that $\mathcal{S}_{\gamma}$ and $\mathcal{U}_{\gamma}$ are linear nonautonomous invariant manifolds of $(\theta, \varphi)$. Given $\gamma \leq \zeta$, the relations $\mathcal{S}_{\gamma} \subset \mathcal{S}_{\zeta}$ and $\mathcal{U}_{\gamma} \supset \mathcal{U}_{\zeta}$ are fulfilled.

We now discuss the important relationship between the projectors of non-hyperbolic exponential dichotomies with growth rate $\gamma$ and the sets $\mathcal{S}_{\gamma}$ and $\mathcal{U}_{\gamma}$.

**Proposition 4.11 (Dynamical properties).** If $(\theta, \varphi)$ admits a nonhyperbolic past exponential dichotomy with growth rate $\gamma$, constants $\alpha > 0, K \geq 1$ and projector $Q$, then we have $\mathcal{N}(Q) = \mathcal{U}_{\gamma}$, and for all $p \in P$, there exists a $\hat{p} \in [p]$ with

$$\|\Phi(t, \theta_{-\tau}\hat{p})\xi\| \leq K\|\xi\|e^{\gamma t} \quad \text{for all } 0 \leq t \leq \tau \text{ and } \xi \in \mathcal{R}(Q(\theta_{-\tau}\hat{p})). \quad (4.1)$$

If $(\theta, \varphi)$ admits a nonhyperbolic future exponential dichotomy with growth rate $\gamma$, constants $\alpha > 0$, $K \geq 1$ and projector $Q$, then we have $\mathcal{R}(Q) = \mathcal{S}_\gamma$, and for all $p \in P$, there exists a $\hat{p} \in [p]$ with

$$\|\Phi(-t, \theta_\tau \hat{p})\xi\| \leq K\|\xi\|e^{-\gamma t} \quad \text{for all } 0 \leq t \leq \tau \text{ and } \xi \in \mathcal{N}(Q(\theta_\tau \hat{p}))\,.$$

If $(\theta, \varphi)$ admits a nonhyperbolic all-time exponential dichotomy with growth rate $\gamma$ and projector $Q$, then $\mathcal{N}(Q) = \mathcal{U}_\gamma$ and $\mathcal{R}(Q) = \mathcal{S}_\gamma$ are fulfilled.

*Proof.* Suppose, $(\theta, \varphi)$ admits a nonhyperbolic past exponential dichotomy with growth rate $\gamma$, constants $\alpha > 0$, $K \geq 1$ and projector $Q$. Due to Proposition 4.10, for given $p \in P$, there exists a $\hat{p} \in [p]$ with

$$\|\Phi(t, \theta_{-\tau}\hat{p})Q(\theta_{-\tau}\hat{p})\| \leq Ke^{(\gamma-\alpha)t} \quad \text{for all } \tau \geq t \geq 0\,,$$
$$\|\Phi(-t, \theta_{-\tau}\hat{p})(\mathbb{1} - Q(\theta_{-\tau}\hat{p}))\| \leq Ke^{-(\gamma+\alpha)t} \quad \text{for all } \tau, t \geq 0\,.$$

The first inequality implies (4.1). Choose $\hat{t} \in \mathbb{T}$ such that $\theta_{\hat{t}} p = \hat{p}$. We now prove the relation $\mathcal{N}(Q) = \mathcal{U}_\gamma$.
($\supseteq$) We choose $(p, \xi) \in \mathcal{U}_\gamma$ arbitrarily. This implies $\|\Phi(-t, \theta_{\hat{t}} p)\Phi(\hat{t}, p)\xi\| \leq Ce^{-\gamma t}$ for all $t \geq 0$ with some real constant $C > 0$. We write $\Phi(\hat{t}, p)\xi = \xi_1 + \xi_2$ with $\xi_1 \in \mathcal{R}(Q(\hat{p}))$ and $\xi_2 \in \mathcal{N}(Q(\hat{p}))$. Hence, for all $t \geq 0$, we get

$$\begin{aligned}
\|\xi_1\| &= \left\|\Phi(t, \theta_{-t}\hat{p})\Phi(-t, \hat{p})Q(\hat{p})\Phi(\hat{t}, p)\xi\right\| \\
&= \left\|\Phi(t, \theta_{-t}\hat{p})Q(\theta_{-t}\hat{p})\Phi(-t, \hat{p})\Phi(\hat{t}, p)\xi\right\| \\
&\leq Ke^{(\gamma-\alpha)t}\left\|\Phi(-t, \hat{p})\Phi(\hat{t}, p)\xi\right\| \leq CKe^{(\gamma-\alpha)t}e^{-\gamma t} = CKe^{-\alpha t}\,.
\end{aligned}$$

The right hand side of this inequality converges to zero in the limit $t \to \infty$. Therefore, $\xi_1 = 0$, and $\Phi(\hat{t}, p)\xi \in \mathcal{N}(Q(\hat{p}))$. Due to the invariance of $\mathcal{N}(Q)$, we finally obtain $(p, \xi) \in \mathcal{N}(Q)$.
($\subseteq$) We choose $(p, \xi) \in \mathcal{N}(Q)$. Thus, for all $t \geq 0$, the relation

$$\left\|\Phi(-t, \hat{p})\Phi(\hat{t}, p)\xi\right\| = \left\|\Phi(-t, \hat{p})(\mathbb{1} - Q(\hat{p}))\Phi(\hat{t}, p)\xi\right\| \leq Ke^{-(\gamma+\alpha)t}\left\|\Phi(\hat{t}, p)\xi\right\|$$

is fulfilled. This means that $\Phi(\cdot, p)\xi$ is $\gamma^-$-quasibounded, i.e., $(p, \xi) \in \mathcal{U}_\gamma$.
The assertions concerning the future exponential dichotomy are treated analogously. In case $(\theta, \varphi)$ admits an all-time exponential dichotomy, Proposition 4.6 (i) yields that $(\theta, \varphi)$ also admits a past exponential dichotomy and a future exponential dichotomy. Hence, we obtain $\mathcal{N}(Q) = \mathcal{U}_\gamma$ and $\mathcal{R}(Q) = \mathcal{S}_\gamma$. $\qquad \square$

*Remark 4.12.* According to this proposition, an invariant projector is uniquely determined only in case of a nonhyperbolic all-time exponential dichotomy. In addition, the null space of a projector of a past exponential dichotomy and the range of a projector of a future exponential dichotomy are uniquely determined. For further information about the kind of nonuniqueness of ranges of projectors of past exponential dichotomies and null spaces of projectors of future exponential dichotomies, we refer to Lemma 4.19.

This section is concluded by pointing out several evidences that the notions of dichotomy are consistent to the concepts of attractivity and repulsivity.

**Theorem 4.13 (Nonhyperbolic dichotomies and the notions of attractivity and repulsivity).** *Suppose, $(\theta, \varphi)$ admits a nonhyperbolic past exponential (future exponential, all-time exponential, $(p, T)$-, respectively) dichotomy with growth rate $\gamma$ and invariant projector $Q$. Then the following statements are fulfilled:*

(i) *If $\gamma \leq 0$ and $\mathrm{rk}\, Q(\hat{p}) \geq 1$ for all $\hat{p} \in P$, then every trivial solution of $(\theta, \varphi)$ is not past (future, all-time, $(p, T)$-, respectively) repulsive.*

(ii) *If $\gamma \geq 0$ and $\mathrm{rk}\, Q(\hat{p}) \leq N - 1$ for all $\hat{p} \in P$, then every trivial solution of $(\theta, \varphi)$ is not past (future, all-time, $(p, T)$-, respectively) attractive.*

(iii) *If $\gamma \leq 0$ and $Q \equiv \mathbb{1}$, then every trivial solution of $(\theta, \varphi)$ is past (future, all-time, $(p, T)$-, respectively) attractive with $\mathfrak{A}_0 = \infty$.*

(iv) *If $\gamma \geq 0$ and $Q \equiv 0$, then every trivial solution of $(\theta, \varphi)$ is past (future, all-time, $(p, T)$-, respectively) repulsive with $\mathfrak{R}_0 = \infty$.*

*Proof.* These assertions are direct consequences of Proposition 4.10 and 4.11.
$\square$

For the rest of this section, the studies are concentrated on the induced nonautonomous dynamical system $(\theta, \mathbb{P}\Phi)$ on the real projective space $\mathbb{P}^{N-1}$ (cf. Section 3.5).

**Lemma 4.14.** *The following statements are fulfilled:*

(i) *We suppose that $(\theta, \varphi)$ admits a nonhyperbolic past exponential dichotomy with invariant projector $Q$. Then there exists a $\beta > 0$ such that for all $p \in P$, there exists a $\hat{p} \in [p]$ with*

$$U_\beta\big(\mathbb{P}\mathcal{R}(Q(\theta_{-t}\hat{p}))\big) \cap U_\beta\big(\mathbb{P}\mathcal{N}(Q(\theta_{-t}\hat{p}))\big) = \emptyset \quad \text{for all } t \geq 0$$

*(i.e., $\mathbb{P}\mathcal{R}(Q)$ and $\mathbb{P}\mathcal{N}(Q)$ are past isolated). Moreover, for all $p \in P$ and compact sets $C \subset \mathbb{S}^{N-1} \setminus \mathcal{N}(Q(p))$, we have*

$$\lim_{t \to \infty} \frac{\sup_{v \in \mathbb{S}^{N-1} \cap \mathcal{N}(Q(p))} \|\Phi(-t, p)v\|}{\inf_{w \in C} \|\Phi(-t, p)w\|} = 0.$$

(ii) *We suppose that $(\theta, \varphi)$ admits a nonhyperbolic future exponential dichotomy with invariant projector $Q$. Then there exists a $\beta > 0$ such that for all $p \in P$, there exists a $\hat{p} \in [p]$ with*

$$U_\beta\big(\mathbb{P}\mathcal{R}(Q(\theta_t\hat{p}))\big) \cap U_\beta\big(\mathbb{P}\mathcal{N}(Q(\theta_t\hat{p}))\big) = \emptyset \quad \text{for all } t \geq 0$$

*(i.e., $\mathbb{P}\mathcal{R}(Q)$ and $\mathbb{P}\mathcal{N}(Q)$ are future isolated). Moreover, for all $p \in P$ and compact sets $C \subset \mathbb{S}^{N-1} \setminus \mathcal{R}(Q(p))$, we have*

$$\lim_{t\to\infty} \frac{\sup_{v\in\mathbb{S}^{N-1}\cap\mathcal{R}(Q(p))}\|\Phi(t,p)v\|}{\inf_{w\in C}\|\Phi(t,p)w\|} = 0\,.$$

(iii)  *We suppose that $(\theta,\varphi)$ admits a nonhyperbolic all-time exponential dichotomy with invariant projector $Q$. Then there exists a $\beta > 0$ with*

$$U_\beta\big(\mathbb{P}\mathcal{R}(Q(p))\big) \cap U_\beta\big(\mathbb{P}\mathcal{N}(Q(p))\big) = \emptyset \quad \text{for all } p \in P$$

*(i.e., $\mathbb{P}\mathcal{R}(Q)$ and $\mathbb{P}\mathcal{N}(Q)$ are all-time isolated),*

$$\lim_{\substack{t\to\infty \\ p\in P}} \sup \frac{\sup_{v\in\mathbb{S}^{N-1}\cap\mathcal{N}(Q(p))}\|\Phi(-t,p)v\|}{\inf_{w\in\mathbb{S}^{N-1}\cap\mathbb{P}^{-1}U_\beta(\mathbb{P}\mathcal{R}(Q(p)))}\|\Phi(-t,p)w\|} = 0$$

*and*

$$\lim_{\substack{t\to\infty \\ p\in P}} \sup \frac{\sup_{v\in\mathbb{S}^{N-1}\cap\mathcal{R}(Q(p))}\|\Phi(t,p)v\|}{\inf_{w\in\mathbb{S}^{N-1}\cap\mathbb{P}^{-1}U_\beta(\mathbb{P}\mathcal{N}(Q(p)))}\|\Phi(t,p)w\|} = 0\,.$$

*Proof.* (i) Suppose that $(\theta,\varphi)$ admits a nonhyperbolic past exponential dichotomy with growth rate $\gamma$, constants $\alpha > 0, K \geq 1$ and projector $Q$. We define $\beta := 1/(3K)$, fix an arbitrary $p \in P$ and choose $\hat{p} \in [p]$ as in Definition 4.4 (i). The remaining proof of (i) is divided into four steps.

*Step 1. The sets $\mathbb{P}\mathcal{R}(Q)$ and $\mathbb{P}\mathcal{N}(Q)$ are past isolated.*
Assume, for some $t \geq 0$, we have $U_\beta\big(\mathbb{P}\mathcal{R}(Q(\theta_{-t}\hat{p}))\big) \cap U_\beta\big(\mathbb{P}\mathcal{N}(Q(\theta_{-t}\hat{p}))\big) \neq \emptyset$. Hence, there exist $x \in \mathbb{P}\mathcal{R}(Q(\theta_{-t}\hat{p}))$ and $y \in \mathbb{P}\mathcal{N}(Q(\theta_{-t}\hat{p}))$ with $d_{\mathbb{P}}(x,y) \leq 2\beta$. Due to the definition of $d_{\mathbb{P}}$ (cf. Appendix A.3), there exist $\tilde{x} \in \mathbb{S}^{N-1}\cap\mathbb{P}^{-1}\{x\}$ and $\tilde{y} \in \mathbb{S}^{N-1}\cap\mathbb{P}^{-1}\{y\}$ such that $\|\tilde{x} - \tilde{y}\| \leq 2\beta$. This yields

$$\frac{\|Q(\theta_{-t}\hat{p})(\tilde{x} - \tilde{y})\|}{\|\tilde{x} - \tilde{y}\|} = \frac{\|\tilde{x}\|}{\|\tilde{x} - \tilde{y}\|} \geq \frac{1}{2\beta} = \frac{3K}{2}\,,$$

and this is a contradiction, since Definition 4.4 (i) implies $\|Q(\theta_{-t}\hat{p})\| \leq K$.

*Step 2. We have*

$$\|\Phi(-t,\theta_{-\tau}\hat{p})x\| \geq \frac{1}{K}e^{-(\gamma-\alpha)t}\|x\| \quad \text{for all } \tau, t \geq 0 \text{ and } x \in \mathcal{R}(Q(\theta_{-\tau}\hat{p}))\,.$$

The assertion follows from

$$\|x\| = \|\Phi(t,\theta_{-\tau-t}\hat{p})\Phi(-t,\theta_{-\tau}\hat{p})Q(\theta_{-\tau}\hat{p})x\|$$
$$\overset{\text{Def. 4.4 (i)}}{\leq} Ke^{(\gamma-\alpha)t}\|\Phi(-t,\theta_{-\tau}\hat{p})x\|\,.$$

*Step 3. Let $M \subset \mathbb{S}^{N-1}\setminus\mathcal{N}(Q(\hat{p}))$ be a compact set. For $w \in M$, we write $w = w_r + w_n$ with $w_r \in \mathcal{R}(Q(\hat{p}))$ and $w_n \in \mathcal{N}(Q(\hat{p}))$. Then*

$$W_r(M) := \{w_r : w \in M\} = Q(\hat{p})M$$

*is bounded away from zero, and the set*

$$W_n(M) := \{w_n : w \in M\} = (\mathbb{1} - Q(\hat{p}))M$$

*is bounded.*

Assume, the set $W_r(M) = Q(\hat{p})M$ is not bounded away from zero. Then it contains 0, since it is compact, and thus, there exists a $w \in M$ with $w \in \mathcal{N}(Q(\hat{p}))$. This is a contradiction. Moreover, the set $W_n(M) = (\mathbb{1} - Q(\hat{p}))M$ is bounded, since it is compact.

*Step 4. For all compact sets $C \subset \mathbb{S}^{N-1} \setminus \mathcal{N}(Q(p))$, we have*

$$\lim_{t \to \infty} \frac{\sup_{v \in \mathbb{S}^{N-1} \cap \mathcal{N}(Q(p))} \|\Phi(-t,p)v\|}{\inf_{w \in C} \|\Phi(-t,p)w\|} = 0.$$

Choose $\tau \in \mathbb{T}$ such that $\hat{p} = \theta_\tau p$. Defining $\hat{C} := \varphi(\tau,p)C$, it is sufficient to show

$$\lim_{t \to \infty} \frac{\sup_{v \in \mathbb{S}^{N-1} \cap \mathcal{N}(Q(\hat{p}))} \|\Phi(-t,\hat{p})v\|}{\inf_{w \in \hat{C}} \|\Phi(-t,\hat{p})w\|} = 0.$$

We have

$$\frac{\sup_{v \in \mathbb{S}^{N-1} \cap \mathcal{N}(Q(\hat{p}))} \|\Phi(-t,\hat{p})v\|}{\inf_{w \in \hat{C}} \|\Phi(-t,\hat{p})w\|}$$

$$\overset{\text{Def. 4.4 (i)}}{\leq} \frac{\sup_{v \in \mathbb{S}^{N-1} \cap \mathcal{N}(Q(\hat{p}))} Ke^{-(\gamma+\alpha)t}\|v\|}{\inf_{w \in \hat{C}} \|\Phi(-t,\hat{p})w_r + \Phi(-t,\hat{p})w_n\|} \leq \sup_{w \in \hat{C}} \frac{\frac{Ke^{-(\gamma+\alpha)t}}{\|\Phi(-t,\hat{p})w_r\|}}{\left|1 - \frac{\|\Phi(-t,\hat{p})w_n\|}{\|\Phi(-t,\hat{p})w_r\|}\right|}.$$

Please note that for the last inequality, we require $w_r \neq 0$ for all $w \in \hat{C}$. This is fulfilled, since $W_r(\hat{C})$ is bounded away from zero (cf. Step 3). Furthermore, using

$$\frac{Ke^{-(\gamma+\alpha)t}}{\|\Phi(-t,\hat{p})w_r\|} \overset{\text{Step 2}}{\leq} \frac{Ke^{-(\gamma+\alpha)t}}{\frac{1}{K}e^{-(\gamma-\alpha)t}\|w_r\|} = \frac{K^2 e^{-2\alpha t}}{\|w_r\|},$$

we obtain

$$\lim_{t \to \infty} \sup_{w \in \hat{C}} \frac{Ke^{-(\gamma+\alpha)t}}{\|\Phi(-t,\hat{p})w_r\|} = 0,$$

since $W_r(\hat{C})$ is bounded away from zero. Moreover, due to

$$\frac{\|\Phi(-t,\hat{p})w_n\|}{\|\Phi(-t,\hat{p})w_r\|} \overset{\text{Def. 4.4 (i), Step 2}}{\leq} \frac{Ke^{-(\gamma+\alpha)t}\|w_n\|}{\frac{1}{K}e^{-(\gamma-\alpha)t}\|w_r\|} = \frac{K^2 e^{-2\alpha t}\|w_n\|}{\|w_r\|},$$

we get

$$\lim_{t \to \infty} \sup_{w \in \hat{C}} \frac{\|\Phi(-t,\hat{p})w_n\|}{\|\Phi(-t,\hat{p})w_r\|} = 0$$

(please note that Step 3 says that $W_n(\hat{C})$ is bounded and $W_r(\hat{C})$ is bounded away from zero). This implies the assertion.

(ii) can be proved similarly to (i).

(iii) Suppose that $(\theta, \varphi)$ admits a nonhyperbolic all-time exponential dichotomy with constants $\alpha > 0, K \geq 1$ and projector $Q$. We define $\beta := 1/(3K)$. The remaining proof of (iii) is divided into five steps.

*Step 1. The sets $\mathbb{PR}(Q)$ and $\mathbb{PN}(Q)$ are all-time isolated.*

Assume, there exists a $p \in P$ such that $U_\beta\big(\mathbb{PR}(Q(p))\big) \cap U_\beta\big(\mathbb{PN}(Q(p))\big) \neq \emptyset$. Hence, there exist $x \in \mathbb{PR}(Q(p))$ and $y \in \mathbb{PN}(Q(p))$ with $d_\mathbb{P}(x, y) \leq 2\beta$. Due to the definition of $d_\mathbb{P}$ (cf. Appendix A.3), there exist $\tilde{x} \in \mathbb{S}^{N-1} \cap \mathbb{P}^{-1}\{x\}$ and $\tilde{y} \in \mathbb{S}^{N-1} \cap \mathbb{P}^{-1}\{y\}$ with $\|\tilde{x} - \tilde{y}\| \leq 2\beta$. This yields

$$\frac{\|Q(p)(\tilde{x} - \tilde{y})\|}{\|\tilde{x} - \tilde{y}\|} = \frac{\|\tilde{x}\|}{\|\tilde{x} - \tilde{y}\|} \geq \frac{1}{2\beta} = \frac{3K}{2}.$$

This is a contradiction, since Definition 4.4 (iii) implies $\|Q(p)\| \leq K$.

*Step 2. We have*

$$\|\Phi(-t, p)x\| \geq \frac{1}{K} e^{-(\gamma - \alpha)t} \|x\| \quad \text{for all } p \in P, t \geq 0 \text{ and } x \in \mathcal{R}(Q(p)).$$

The assertion follows from

$$\|x\| = \|\Phi(t, \theta_{-t}p)\Phi(-t, p)Q(p)x\| \overset{\text{Def. 4.4 (iii)}}{\leq} Ke^{(\gamma-\alpha)t}\|\Phi(-t, p)x\|.$$

*Step 3. For $p \in P$ and $w \in \mathbb{S}^{N-1} \cap \mathbb{P}^{-1}U_\beta\big(\mathbb{PR}(Q(p))\big)$, we write $w = w_r^p + w_n^p$ with $w_r^p \in \mathcal{R}(Q(p))$ and $w_n^p \in \mathcal{N}(Q(p))$. Then*

$$W_r := \big\{w_r^p : p \in P, \, w \in \mathbb{S}^{N-1} \cap \mathbb{P}^{-1}U_\beta\big(\mathbb{PR}(Q(p))\big)\big\}$$

*is bounded away from zero, and*

$$W_n := \big\{w_n^p : p \in P, \, w \in \mathbb{S}^{N-1} \cap \mathbb{P}^{-1}U_\beta\big(\mathbb{PR}(Q(p))\big)\big\}$$

*is bounded.*

To show that $W_r$ is bounded away from zero, assume for contradiction, there exist sequences $\{p_n\}_{n \in \mathbb{N}}$ in $P$ and $\big\{w^{(n)}\big\}_{n \in \mathbb{N}}$ in $\mathbb{S}^{N-1}$ such that

$$w^{(n)} \in \mathbb{S}^{N-1} \cap \mathbb{P}^{-1}U_\beta\big(\mathbb{PR}(Q(p_n))\big) \quad \text{for all } n \in \mathbb{N}$$

and $\lim_{n\to\infty} w_r^{(n)p_n} = 0$. Hence, we have $\lim_{n\to\infty} d_\mathbb{P}\big(w^{(n)}, \mathbb{PN}(Q(p_n))\big) = 0$, and this is a contradiction to Step 1. Furthermore, because we have for all $p \in P$ and $w \in \mathbb{S}^{N-1} \cap \mathbb{P}^{-1}U_\beta\big(\mathbb{PR}(Q(p))\big)$ the relation

$$\|w_n^p\| = \|(\mathbb{1} - Q(p))w\| \overset{\text{Def. 4.4 (iii)}}{\leq} K,$$

the set $W_n$ is bounded.

*Step 4. The relation*

$$\lim_{t \to \infty} \sup_{p \in P} \frac{\sup_{v \in \mathbb{S}^{N-1} \cap \mathcal{N}(Q(p))} \|\Phi(-t, p)v\|}{\inf_{w \in \mathbb{S}^{N-1} \cap \mathbb{P}^{-1}U_\beta(\mathbb{PR}(Q(p)))} \|\Phi(-t, p)w\|} = 0$$

*is fulfilled.*

For $p \in P$, we have

$$\frac{\sup_{v \in \mathbb{S}^{N-1} \cap \mathcal{N}(Q(p))} \|\Phi(-t,p)v\|}{\inf_{w \in \mathbb{S}^{N-1} \cap \mathbb{P}^{-1}U_\beta(\mathbb{P}\mathcal{R}(Q(p)))} \|\Phi(-t,p)w\|}$$

$$\overset{\text{Def. 4.4 (iii)}}{\leq} \frac{\sup_{v \in \mathbb{S}^{N-1} \cap \mathcal{N}(Q(p))} Ke^{-(\gamma+\alpha)t}\|v\|}{\inf_{w \in \mathbb{S}^{N-1} \cap \mathbb{P}^{-1}U_\beta(\mathbb{P}\mathcal{R}(Q(p)))} \|\Phi(-t,p)w_r^p + \Phi(-t,p)w_n^p\|}$$

$$\leq \sup_{w \in \mathbb{S}^{N-1} \cap \mathbb{P}^{-1}U_\beta(\mathbb{P}\mathcal{R}(Q(p)))} \frac{\frac{Ke^{-(\gamma+\alpha)t}}{\|\Phi(-t,p)w_r^p\|}}{\left|1 - \frac{\|\Phi(-t,p)w_n^p\|}{\|\Phi(-t,p)w_r^p\|}\right|}.$$

Please note that for the last inequality, we require that $w_r^p \neq 0$ holds for all $w \in \mathbb{S}^{N-1} \cap \mathbb{P}^{-1}U_\beta(\mathbb{P}\mathcal{R}(Q(p)))$. This is fulfilled, since $W_r$ is bounded away from zero (cf. Step 3). Furthermore, using

$$\frac{Ke^{-(\gamma+\alpha)t}}{\|\Phi(-t,p)w_r^p\|} \overset{\text{Step 2}}{\leq} \frac{Ke^{-(\gamma+\alpha)t}}{\frac{1}{K}e^{-(\gamma-\alpha)t}\|w_r^p\|} = \frac{K^2 e^{-2\alpha t}}{\|w_r^p\|},$$

we obtain

$$\limsup_{t\to\infty} \sup_{p \in P} \sup_{w \in \mathbb{S}^{N-1} \cap \mathbb{P}^{-1}U_\beta(\mathbb{P}\mathcal{R}(Q(p)))} \frac{Ke^{-(\gamma+\alpha)t}}{\|\Phi(-t,p)w_r^p\|} = 0,$$

since $W_r$ is bounded away from zero. Moreover, due to

$$\frac{\|\Phi(-t,p)w_n^p\|}{\|\Phi(-t,p)w_r^p\|} \overset{\text{Def. 4.4 (iii), Step 2}}{\leq} \frac{Ke^{-(\gamma+\alpha)t}\|w_n^p\|}{\frac{1}{K}e^{-(\gamma-\alpha)t}\|w_r^p\|} = \frac{K^2 e^{-2\alpha t}\|w_n^p\|}{\|w_r^p\|},$$

we get

$$\limsup_{t\to\infty} \sup_{p \in P} \sup_{w \in \mathbb{S}^{N-1} \cap \mathbb{P}^{-1}U_\beta(\mathbb{P}\mathcal{R}(Q(p)))} \frac{\|\Phi(-t,p)w_n^p\|}{\|\Phi(-t,p)w_r^p\|} = 0$$

(please note that Step 3 says that $W_n$ is bounded and $W_r$ is bounded away from zero). This implies the assertion.

*Step 5. The relation*

$$\limsup_{t\to\infty} \sup_{p \in P} \frac{\sup_{v \in \mathbb{S}^{N-1} \cap \mathcal{R}(Q(p))} \|\Phi(t,p)v\|}{\inf_{w \in \mathbb{S}^{N-1} \cap \mathbb{P}^{-1}U_\beta(\mathbb{P}\mathcal{N}(Q(p)))} \|\Phi(t,p)w\|} = 0$$

*is fulfilled.*

See proof of Step 4. $\qquad\qquad\qquad\qquad\qquad\qquad\qquad\qquad\qquad\qquad\square$

The following theorem says that ranges and null spaces of invariant projectors give rise to nonautonomous repellers and attractors. Similar questions are treated in PALMER & SIEGMUND [128, Proposition 3.1], where so-called generalized attractor-repeller pairs on the projective space are examined.

**Theorem 4.15 (Ranges and null spaces of invariant projectors as nonautonomous repellers and attractors).** *We suppose that $(\theta, \varphi)$ admits a nonhyperbolic past (future, all-time, respectively) exponential dichotomy with projector $Q$ and consider the nonautonomous dynamical system $(\theta, \mathbb{P}\Phi)$ on the real projective space $\mathbb{P}^{N-1}$. Then the following statements are fulfilled:*

(i) $\mathbb{P}\mathcal{R}(Q)$ *is a past (future, all-time, respectively) repeller,*

(ii) $\mathbb{P}\mathcal{N}(Q)$ *is a past (future, all-time, respectively) attractor,*

(iii) *in case of a nonhyperbolic past exponential dichotomy, we have $\mathbb{P}\mathcal{N}(Q) = \mathbb{P}\mathcal{R}(Q)^*$, and in case of a nonhyperbolic future exponential dichotomy, $\mathbb{P}\mathcal{R}(Q) = \mathbb{P}\mathcal{N}(Q)^*$ is fulfilled.*

*Proof.* In case of a nonhyperbolic past exponential dichotomy, the fact that $\mathbb{P}\mathcal{R}(Q)$ is a past repeller can be proved as in Proposition 3.20 (Step 1 to Step 4), where instead of Proposition 3.19 and Step 3 one should use Lemma 4.14 (i). Moreover, the proof that $\mathbb{P}\mathcal{N}(Q)$ is a past attractor and $\mathbb{P}\mathcal{N}(Q) = \mathbb{P}\mathcal{R}(Q)^*$ is analogous to Step 5 of the proof of Proposition 3.20.

The assertions concerning the case of a nonhyperbolic future exponential dichotomy are now easily obtained by using Proposition 2.32.

In case $(\theta, \varphi)$ admits a nonhyperbolic all-time exponential dichotomy, we now prove that $\mathbb{P}\mathcal{R}(Q)$ is an all-time repeller. First, we choose $\beta > 0$ from Lemma 4.14 (iii). The remaining proof is divided into two steps.

*Step 1.* We have

$$
\begin{aligned}
1 &= \lim_{t \to \infty} \inf_{p \in P} \inf_{0 \neq v \in \mathbb{P}^{-1} U_\beta(\mathbb{P}\mathcal{R}(Q(p)))} \frac{\left\|\Phi(-t,p)v_r^p\right\|}{\left\|\Phi(-t,p)v\right\|} \\
&= \lim_{t \to \infty} \sup_{p \in P} \sup_{0 \neq v \in \mathbb{P}^{-1} U_\beta(\mathbb{P}\mathcal{R}(Q(p)))} \frac{\left\|\Phi(-t,p)v_r^p\right\|}{\left\|\Phi(-t,p)v\right\|} \,,
\end{aligned}
$$

*where $v = v_r^p + v_n^p$ with $v_r^p \in \mathcal{R}(Q(p))$ and $v_n^p \in \mathcal{N}(Q(p))$.*
The first assertion follows from

$$
\begin{aligned}
&\lim_{t \to \infty} \inf_{p \in P} \inf_{0 \neq v \in \mathbb{P}^{-1} U_\beta(\mathbb{P}\mathcal{R}(Q(p)))} \frac{\left\|\Phi(-t,p)v_r^p\right\|}{\left\|\Phi(-t,p)v\right\|} \\
&\geq \left( \lim_{t \to \infty} \sup_{p \in P} \sup_{0 \neq v \in \mathbb{P}^{-1} U_\beta(\mathbb{P}\mathcal{R}(Q(p)))} \frac{\left\|\Phi(-t,p)v_n^p\right\|}{\left\|\Phi(-t,p)v_r^p\right\|} + 1 \right)^{-1} \\
&= \left( \lim_{t \to \infty} \sup_{p \in P} \sup_{v \in \mathbb{P}^{-1} U_\beta(\mathbb{P}\mathcal{R}(Q(p))), v_n^p \neq 0} \frac{\left\|v_n^p\right\| \left\|\Phi(-t,p)\frac{v_n^p}{\|v_n^p\|}\right\|}{\left\|v_r^p\right\| \left\|\Phi(-t,p)\frac{v_r^p}{\|v_r^p\|}\right\|} + 1 \right)^{-1} \\
&\overset{\text{L. 4.14(iii)}}{=} 1
\end{aligned}
$$

and

$$\lim_{t\to\infty} \inf_{p\in P} \inf_{0\neq v\in \mathbb{P}^{-1}U_\beta(\mathbb{PR}(Q(p)))} \frac{\left\|\Phi(-t,p)v_r^p\right\|}{\left\|\Phi(-t,p)v\right\|}$$

$$\leq \left(\lim_{t\to\infty} \sup_{p\in P} \sup_{0\neq v\in \mathbb{P}^{-1}U_\beta(\mathbb{PR}(Q(p)))} \left|1 - \frac{\left\|\Phi(-t,p)v_n^p\right\|}{\left\|\Phi(-t,p)v_r^p\right\|}\right|\right)^{-1}$$

$$= \left(\lim_{t\to\infty} \sup_{p\in P} \sup_{v\in \mathbb{P}^{-1}U_\beta(\mathbb{PR}(Q(p))),v_n^p\neq 0} \left|1 - \frac{\left\|v_n^p\right\|\left\|\Phi(-t,p)\frac{v_n^p}{\|v_n^p\|}\right\|}{\left\|v_r^p\right\|\left\|\Phi(-t,p)\frac{v_r^p}{\|v_r^p\|}\right\|}\right|\right)^{-1}$$

$$\overset{\text{L. 4.14(iii)}}{=} 1. \quad \bullet$$

In both relations, the last equality holds, because the sets

$$\left\{v_n^p : v \in \mathbb{P}^{-1}U_\beta\big(\mathbb{PR}(Q(p))\big)\right\} \quad \text{for all } p \in P$$

are compact and the sets

$$\left\{v_r^p : v \in \mathbb{P}^{-1}U_\beta\big(\mathbb{PR}(Q(p))\big)\right\} \quad \text{for all } p \in P$$

are bounded away from zero (cf. also Step 3 of Lemma 4.14 and Step 1 of Proposition 3.20). The assertion

$$1 = \lim_{t\to\infty} \sup_{p\in P} \sup_{0\neq v\in \mathbb{P}^{-1}U_\beta(\mathbb{PR}(Q(p)))} \frac{\left\|\Phi(-t,p)v_r^p\right\|}{\left\|\Phi(-t,p)v\right\|}$$

follows analogously.

*Step 2. We have*

$$\lim_{t\to\infty} \sup_{p\in P} d_{\mathbb{P}}\big(\mathbb{P}\Phi(-t,p)U_\beta\big(\mathbb{PR}(Q(p))\big)\big|\mathbb{PR}(Q(\theta_{-t}p))\big) = 0,$$

*i.e., $\mathbb{PR}(Q)$ is an all-time repeller.*

With $v_r^p$ and $v_n^p$ defined as in Step 1, the relation

$$\frac{\left\langle\Phi(-t,p)v,\Phi(-t,p)v_r^p\right\rangle^2}{\left\|\Phi(-t,p)v\right\|^2\left\|\Phi(-t,p)v_r^p\right\|^2}$$

$$= \frac{\left\langle\Phi(-t,p)v_n^p,\Phi(-t,p)v_r^p\right\rangle^2}{\left\|\Phi(-t,p)v\right\|^2\left\|\Phi(-t,p)v_r^p\right\|^2} + \frac{\left\|\Phi(-t,p)v_r^p\right\|^2}{\left\|\Phi(-t,p)v\right\|^2} + \frac{2\left\langle\Phi(-t,p)v_n^p,\Phi(-t,p)v_r^p\right\rangle}{\left\|\Phi(-t,p)v\right\|^2}$$

holds for all $t \geq 0, p \in P$ and $v \in \mathbb{S}^{N-1}\cap\mathbb{P}^{-1}U_\beta\big(\mathbb{PR}(Q(p))\big)$ (cf. Step 2 of the proof of Proposition 3.20). Using the Cauchy-Schwartz inequality, we obtain the following relations:

$$0 \leq \limsup_{t \to \infty} \sup_{p \in P} \sup_{v \in \mathbb{S}^{N-1} \cap \mathbb{P}^{-1} U_\beta(\mathbb{PR}(Q(p)))} \frac{\langle \Phi(-t,p)v_n^p, \Phi(-t,p)v_r^p \rangle^2}{\|\Phi(-t,p)v\|^2 \|\Phi(-t,p)v_r^p\|^2}$$

$$\leq \limsup_{t \to \infty} \sup_{p \in P} \sup_{v \in \mathbb{S}^{N-1} \cap \mathbb{P}^{-1} U_\beta(\mathbb{PR}(Q(p)))} \frac{\|\Phi(-t,p)v_n^p\|^2}{\|\Phi(-t,p)v\|^2}$$

$$\overset{\text{L. } 4.14\text{(iii)}}{=} 0$$

and

$$0 \leq \limsup_{t \to \infty} \sup_{p \in P} \sup_{v \in \mathbb{S}^{N-1} \cap \mathbb{P}^{-1} U_\beta(\mathbb{PR}(Q(p)))} \frac{2 |\langle \Phi(-t,p)v_n^p, \Phi(-t,p)v_r^p \rangle|}{\|\Phi(-t,p)v\|^2}$$

$$\leq \limsup_{t \to \infty} \sup_{p \in P} \sup_{v \in \mathbb{S}^{N-1} \cap \mathbb{P}^{-1} U_\beta(\mathbb{PR}(Q(p)))} 2 \frac{\|\Phi(-t,p)v_n^p\|}{\|\Phi(-t,p)v\|} \frac{\|\Phi(-t,p)v_r^p\|}{\|\Phi(-t,p)v\|}$$

$$\overset{\text{Step 1}}{=} \limsup_{t \to \infty} \sup_{p \in P} \sup_{v \in \mathbb{S}^{N-1} \cap \mathbb{P}^{-1} U_\beta(\mathbb{PR}(Q(p)))} \frac{2 \|\Phi(-t,p)v_n^p\|}{\|\Phi(-t,p)v\|}$$

$$\overset{\text{L. } 4.14\text{(iii)}}{=} 0 .$$

Hence, due to Step 1, we have

$$\liminf_{t \to \infty} \inf_{p \in P} \inf_{v \in \mathbb{S}^{N-1} \cap \mathbb{P}^{-1} U_\beta(\mathbb{PR}(Q(p)))} \frac{\langle \Phi(-t,p)v, \Phi(-t,p)v_r^p \rangle^2}{\|\Phi(-t,p)v\|^2 \|\Phi(-t,p)v_r^p\|^2}$$

$$= \liminf_{t \to \infty} \inf_{p \in P} \inf_{v \in \mathbb{S}^{N-1} \cap \mathbb{P}^{-1} U_\beta(\mathbb{PR}(Q(p)))} \left( \frac{\langle \Phi(-t,p)v_n^p, \Phi(-t,p)v_r^p \rangle^2}{\|\Phi(-t,p)v\|^2 \|\Phi(-t,p)v_r^p\|^2} + \right.$$

$$\left. + \frac{\|\Phi(-t,p)v_r^p\|^2}{\|\Phi(-t,p)v\|^2} + \frac{2 \langle \Phi(-t,p)v_n^p, \Phi(-t,p)v_r^p \rangle}{\|\Phi(-t,p)v\|^2} \right)$$

$$= 1 .$$

Using Lemma A.11, this implies that $\mathbb{PR}(Q)$ is an all-time repeller. Moreover, with Proposition 2.32, it is easy to show that $\mathbb{PN}(Q)$ is an all-time attractor. This finishes the proof of this theorem. $\qquad \square$

## 4.2 Dichotomy Spectra

In the previous section, notions of dichotomy have been introduced by localizing attractive and repulsive directions. To classify the strength of attractivity and repulsivity of linear systems, the concept of the dichotomy spectrum is essential. For linear skew product flows with compact base sets, the so-called Sacker-Sell spectrum (see SACKER & SELL [154]) has become widely accepted. In SIEGMUND [172] and AULBACH & SIEGMUND [19], this spectrum has been adapted for arbitrary classes of linear differential and difference equations, respectively (for the noninvertible case, see AULBACH & SIEGMUND [20]). In addition to this dichotomy spectrum, in this section, three other kinds of spectra are introduced with respect to the notions of past, future and finite-time attractivity and repulsivity. Thereby, attention is restricted to the following situation.

**Standing Hypothesis.** We suppose that $(\theta, \varphi)$ is generated by the nonautonomous differential equation

$$\dot{x} = A(t)x, \tag{4.2}$$

$P = \mathbb{T} = \mathbb{R}$, or the nonautonomous difference equation

$$x_{n+1} = A(n)x_n, \tag{4.3}$$

$P = \mathbb{T} = \mathbb{Z}$, where $A : \mathbb{T} \to \mathbb{R}^{N \times N}$ is a continuous function. The base flow fulfills the relation $\theta(t, \tau) = t + \tau$ for all $t, \tau \in \mathbb{T}$ (cf. Section 2.2).

This restriction is necessary, since we do not want to make assumptions concerning the structure of the base flow $\theta$ and the base set $P$ such as compactness, minimality, chain recurrence or invariant connectedness.

We consider unbounded and closed $\mathbb{T}$-intervals $\mathbb{I}$, i.e., $\mathbb{I}$ is of the form $\mathbb{T}_\kappa^-$, $\mathbb{T}_\kappa^+$ or $\mathbb{T}$ for some $\kappa \in \mathbb{T}$. We say, $(\theta, \varphi)$ admits a *nonhyperbolic exponential dichotomy on* $\mathbb{I}$ with growth rate $\gamma \in \mathbb{R}$, constants $\alpha > 0, K \geq 1$ and invariant projector $Q$ if

$$\|\Phi(t, \tau)Q(\tau)\| \leq Ke^{(\gamma - \alpha)t} \quad \text{for all } \tau \in \mathbb{I}, t \geq 0 \text{ with } \tau + t \in \mathbb{I},$$
$$\|\Phi(-t, \tau)(\mathbb{1} - Q(\tau))\| \leq Ke^{-(\gamma + \alpha)t} \quad \text{for all } \tau \in \mathbb{I}, t \geq 0 \text{ with } \tau - t \in \mathbb{I}.$$

The following proposition says that this definition coincides with the notions of nonhyperbolic exponential dichotomy from in the previous section.

**Proposition 4.16.** *Let $\kappa \in \mathbb{T}$ and $\gamma \in \mathbb{R}$. Then the following statements are fulfilled:*

(i) *$(\theta, \varphi)$ admits a nonhyperbolic exponential dichotomy on $\mathbb{T}_\kappa^-$ with growth rate $\gamma$ if and only if $(\theta, \varphi)$ admits a nonhyperbolic past exponential dichotomy with growth rate $\gamma$,*

(ii) *$(\theta, \varphi)$ admits a nonhyperbolic exponential dichotomy on $\mathbb{T}_\kappa^+$ with growth rate $\gamma$ if and only if $(\theta, \varphi)$ admits a nonhyperbolic future exponential dichotomy with growth rate $\gamma$,*

(iii) *$(\theta, \varphi)$ admits a nonhyperbolic exponential dichotomy on $\mathbb{T}$ with growth rate $\gamma$ if and only if $(\theta, \varphi)$ admits a nonhyperbolic all-time exponential dichotomy with growth rate $\gamma$.*

*Proof.* (i) ($\Rightarrow$) The conditions of Proposition 4.10 (i) are fulfilled by choosing $\hat{p} := \kappa$.

($\Leftarrow$) Suppose, $(\theta, \varphi)$ admits a nonhyperbolic past exponential dichotomy with growth rate $\gamma$, constants $\alpha, K$ and projector $Q$. Thus, there exists a $\hat{p} \in \mathbb{T}$ such that the two conditions in Proposition 4.10 (i) are fulfilled. In case $\hat{p} \geq \kappa$, the assertion follows immediately. Otherwise, we define

$$\tilde{K} := \max \Big\{ \max \big\{ \|\Phi(t, \tau)Q(\tau)\| : t, \tau \in [\hat{p}, \kappa] \cap \mathbb{T} \text{ with } t \geq \tau \big\},$$
$$\max \big\{ \|\Phi(t, \tau)(\mathbb{1} - Q(\tau))\| : t, \tau \in [\hat{p}, \kappa] \cap \mathbb{T} \text{ with } t \leq \tau \big\} \Big\}.$$

Then $(\theta, \varphi)$ admits a nonhyperbolic exponential dichotomy on $\mathbb{T}_\kappa^-$ with growth rate $\gamma$ and constants $\alpha, K\tilde{K}$.

(ii) can be shown analogously to (i), and (iii) is obviously fulfilled.    $\square$

In this section, we make use of these alternative characterizations instead of Definition 4.7.

*Remark 4.17.* In case the function $A$ of the differential equation (4.2) or difference equation (4.3) is only defined on an interval of the form $\mathbb{T}_\kappa^-$ or $\mathbb{T}_\kappa^+$ for some $\kappa \in \mathbb{T}$, respectively, the nonautonomous dynamical system generated by this equation does not fulfill the hypotheses of this chapter. Due to Proposition 4.16, however, we are able to use the notions of past or future exponential dichotomy and the notions of past and future dichotomy spectrum (see Definition 4.20 below) also for these types of equations.

Given an invariant projector $Q$, the fibres of $\mathcal{R}(Q)$ and $\mathcal{N}(Q)$, respectively, have the same dimension, since the base set $P$ is a trajectory of base flow $\theta$. We therefore define the rank of $Q$ by

$$\operatorname{rk} Q := \dim \mathcal{R}(Q) := \dim \mathcal{R}(Q(t)) \quad \text{for all } t \in \mathbb{T},$$

and we set

$$\dim \mathcal{N}(Q) := \dim \mathcal{N}(Q(t)) \quad \text{for all } t \in \mathbb{T}.$$

**Proposition 4.18.** *Suppose, both $Q$ and $\hat{Q}$ are invariant projectors of a nonhyperbolic past exponential (future exponential, all-time exponential, $(\tau, T)$-, respectively) dichotomy with growth rate $\gamma$. Then $\operatorname{rk} Q = \operatorname{rk} \hat{Q}$ is fulfilled.*

*Proof.* In case of a nonhyperbolic past (future, all-time, respectively) exponential dichotomy, the assertion follows directly from Proposition 4.11. Arguing negatively, we suppose that $(\theta, \varphi)$ admits a $(\tau, T)$-dichotomy with two invariant projectors $Q$ and $\hat{Q}$ such that $\operatorname{rk} Q < \operatorname{rk} \hat{Q}$. Thus,

$$\dim \big( \mathcal{N}(Q) \cap \mathcal{R}(\hat{Q}) \big) = \dim \mathcal{N}(Q) + \dim \mathcal{R}(\hat{Q}) - \dim \big( \mathcal{N}(Q) + \mathcal{R}(\hat{Q}) \big)$$
$$> \dim \mathcal{N}(Q) + \dim \mathcal{R}(Q) - \dim \big( \mathcal{N}(Q) + \mathcal{R}(\hat{Q}) \big) \geq 0.$$

Hence, there exists a nonzero element $\xi \in \mathcal{N}(Q(\tau)) \cap \mathcal{R}(\hat{Q}(\tau))$. We obtain

$$\|\Phi(T,\tau)\xi\| < \|\xi\| = \|\Phi(-T,\tau+T)\Phi(T,\tau)\xi\| < \|\Phi(T,\tau)\xi\|,$$

since $0 \neq \xi \in \mathcal{R}(\hat{Q}(\tau))$ and $0 \neq \Phi(T,\tau)\xi \in \mathcal{N}(Q(\tau+T))$. This contradiction finishes the proof of this proposition. $\qquad\square$

As indicated in Remark 4.12, an invariant projector is uniquely determined only in case of a nonhyperbolic all-time exponential dichotomy. The degree of nonuniqueness of projectors of past and future exponential dichotomies is described in the following lemma, which is adapted from AULBACH & SIEG-MUND [20, Lemma 2.4].

**Lemma 4.19.** *The following statements are fulfilled:*

(i) *We assume that $(\theta, \varphi)$ admits a nonhyperbolic past exponential dichotomy with growth rate $\gamma$ and projector $Q$, and $\hat{Q}$ is another invariant projector with*

$$\sup_{t \in \mathbb{T}_0^-} \|\hat{Q}(t)\| < \infty \quad and \quad \mathcal{N}(Q) = \mathcal{N}(\hat{Q}).$$

*Then $(\theta, \varphi)$ also admits a nonhyperbolic past exponential dichotomy with growth rate $\gamma$ and projector $\hat{Q}$.*

(ii) *We assume that $(\theta, \varphi)$ admits a nonhyperbolic future exponential dichotomy with growth rate $\gamma$ and projector $Q$, and $\hat{Q}$ is another invariant projector with*

$$\sup_{t \in \mathbb{T}_0^+} \|\hat{Q}(t)\| < \infty \quad and \quad \mathcal{R}(Q) = \mathcal{R}(\hat{Q}).$$

*Then $(\theta, \varphi)$ also admits a nonhyperbolic future exponential dichotomy with growth rate $\gamma$ and projector $\hat{Q}$.*

*Proof.* (i) Suppose, $(\theta, \varphi)$ admits a nonhyperbolic exponential dichotomy on $\mathbb{T}_0^-$ with growth rate $\gamma$, constants $\alpha > 0$, $K \geq 1$ and projector $Q$, and let $\hat{Q}$ be given as above. First, we observe that $\sup_{t \in \mathbb{T}_0^-} \|Q(t)\| \leq K$, and we define

$$M := \sup_{t \in \mathbb{T}_0^-} \|\hat{Q}(t)\|.$$

The relation $\mathcal{N}(Q) = \mathcal{N}(\hat{Q})$ implies the two equations

$$(\mathbf{1} - \hat{Q}) = (\mathbf{1} - Q)(\mathbf{1} - \hat{Q}) \quad and \quad \hat{Q} = (\mathbf{1} - Q + \hat{Q})Q.$$

The first equation yields for all $\tau \in \mathbb{T}_0^-$ and $t \geq 0$,

$$\left\|\Phi(-t,\tau)\big(\mathbb{1}-\hat{Q}(\tau)\big)\right\| = \left\|\Phi(-t,\tau)\big(\mathbb{1}-Q(\tau)\big)\big(\mathbb{1}-\hat{Q}(\tau)\big)\right\|$$
$$\leq \left\|\Phi(-t,\tau)\big(\mathbb{1}-Q(\tau)\big)\right\|\left\|\mathbb{1}-\hat{Q}(\tau)\right\|$$
$$\leq K(1+M)e^{-(\gamma+\alpha)t}.$$

Using the invariance of $Q$ and $\hat{Q}$, the second equation implies

$$\left\|\Phi(t,\tau)\hat{Q}(\tau)\right\| = \left\|\Phi(t,\tau)\big(\mathbb{1}-Q(\tau)+\hat{Q}(\tau)\big)Q(\tau)\right\|$$
$$\leq \left\|\big(\mathbb{1}-Q(\tau+t)+\hat{Q}(\tau+t)\big)\right\|\left\|\Phi(t,\tau)Q(\tau)\right\|$$
$$\leq K(1+K+M)e^{(\gamma-\alpha)t}$$

for all $\tau \in \mathbb{T}_0^-$ and $t \geq 0$ with $\tau + t \in \mathbb{T}_0^-$.
The assertion (ii) can be proved similarly.                    $\square$

It is crucial for the definition of the dichotomy spectra, for which growth rates, the linear nonautonomous dynamical systems $(\theta,\varphi)$ admits a nonhyperbolic dichotomy. In case of a past (future, all-time, respectively) exponential dichotomy, we will not exclude growth rates $\gamma = \pm\infty$ from our considerations. We say, $(\theta,\varphi)$ admits a nonhyperbolic dichotomy with growth rate $\infty$ if there exists a $\gamma \in \mathbb{R}$ such that $(\theta,\varphi)$ admits a nonhyperbolic dichotomy with growth rate $\gamma$ and projector $P_\gamma \equiv \mathbb{1}$. Accordingly, we say that $(\theta,\varphi)$ admits a nonhyperbolic dichotomy with growth rate $-\infty$ if there exists a $\gamma \in \mathbb{R}$ such that $(\theta,\varphi)$ admits a nonhyperbolic dichotomy with growth rate $\gamma$ and projector $P_\gamma \equiv 0$.

**Definition 4.20 (Dichotomy spectra).**

(i) *The* past dichotomy spectrum *of $(\theta,\varphi)$ is defined by*

$$\Sigma_\Phi^\leftarrow := \big\{\gamma \in \overline{\mathbb{R}} : (\theta,\varphi) \text{ does not admit a nonhyperbolic past}$$
$$\text{exponential dichotomy with growth rate } \gamma\big\}.$$

(ii) *The* future dichotomy spectrum *of $(\theta,\varphi)$ is defined by*

$$\Sigma_\Phi^\rightarrow := \big\{\gamma \in \overline{\mathbb{R}} : (\theta,\varphi) \text{ does not admit a nonhyperbolic future}$$
$$\text{exponential dichotomy with growth rate } \gamma\big\}.$$

(iii) *The* all-time dichotomy spectrum *of $(\theta,\varphi)$ is defined by*

$$\Sigma_\Phi^\leftrightarrow := \big\{\gamma \in \overline{\mathbb{R}} : (\theta,\varphi) \text{ does not admit a nonhyperbolic all-time}$$
$$\text{exponential dichotomy with growth rate } \gamma\big\}.$$

(iv) *Given $\tau \in \mathbb{T}$ and $T \in \mathbb{T}^+$, the* $(\tau,T)$-dichotomy spectrum *of $(\theta,\varphi)$ is defined by*

$$\Sigma_\Phi^{(\tau,T)} := \big\{\gamma \in \mathbb{R} : (\theta,\varphi) \text{ does not admit a nonhyperbolic}$$
$$(\tau,T)\text{-dichotomy with growth rate } \gamma\big\}.$$

*The corresponding* resolvent sets *are defined as follows:*

$$\rho_\Phi^\leftarrow := \overline{\mathbb{R}} \setminus \Sigma_\Phi^\leftarrow, \qquad \rho_\Phi^\rightarrow := \overline{\mathbb{R}} \setminus \Sigma_\Phi^\rightarrow,$$

$$\rho_\Phi^\leftrightarrow := \overline{\mathbb{R}} \setminus \Sigma_\Phi^\leftrightarrow \quad and \quad \rho_\Phi^{(\tau,T)} := \overline{\mathbb{R}} \setminus \Sigma_\Phi^{(p,T)}.$$

In regard to the Standing Hypothesis, also the notation $\Sigma_A$ and $\rho_A$ is used for the dichotomy spectra and resolvent sets of $\dot{x} = A(t)x$ and $x_{n+1} = A(n)x_n$, respectively.

*Remark 4.21.*

(i) The all-time dichotomy spectrum without $\{-\infty, \infty\}$, i.e., $\Sigma_\Phi^\leftrightarrow \cap \mathbb{R}$, coincides with the dichotomy spectrum for differential equations introduced in SIEGMUND [172] (see also SIEGMUND [171]). In case of linear difference equations, $\exp\left(\Sigma_\Phi^\leftrightarrow \cap \mathbb{R}\right)$ is the dichotomy spectrum introduced in AULBACH & SIEGMUND [19].

(ii) In contrast to the past, future or all-time dichotomy spectrum, the notion of $(\tau, T)$-dichotomy spectrum is not invariant with respect to a change of the norm to an equivalent norm (cf. also Remark 4.5 (iii) and Remark 2.18).

(iii) From Proposition 4.6, we obtain directly $\Sigma_\Phi^\leftarrow \subset \Sigma_\Phi^\leftrightarrow$ and $\Sigma_\Phi^\rightarrow \subset \Sigma_\Phi^\leftrightarrow$.

The aim of the following lemma is to analyze the topological structure of the resolvent sets.

**Lemma 4.22.** *We suppose that* $\rho_\Phi := \rho_\Phi^\leftarrow, \rho_\Phi^\rightarrow, \rho_\Phi^\leftrightarrow, \rho_\Phi^{(\tau,T)}$, *respectively. Then* $\rho_\Phi \cap \mathbb{R}$ *is open, more precisely, for all* $\gamma \in \rho_\Phi \cap \mathbb{R}$, *there exists an* $\varepsilon > 0$ *such that* $U_\varepsilon(\gamma) \subset \rho_\Phi$. *Furthermore, the relation* $\operatorname{rk} Q_\zeta = \operatorname{rk} Q_\gamma$ *is fulfilled for all* $\zeta \in U_\varepsilon(\gamma)$ *and every invariant projector* $Q_\gamma$ *and* $Q_\zeta$ *of the nonhyperbolic dichotomies of* $(\theta, \varphi)$ *with growth rates* $\gamma$ *and* $\zeta$, *respectively.*

*Proof.* We first treat the case $\rho_\Phi = \rho_\Phi^\leftarrow, \rho_\Phi^\rightarrow, \rho_\Phi^\leftrightarrow$ and choose $\gamma \in \rho_\Phi$ arbitrarily. Since $(\theta, \varphi)$ admits a nonhyperbolic exponential dichotomy on $\mathbb{I} = \mathbb{T}_0^-, \mathbb{T}_0^+, \mathbb{T}$ with growth rate $\gamma$, respectively, there exists an invariant projector $Q_\gamma$ and constants $\alpha > 0$, $K \geq 1$ such that

$$\|\Phi(t,\tau)Q_\gamma(\tau)\| \leq Ke^{(\gamma-\alpha)t} \quad \text{for all } \tau \in \mathbb{I}, t \geq 0 \text{ with } \tau + t \in \mathbb{I},$$

$$\|\Phi(-t,\tau)(1 - Q_\gamma(\tau))\| \leq Ke^{-(\gamma+\alpha)t} \quad \text{for all } \tau \in \mathbb{I}, t \geq 0 \text{ with } \tau - t \in \mathbb{I}.$$

We set $\varepsilon := \alpha/2$ and choose $\zeta \in U_\varepsilon(\gamma)$. Thus,

$$\|\Phi(t,\tau)Q_\gamma(\tau)\| \leq Ke^{(\zeta-\alpha/2)t} \quad \text{for all } \tau \in \mathbb{I}, t \geq 0 \text{ with } \tau + t \in \mathbb{I},$$

$$\|\Phi(-t,\tau)(1 - Q_\gamma(\tau))\| \leq Ke^{-(\zeta+\alpha/2)t} \quad \text{for all } \tau \in \mathbb{I}, t \geq 0 \text{ with } \tau - t \in \mathbb{I}.$$

This yields $\zeta \in \rho_\Phi$. Since Proposition 4.18 says that the ranks of the projectors of nonhyperbolic exponential dichotomies on $\mathbb{I}$ with the same growth rate are equal, we have $\operatorname{rk} Q_\zeta = \operatorname{rk} Q_\gamma$ for any projector $Q_\zeta$ of the nonhyperbolic exponential dichotomy on $\mathbb{I}$ with growth rate $\zeta$. In case of finite-time intervals and $\gamma \in \rho_\Phi$, there exists an invariant projector $Q_\gamma$ such that

$$\|\Phi(T,\tau)\xi\| < e^{\gamma T}\|\xi\| \qquad \text{for all } 0 \neq \xi \in \mathcal{R}(Q_\gamma(\tau)),$$
$$\|\Phi(-T,\tau+T)\xi\| < e^{-\gamma T}\|\xi\| \qquad \text{for all } 0 \neq \xi \in \mathcal{N}(Q_\gamma(\tau+T)).$$

We define

$$\beta := \max \left\{ \max_{0 \neq \xi \in \mathcal{R}(Q_\gamma(\tau))} \frac{\|\Phi(T,\tau)\xi\|}{e^{\gamma T}\|\xi\|}, \max_{0 \neq \xi \in \mathcal{N}(Q_\gamma(\tau+T))} \frac{\|\Phi(T,\tau)\xi\|}{e^{\gamma T}\|\xi\|} \right\} < 1$$

and set $\varepsilon := \ln\beta/(2T)$. Thus, for all $\zeta \in U_\varepsilon(\gamma)$, we have

$$\|\Phi(T,\tau)\xi\| < e^{\zeta T}\|\xi\| \qquad \text{for all } 0 \neq \xi \in \mathcal{R}(Q_\gamma(\tau)),$$
$$\|\Phi(-T,\tau+T)\xi\| < e^{-\zeta T}\|\xi\| \qquad \text{for all } 0 \neq \xi \in \mathcal{N}(Q_\gamma(\tau+T)).$$

This implies $\zeta \in \rho_\Phi$. The equality of the ranks of the invariant projectors follows from Proposition 4.18. $\qquad\square$

**Lemma 4.23.** *Assume, the resolvent is of the form $\rho_\Phi := \rho_\Phi^\leftarrow, \rho_\Phi^\rightarrow, \rho_\Phi^\leftrightarrow, \rho_\Phi^{(\tau,T)}$, respectively, let $\gamma_1, \gamma_2 \in \rho_\Phi \cap \mathbb{R}$ with $\gamma_1 < \gamma_2$, and choose invariant projectors $Q_{\gamma_1}$ and $Q_{\gamma_2}$ for the corresponding nonhyperbolic dichotomies with growth rates $\gamma_1$ and $\gamma_2$. Then we have $\operatorname{rk} Q_{\gamma_1} \leq \operatorname{rk} Q_{\gamma_2}$. Moreover, $[\gamma_1, \gamma_2] \subset \rho_\Phi$ is fulfilled if and only if $\operatorname{rk} Q_{\gamma_1} = \operatorname{rk} Q_{\gamma_2}$.*

*Proof.* We first prove the relation $\operatorname{rk} Q_{\gamma_1} \leq \operatorname{rk} Q_{\gamma_2}$. In case $\rho_\Phi = \rho_\Phi^\leftarrow, \rho_\Phi^\rightarrow, \rho_\Phi^\leftrightarrow$, respectively, this is a direct consequence of Proposition 4.11, since $\mathcal{S}_{\gamma_1} \subset \mathcal{S}_{\gamma_2}$ and $\mathcal{U}_{\gamma_1} \supset \mathcal{U}_{\gamma_2}$. In case of finite time intervals, we observe that the relation $\mathcal{R}(Q_{\gamma_1}) \cap \mathcal{N}(Q_{\gamma_2}) = \mathbb{T} \times \{0\}$ holds, because $0 \neq \xi \in \mathcal{R}(Q_{\gamma_1}(\tau)) \cap \mathcal{N}(Q_{\gamma_2}(\tau))$ would satisfy

$$\|\xi\| = \|\Phi(-T,\tau+T)\Phi(T,\tau)\xi\| < e^{-\gamma_2 T}\|\Phi(T,\tau)\xi\| < e^{-\gamma_2 T}e^{+\gamma_1 T}\|\xi\| < \|\xi\|.$$

This yields

$$0 = \dim\left(\mathcal{R}(Q_{\gamma_1}) \cap \mathcal{N}(Q_{\gamma_2})\right)$$
$$= \operatorname{rk} Q_{\gamma_1} + \dim\mathcal{N}(Q_{\gamma_2}) - \dim\left(\mathcal{R}(Q_{\gamma_1}) + \mathcal{N}(Q_{\gamma_2})\right),$$

and therefore,

$$\operatorname{rk} Q_{\gamma_2} = \operatorname{rk} Q_{\gamma_1} + N - \dim\left(\mathcal{R}(Q_{\gamma_1}) + \mathcal{N}(Q_{\gamma_2})\right) \geq \operatorname{rk} Q_{\gamma_1}.$$

Assume that $[\gamma_1, \gamma_2] \subset \rho_\Phi$. Arguing negatively, suppose that $\operatorname{rk} Q_{\gamma_1} \neq \operatorname{rk} Q_{\gamma_2}$. We choose invariant projectors $Q_\gamma$ for the nonhyperbolic dichotomies of $(\theta, \varphi)$ with growth rate $\gamma$ for all $\gamma \in (\gamma_1, \gamma_2)$ and define

$$\zeta_0 := \sup\left\{\zeta \in [\gamma_1, \gamma_2] : \mathrm{rk}\, Q_\zeta \neq \mathrm{rk}\, Q_{\gamma_2}\right\}.$$

Due to Lemma 4.22, there exists an $\varepsilon > 0$ such that $\mathrm{rk}\, Q_{\zeta_0} = \mathrm{rk}\, Q_\zeta$ for all $\zeta \in U_\varepsilon(\zeta_0)$. This is a contradiction to the definition of $\zeta_0$. Conversely, let $\mathrm{rk}\, Q_{\gamma_1} = \mathrm{rk}\, Q_{\gamma_2}$. We first treat the case $\rho_\Phi = \rho_\Phi^\leftarrow$. Because of $\mathrm{rk}\, Q_{\gamma_1} = \mathrm{rk}\, Q_{\gamma_2}$, Proposition 4.11 yields that $\mathcal{N}(Q_{\gamma_1}) = \mathcal{N}(Q_{\gamma_2})$. Due to Lemma 4.19, $Q_{\gamma_2}$ is an invariant projector of the nonhyperbolic past exponential dichotomy with growth rate $\gamma_1$. Thus, we have

$$\|\Phi(t, \tau)Q_{\gamma_2}(\tau)\| \leq K_1 e^{(\gamma_1 - \alpha_1)t} \quad \text{for all } t \geq 0, \tau \leq 0 \text{ with } t + \tau \leq 0$$

for some $K_1 \geq 1$ and $\alpha_1 > 0$. $Q_{\gamma_2}$ is also projector of the nonhyperbolic past exponential dichotomy with growth rate $\gamma_2$. Hence,

$$\left\|\Phi(-t, \tau)(\mathbb{1} - Q_{\gamma_2}(\tau))\right\| \leq K_2 e^{-(\gamma_2 + \alpha_2)t} \quad \text{for all } t \geq 0, \tau \leq 0$$

is fulfilled for some $K_2 \geq 1$ and $\alpha_2 > 0$. For all $\gamma \in [\gamma_1, \gamma_2]$, these two inequalities imply by setting $K := \max\{K_1, K_2\}$ and $\alpha := \min\{\alpha_1, \alpha_2\}$ that

$$\|\Phi(t, \tau)Q_{\gamma_2}(\tau)\| \leq K e^{(\gamma - \alpha)t} \quad \text{for all } t \geq 0, \tau \leq 0 \text{ with } t + \tau \leq 0,$$
$$\left\|\Phi(-t, \tau)(\mathbb{1} - Q_{\gamma_2}(\tau))\right\| \leq K e^{-(\gamma + \alpha)t} \quad \text{for all } t \geq 0, \tau \leq 0.$$

This means $\gamma \in \rho_\Phi$, and therefore, $[\gamma_1, \gamma_2] \subset \rho_\Phi$. The case $\rho_\Phi = \rho_\Phi^\rightarrow, \rho_\Phi^\leftrightarrow$ is treated analogously. It remains to show the implication for the $(\tau, T)$-resolvent set. We have already seen at the beginning of this proof that $\mathcal{R}(Q_{\gamma_1}) \cap \mathcal{N}(Q_{\gamma_2}) = \mathbb{T} \times \{0\}$. Since $\mathrm{rk}\, Q_{\gamma_1} = \mathrm{rk}\, Q_{\gamma_2}$, this implies the existence of an invariant projector $Q$ with $\mathcal{N}(Q) = \mathcal{N}(Q_{\gamma_2})$ and $\mathcal{R}(Q) = \mathcal{R}(Q_{\gamma_1})$. Thus, for all $\gamma \in [\gamma_1, \gamma_2]$, we have

$$\|\Phi(T, \tau)\xi\| < e^{\gamma T}\|\xi\| \quad \text{for all } 0 \neq \xi \in \mathcal{R}(Q_{(\gamma_1)}(\tau)),$$
$$\|\Phi(-T, \tau + T)\xi\| < e^{-\gamma T}\|\xi\| \quad \text{for all } 0 \neq \xi \in \mathcal{N}(Q_{(\gamma_2)}(\tau + T)).$$

This implies $[\gamma_1, \gamma_2] \subset \rho_\Phi$ and finishes the proof of this lemma. □

For arbitrarily chosen $a \in \mathbb{R}$, we define

$$[-\infty, a] := (-\infty, a] \cup \{-\infty\}, \qquad [a, \infty] := [a, \infty) \cup \{\infty\},$$
$$[-\infty, -\infty] := \{-\infty\} \qquad [\infty, \infty] := \{\infty\},$$
$$\text{and} \qquad [-\infty, \infty] = \overline{\mathbb{R}}.$$

We now state the main result of this section.

**Theorem 4.24 (Spectral Theorem).** *We consider dichotomy spectra of the form $\Sigma_\Phi := \Sigma_\Phi^\leftarrow, \Sigma_\Phi^\rightarrow, \Sigma_\Phi^\leftrightarrow, \Sigma_\Phi^{(\tau,T)}$. Then there exists an $n \in \{1, \ldots, N\}$ such that*

$$\Sigma_\Phi = [a_1, b_1] \cup \cdots \cup [a_n, b_n]$$

*with $-\infty \leq a_1 \leq b_1 < a_2 \leq b_2 < \cdots < a_n \leq b_n \leq \infty$. In case $\Sigma_\Phi = \Sigma_\Phi^{(\tau,T)}$, we have $-\infty < a_1$ and $b_n < \infty$.*

*Proof.* Due to Lemma 4.22, the set $\rho_\Phi \cap \mathbb{R}$ is open. Therefore, $\Sigma_\Phi \cap \mathbb{R}$ is the disjoint union of closed intervals. In case $\Sigma_\Phi = \Sigma_\Phi^{(\tau,T)}$, the boundedness of $\Sigma_\Phi$ is obvious, and if $\Sigma_\Phi = \Sigma_\Phi^\leftarrow, \Sigma_\Phi^\rightarrow, \Sigma_\Phi^\leftrightarrow$, respectively, then the relation $(-\infty, b_1] \subset \Sigma_\Phi$ implies $[-\infty, b_1] \subset \Sigma_\Phi$, because the assumption of the existence of a $\gamma \in \mathbb{R}$ such that $(\theta, \varphi)$ admits a nonhyperbolic dichotomy with growth rate $\gamma$ and projector $Q_\gamma \equiv 0$ leads to $(-\infty, \gamma] \subset \rho_\Phi$ using Lemma 4.9. This is a contradiction. Analogously, it follows from $[a_n, \infty) \subset \Sigma_\Phi$ that $[a_n, \infty] \subset \Sigma_\Phi$. To show the relation $n \leq N$, we assume that $n \geq N + 1$. Thus, there exist

$$\zeta_1 < \zeta_2 < \cdots < \zeta_N \in \rho_\Phi$$

such that the $N + 1$ intervals

$$(-\infty, \zeta_1), (\zeta_1, \zeta_2), \ldots, (\zeta_N, \infty)$$

have nonempty intersection with the spectrum $\Sigma_\Phi$. It follows from Lemma 4.23 that

$$0 \leq \operatorname{rk} Q_{\zeta_1} < \operatorname{rk} Q_{\zeta_2} < \cdots < \operatorname{rk} Q_{\zeta_N} \leq N$$

is fulfilled for invariant projectors $Q_{\zeta_i}$ of the nonhyperbolic dichotomy with growth rate $\zeta_i$, $i \in \{1, \ldots, n\}$. This implies either $\operatorname{rk} Q_{\zeta_1} = 0$ or $\operatorname{rk} Q_{\zeta_N} = N$. Thus,

$$[-\infty, \zeta_1] \cap \Sigma_\Phi = \emptyset \quad \text{or} \quad [\zeta_N, \infty] \cap \Sigma_\Phi = \emptyset,$$

and this is a contradiction. To show $n \geq 1$, we assume that $\Sigma_\Phi = \emptyset$. This implies $\{-\infty, \infty\} \subset \rho_\Phi$. Thus, there exist $\zeta_1, \zeta_2 \in \mathbb{R}$ such that $(\theta, \varphi)$ admits a nonhyperbolic dichotomy with growth rate $\zeta_1$ and projector $Q_{\zeta_1} \equiv 0$ and a nonhyperbolic dichotomy with growth rate $\zeta_2$ and projector $Q_{\zeta_2} \equiv \mathbb{1}$. Applying Lemma 4.23, we get $(\zeta_1, \zeta_2) \cap \Sigma_\Phi \neq \emptyset$. This contradiction yields $n \geq 1$ and finishes the proof of this theorem. □

In the following example, spectra of scalar linear differential equations are studied.

*Example 4.25.* We consider scalar differential equations of the form

$$\dot{x} = a(t)x,$$

where $a : \mathbb{R} \to \mathbb{R}$ is a continuous function. We have

$$\Phi(t, \tau) = \exp\left(\int_\tau^{\tau+t} a(s)\, ds\right) \quad \text{for all } t, \tau \in \mathbb{R}.$$

The Spectral Theorem says that the past, future, all-time and $(\tau, T)$-dichotomy spectra consist of exactly one closed interval. Furthermore, due to Remark 4.5 (iv), the $(\tau, T)$-dichotomy spectrum fulfills

$$\Sigma_\Phi^{(\tau,T)} = \left\{|\Phi(T, \tau)|\right\}.$$

The following examples show that the past, future and all-time dichotomy spectra can be more complicated.

(i) $\Sigma_{\overleftarrow{\Phi}} = \Sigma_{\overrightarrow{\Phi}} = \Sigma_{\overleftrightarrow{\Phi}} = \{\infty\}$ for $a(t) := |t|$ for all $t \in \mathbb{R}$.
    *Proof.* For $\gamma \in \mathbb{R}$, we have

$$\Phi_\gamma(t,\tau) = \exp\left(\int_\tau^{\tau+t} (|s| - \gamma)\, ds\right) \qquad \text{for all } t, \tau \in \mathbb{R}.$$

Since for all $s \in \mathbb{R}$ with $|s| \geq \gamma + 1$, the relation $|s| - \gamma \geq 1$ is fulfilled, $(\theta, \Phi_\gamma)$ admits a nonhyperbolic exponential dichotomy on $\mathbb{R}_{-|\gamma|-1}^-$ and $\mathbb{R}_{|\gamma|+1}^+$ with growth rate 0, constants $\alpha = 1$, $K = 1$ and invariant projector 0. Moreover, Proposition 4.16 (i), (ii) implies that $\Sigma_{\overleftarrow{\Phi}} = \Sigma_{\overrightarrow{\Phi}} = \{\infty\}$. The remaining assertion $\Sigma_{\overleftrightarrow{\Phi}} = \{\infty\}$ is a consequence of Proposition 4.6 (ii).

(ii) $\Sigma_{\overleftarrow{\Phi}} = \{-\infty\}$, $\Sigma_{\overrightarrow{\Phi}} = \{\infty\}$ and $\Sigma_{\overleftrightarrow{\Phi}} = \overline{\mathbb{R}}$ for $a(t) := t$ for all $t \in \mathbb{R}$.
    *Proof.* The assertions concerning the past and future dichotomy spectrum are proved analogously to (i). Concerning the all-time dichotomy spectrum, we assume to the contrary that there exists a $\gamma \in \mathbb{R}$ such that $\Phi_\gamma$ admits an all-time exponential dichotomy. Please note that the relation

$$\Phi_\gamma(t,\tau) = \exp\left(\frac{1}{2}t^2 + \tau t + \gamma t\right) \qquad \text{for all } t, \tau \in \mathbb{R}$$

holds. For the corresponding invariant projector $Q_\gamma$, there are only the possibilities $Q_\gamma \equiv 0$ or $Q_\gamma \equiv 1$. In case $Q_\gamma \equiv 1$, the dichotomy estimate

$$\Phi_\gamma(t,0) = \exp\left(\frac{1}{2}t^2 + \gamma t\right) \leq Ke^{-\alpha t} \quad \text{for all } t \geq 0$$

yields a contradiction in the limit $t \to \infty$. Analogously, the case $Q_\gamma \equiv 0$ is treated.

(iii) $\Sigma_{\overleftarrow{\Phi}} = [-\infty, \beta]$, $\Sigma_{\overrightarrow{\Phi}} = \{\beta\}$ and $\Sigma_{\overleftrightarrow{\Phi}} = [-\infty, \beta]$ for

$$a(t) := \begin{cases} \beta & : \ t \geq -1 \\ \beta - n + n(t + 2^{2n} + 1) & : \ t \in \left[-2^{2n} - 1, -2^{2n}\right] \\ \beta - n & : \ t \in \left[-2^{2n+1}, -2^{2n} - 1\right] \\ \beta - n(t + 2^{2n+1} + 1) & : \ t \in \left[-2^{2n+1} - 1, -2^{2n+1}\right] \\ \beta & : \ t \in \left[-2^{2(n+1)}, -2^{2n+1} - 1\right] \end{cases}.$$

In all cases above, $n \in \mathbb{N}_0$.
    *Proof.* The statement concerning $\Sigma_{\overrightarrow{\Phi}}$ is clear. To compute $\Sigma_{\overleftarrow{\Phi}}$, assume to the contrary that for some $\gamma \leq \beta$, the linear nonautonomous dynamical system $(\theta, \Phi_\gamma)$ admits a past exponential dichotomy with projector $Q_\gamma$. In the one-dimensional context, there are only the possibilities $Q_\gamma \equiv 0$ or $Q_\gamma \equiv 1$. In case $Q_\gamma \equiv 1$, we have the dichotomy estimate

$$\Phi_\gamma(t,\tau) = \exp\left(\int_\tau^{\tau+t} (a(s) - \gamma)\, ds\right) \leq Ke^{-\alpha t} \quad \text{for all } 0 \leq t \leq -\tau$$

for some $K \geq 1$ and $\alpha > 0$. Let $n \in \mathbb{N}_0$ with $K \exp\left(-\alpha\left(2^{2n+1} - 1\right)\right) < 1$. Then

$$\Phi_\gamma\left(2^{2n+1} - 1, -2^{2(n+1)}\right) = \exp\left(\int_{-2^{2(n+1)}}^{-2^{2n+1}-1} \underbrace{(\beta - \gamma)}_{\geq 0}\, ds\right) \geq 1\,.$$

This is a contradiction. In case $Q_\gamma \equiv 0$, we have the dichotomy estimate

$$\Phi_\gamma(-t, \tau) = \exp\left(\int_\tau^{\tau-t} (a(s) - \gamma)\, ds\right) \leq K e^{-\alpha t} \quad \text{for all } \tau \leq 0, t \geq 0$$

for some $K \geq 1$ and $\alpha > 0$. We choose an integer $n \in \mathbb{N}_0$ such that $K \exp\left(-\alpha\left(2^{2n} - 1\right)\right) < 1$ and $\beta - n - \gamma \leq 0$. Then

$$\Phi_\gamma\left(-2^{2n+1}, -2^{2n} - 1\right) = \exp\left(\int_{-2^{2n}-1}^{-2^{2n+1}} \underbrace{(\beta - n - \gamma)}_{\leq 0}\, ds\right) \geq 1\,.$$

This is also a contradiction. It is easy to see that for $\gamma > \beta$, the linear nonautonomous dynamical system $(\theta, \Phi_\gamma)$ admits a past exponential dichotomy with projector $Q_\gamma \equiv \mathbb{1}$. Hence, we have $\Sigma_{\Phi}^{\leftarrow} = [-\infty, \beta]$. Due to Remark 4.21 (iii), $\Sigma_{\Phi}^{\leftrightarrow} \supset \Sigma_{\Phi}^{\leftarrow} \cup \Sigma_{\Phi}^{\rightarrow} = [-\infty, \beta]$ is fulfilled. It is also easily shown that for $\gamma > \beta$, $(\theta, \Phi_\gamma)$ admits an all-time exponential dichotomy with projector $Q_\gamma \equiv \mathbb{1}$. Thus, we obtain $\Sigma_{\Phi}^{\leftrightarrow} = [-\infty, \beta]$.

(iv) $\Sigma_{\Phi}^{\leftarrow} = \{\beta\}$, $\Sigma_{\Phi}^{\rightarrow} = [\beta, \infty]$ and $\Sigma_{\Phi}^{\leftrightarrow} = [\beta, \infty]$ for

$$a(t) := \begin{cases} \beta & : \ t \leq 1 \\ \beta + n\left(t - 2^{2n}\right) & : \ t \in \left[2^{2n}, 2^{2n} + 1\right] \\ \beta + n & : \ t \in \left[2^{2n} + 1, 2^{2n+1}\right] \\ \beta + n - n\left(t - 2^{2n+1}\right) & : \ t \in \left[2^{2n+1}, 2^{2n+1} + 1\right] \\ \beta & : \ t \in \left[2^{2n+1} + 1, 2^{2(n+1)}\right] \end{cases}.$$

In all cases above, $n \in \mathbb{N}_0$.
*Proof.* See proof of (iii).

(v) $\Sigma_{\Phi}^{\leftarrow} = \{\beta\}$, $\Sigma_{\Phi}^{\rightarrow} = [\beta, \delta]$ and $\Sigma_{\Phi}^{\leftrightarrow} = [\beta, \delta]$ for

$$a(t) := \begin{cases} \beta & : \ t \leq 1 \\ \beta + \left(t - 2^{2n}\right)(\delta - \beta) & : \ t \in \left[2^{2n}, 2^{2n} + 1\right] \\ \delta & : \ t \in \left[2^{2n} + 1, 2^{2n+1}\right] \\ \delta + \left(t - 2^{2n+1}\right)(\beta - \delta) & : \ t \in \left[2^{2n+1}, 2^{2n+1} + 1\right] \\ \beta & : \ t \in \left[2^{2n+1} + 1, 2^{2(n+1)}\right] \end{cases}.$$

In all cases above, $n \in \mathbb{N}_0$.
*Proof.* See proof of (iii).

The following theorem says that each interval of the past (future, all-time, respectively) dichotomy spectrum corresponds to a linear nonautonomous invariant manifold.

**Theorem 4.26 (Spectral manifolds).** *Let*

$$\Sigma_\Phi := \Sigma_\Phi^\leftarrow, \Sigma_\Phi^\rightarrow, \Sigma_\Phi^\leftrightarrow = [a_1, b_1] \cup \cdots \cup [a_n, b_n],$$

*respectively, define the invariant projectors* $Q_{\gamma_0} := 0$, $Q_{\gamma_n} := 1$, *and for integers* $i \in \{1, \ldots, n-1\}$, *choose* $\gamma_i \in (b_i, a_{i+1})$ *and projectors* $Q_{\gamma_i}$ *of the nonhyperbolic dichotomy of* $(\theta, \varphi)$ *with growth rate* $\gamma_i$. *Then the sets*

$$\mathcal{W}_i := \mathcal{R}(Q_{\gamma_i}) \cap \mathcal{N}(Q_{\gamma_{i-1}}) \quad \textit{for all } i \in \{1, \ldots, n\}$$

*are linear nonautonomous invariant manifolds, the so-called* spectral manifolds, *such that*

$$\mathcal{W}_1 \oplus \cdots \oplus \mathcal{W}_n = \mathbb{T} \times \mathbb{R}^N$$

*and* $\mathcal{W}_i \neq \mathbb{T} \times \{0\}$ *for* $i \in \{1, \ldots, n\}$.

*Proof.* The sets $\mathcal{W}_1, \ldots, \mathcal{W}_n$ are obviously linear nonautonomous invariant manifolds. We suppose that there exists an $i \in \{1, \ldots, n\}$ with $\mathcal{W}_i = \mathbb{T} \times \{0\}$. In case $i = 1$ or $i = n$, Lemma 4.9 implies either $[-\infty, \gamma_1] \cap \Sigma_\Phi = \emptyset$ or $[\gamma_{n-1}, \infty] \cap \Sigma_\Phi = \emptyset$, and this is a contradiction. In case $1 < i < n$, due to Lemma 4.23, we obtain

$$\dim \mathcal{W}_i = \dim \left( \mathcal{R}(Q_{\gamma_i}) \cap \mathcal{N}(Q_{\gamma_{i-1}}) \right)$$
$$= \operatorname{rk} Q_{\gamma_i} + N - \operatorname{rk} Q_{\gamma_{i-1}} - \dim \left( \mathcal{R}(Q_{\gamma_i}) + \mathcal{N}(Q_{\gamma_{i-1}}) \right) \geq 1,$$

and this is also a contradiction. We now prove $\mathcal{W}_1 \oplus \cdots \oplus \mathcal{W}_n = \mathbb{T} \times \mathbb{R}^N$. W.l.o.g., we assume $\Sigma_\Phi = \Sigma_\Phi^\leftarrow, \Sigma_\Phi^\leftrightarrow$. For $1 \leq i < j \leq n$, due to Proposition 4.11, the relations $\mathcal{W}_i \subset \mathcal{R}(Q_{\gamma_i})$ and $\mathcal{W}_j \subset \mathcal{N}(Q_{\gamma_{j-1}}) \subset \mathcal{N}(Q_{\gamma_i})$ are fulfilled. This yields

$$\mathcal{W}_i \cap \mathcal{W}_j \subset \mathcal{R}(Q_{\gamma_i}) \cap \mathcal{N}(Q_{\gamma_i}) = \mathbb{T} \times \{0\}.$$

Moreover, Lemma A.10 implies that

$$\mathbb{T} \times \mathbb{R}^N = \mathcal{W}_1 + \mathcal{N}(Q_{\gamma_1}) = \mathcal{W}_1 + \mathcal{N}(Q_{\gamma_1}) \cap \left( \mathcal{R}(Q_{\gamma_2}) + \mathcal{N}(Q_{\gamma_2}) \right)$$
$$= \mathcal{W}_1 + \mathcal{N}(Q_{\gamma_1}) \cap \mathcal{R}(Q_{\gamma_2}) + \mathcal{N}(Q_{\gamma_2}) = \mathcal{W}_1 + \mathcal{W}_2 + \mathcal{N}(Q_{\gamma_2})$$

holds. It follows inductively that

$$\mathbb{T} \times \mathbb{R}^N = \mathcal{W}_1 + \cdots + \mathcal{W}_n + \mathcal{N}(Q_{\gamma_n}) = \mathcal{W}_1 + \cdots + \mathcal{W}_n.$$

This finishes the proof of this theorem. □

In case of the past and future dichotomy spectrum, the spectral manifolds give rise to a Morse decomposition on the projective space.

**Theorem 4.27 (Spectral manifolds and Morse decompositions).** *Let*

$$\Sigma_\Phi = \Sigma_\Phi^\leftarrow, \Sigma_\Phi^\rightarrow = [a_1, b_1] \cup \cdots \cup [a_n, b_n],$$

*respectively, define the invariant projectors $Q_{\gamma_0} := 0$, $Q_{\gamma_n} := 1$, and for integers $i \in \{1, \ldots, n-1\}$, choose $\gamma_i \in (b_i, a_{i+1})$ and projectors $Q_{\gamma_i}$ of the nonhyperbolic dichotomy of $(\theta, \varphi)$ with growth rate $\gamma_i$. Then the sets*

$$M_i := \mathbb{P}\big(\mathcal{R}(Q_{\gamma_i}) \cap \mathcal{N}(Q_{\gamma_{i-1}})\big) \quad \text{for all } i \in \{1, \ldots, n\}$$

*are the Morse sets of a past (future, respectively) Morse decomposition of $(\theta, \mathbb{P}\varphi)$.*

*Proof.* This is a direct consequence of Theorem 4.15.    □

*Remark 4.28.* It is possible that the above Morse decomposition defined by the spectral intervals is coarser than the finest Morse decomposition of Theorem 3.24 (see also COLONIUS & KLIEMANN [49]).

## 4.3 Lyapunov Spectra

In this section, the so-called Lyapunov spectra are introduced, and their relationship to the past and future dichotomy spectrum is examined. As in the previous section, we restrict to the case that $(\theta, \varphi)$ is generated by a nonautonomous differential or difference equation, i.e., $P = \mathbb{T} = \mathbb{R}, \mathbb{Z}$ and $\theta(t, \tau) = t + \tau$ for all $t, \tau \in \mathbb{T}$.

**Definition 4.29 (Lyapunov exponents and spectra).** *For $0 \neq \xi \in \mathbb{R}^N$, the numbers*

$$\lambda_+^\leftarrow(\xi) = \limsup_{t \to \infty} \frac{1}{t} \ln \|\Phi(-t, 0)\xi\| \quad \text{and} \quad \lambda_-^\leftarrow(\xi) = \liminf_{t \to \infty} \frac{1}{t} \ln \|\Phi(-t, 0)\xi\|$$

*are called* upper and lower Lyapunov exponent for $t \to -\infty$. *Considering the future, the numbers*

$$\lambda_+^\rightarrow(\xi) = \limsup_{t \to \infty} \frac{1}{t} \ln \|\Phi(t, 0)\xi\| \quad \text{and} \quad \lambda_-^\rightarrow(\xi) = \liminf_{t \to \infty} \frac{1}{t} \ln \|\Phi(t, 0)\xi\|$$

*are called* upper and lower Lyapunov exponent for $t \to \infty$. *The Lyapunov spectrum for $t \to -\infty$ is defined by*

$$\sigma_{\varPhi}^{\leftarrow} := \bigcup_{0 \neq \xi \in \mathbb{R}^N} \left[ \lambda_-^{\leftarrow}(\xi), \lambda_+^{\leftarrow}(\xi) \right],$$

*and the* Lyapunov spectrum for $t \to \infty$ *is defined by*

$$\sigma_{\varPhi}^{\rightarrow} := \bigcup_{0 \neq \xi \in \mathbb{R}^N} \left[ \lambda_-^{\rightarrow}(\xi), \lambda_+^{\rightarrow}(\xi) \right].$$

It is well-known that there exist numbers $n^-, n^+ \in \{1, \dots, N\}$ and $\xi_1^-, \dots, \xi_{n^-}^-, \xi_1^+, \dots, \xi_{n^+}^+ \in \mathbb{R}^N$ such that

$$\sigma_{\varPhi}^{\leftarrow} = \left[ \lambda_-^{\leftarrow}(\xi_1), \lambda_+^{\leftarrow}(\xi_1) \right] \cup \cdots \cup \left[ \lambda_-^{\leftarrow}(\xi_{n^-}), \lambda_+^{\leftarrow}(\xi_{n^-}) \right]$$

and

$$\sigma_{\varPhi}^{\rightarrow} = \left[ \lambda_-^{\rightarrow}(\xi_1), \lambda_+^{\rightarrow}(\xi_1) \right] \cup \cdots \cup \left[ \lambda_-^{\rightarrow}(\xi_{n^+}), \lambda_+^{\rightarrow}(\xi_{n^+}) \right]$$

(see, e.g., BARREIRA & PESIN [24] and DIECI & VAN VLECK [62, 63]).

In the following, the relationship between the dichotomy spectra and the Lyapunov spectra is discussed.

**Theorem 4.30 (Relationship to the past and future dichotomy spectrum).** *The relations* $-\sigma_{\varPhi}^{\leftarrow} \subset \Sigma_{\varPhi}^{\leftarrow}$ *and* $\sigma_{\varPhi}^{\rightarrow} \subset \Sigma_{\varPhi}^{\rightarrow}$ *hold.*

*Proof.* Let $\lambda \in -\sigma_{\varPhi}^{\leftarrow}$. Thus, there exists a $\xi \in \mathbb{R}^N$ with $-\lambda \in \left[ \lambda_-^{\leftarrow}(\xi), \lambda_+^{\leftarrow}(\xi) \right]$. Initially, we suppose that $\lambda \in \mathbb{R}$. Arguing negatively, we assume, $(\theta, \varphi)$ admits a nonhyperbolic past exponential dichotomy with growth rate $\gamma := \lambda$, constants $K \geq 1, \alpha > 0$ and invariant projector $Q_\gamma$, i.e.,

$$\|\varPhi(t, \tau) Q_\gamma(\tau)\| \leq K e^{(\gamma - \alpha)t} \quad \text{for all } \tau \leq 0 \leq t \text{ with } \tau + t \leq 0,$$

$$\left\| \varPhi(-t, \tau)(\mathbb{1} - Q_\gamma(\tau)) \right\| \leq K e^{-(\gamma + \alpha)t} \quad \text{for all } \tau \leq 0 \leq t.$$

We write $\xi = \xi_1 + \xi_2$ with $\xi_1 \in \mathcal{R}(Q_\gamma(0))$ and $\xi_2 \in \mathcal{N}(Q_\gamma(0))$. In case $\xi_1 = 0$, we have

$$\lambda_+^{\leftarrow}(\xi) = \limsup_{t \to \infty} \frac{1}{t} \ln \|\varPhi(-t, 0)\xi\| \leq \limsup_{t \to \infty} \frac{1}{t} \ln \left( K e^{-(\gamma + \alpha)t} \right)$$

$$= -\gamma - \alpha = -\lambda - \alpha \leq \lambda_+^{\leftarrow}(\xi) - \alpha.$$

This is a contradiction. Otherwise ($\xi_1 \neq 0$), we observe that for all $t \geq 0$,

$$\|\xi_1\| = \|Q_\gamma(0)\xi\| = \left\| \varPhi(t, -t)\varPhi(-t, 0)Q_\gamma(0)\xi \right\| = \left\| \varPhi(t, -t)Q_\gamma(-t)\varPhi(-t, 0)\xi \right\|$$

$$\leq K e^{(\gamma - \alpha)t} \|\varPhi(-t, 0)\xi\|$$

is fulfilled. Thus, $\|\varPhi(-t, 0)\xi\| \geq K^{-1} e^{-(\gamma - \alpha)t} \|\xi_1\|$ for all $t \geq 0$, and therefore,

$$\lambda_-^{\leftarrow}(\xi) = \liminf_{t \to \infty} \frac{1}{t} \ln \|\varPhi(-t, 0)\xi\| \geq \liminf_{t \to \infty} \frac{1}{t} \ln \left( K^{-1} e^{-(\gamma - \alpha)t} \|\xi_1\| \right)$$

$$= \alpha - \gamma = \alpha - \lambda \geq \alpha + \lambda_-^{\leftarrow}(\xi).$$

This is also a contradiction, and hence, $\lambda \in \Sigma_{\Phi}^{\leftarrow}$. We now treat the case $\lambda \notin \mathbb{R}$, w.l.o.g., $\lambda = \infty$. Assume that $-\infty \notin \Sigma_{\Phi}^{\leftarrow}$. Thus, there exist $\gamma \in \mathbb{R}$, $K \geq 1$ and $\alpha > 0$ with

$$\|\Phi(-t,0)\| \leq Ke^{-(\gamma+\alpha)t} \quad \text{for all } t \geq 0,$$

and this relation implies the contradiction

$$\limsup_{t\to\infty} \frac{1}{t} \ln\left(\|\Phi(-t,0)\|\right) \leq -\gamma - \alpha < \infty.$$

Hence, $\lambda \in \Sigma_{\Phi}^{\leftarrow}$. The remaining assertion $\sigma_{\Phi}^{\rightarrow} \subset \Sigma_{\Phi}^{\rightarrow}$ can be proved analogously. $\qquad\square$

The following example shows that the Lyapunov spectra do not coincide with the dichotomy spectra.

*Example 4.31.* We consider the scalar linear nonautonomous differential equation

$$\dot{x} = \left(t\sin(t) - \cos(t)\right)x,$$

which generates a linear nonautonomous dynamical system with $P = \mathbb{R}$. We have

$$\Phi(t,\tau) = \exp\left(-(t+\tau)\cos(t+\tau) + \tau\cos(\tau)\right) \quad \text{for all } t,\tau \in \mathbb{R}.$$

An easy calculation yields $\sigma_{\Phi}^{\leftarrow} = \sigma_{\Phi}^{\rightarrow} = [-1,1]$. Choosing

$$t_1 := \frac{\pi}{2}, \, t_2 := -\frac{\pi}{2}, \, \tau_1 := 2k\pi - \frac{\pi}{2}, \, \tau_2 := 2k\pi + \frac{\pi}{2} \quad \text{for all } k \in \mathbb{Z},$$

we obtain

$$\Phi(t_1,\tau_1) = \Phi(t_2,\tau_2) = \exp(2k\pi).$$

Hence, for any $\gamma \in \mathbb{R}$, this system does not admit a nonhyperbolic past exponential dichotomy with growth rate $\gamma$. This implies $\Sigma_{\Phi}^{\leftarrow} = \overline{\mathbb{R}}$. Analogously, one can show that $\Sigma_{\Phi}^{\rightarrow} = \overline{\mathbb{R}}$ is fulfilled.

## 4.4 Spectra of Autonomous Linear Systems

It is well-known that an autonomous linear differential equation

$$\dot{x} = Ax \tag{4.4}$$

with a matrix $A \in \mathbb{R}^{N \times N}$ admits an exponential dichotomy on $\mathbb{I} = \mathbb{R}_0^-$, $\mathbb{R}_0^+$, $\mathbb{R}$, respectively, if and only if the real part of every eigenvalue of $A$ is unequal to zero (see, e.g., KALKBRENNER [88, Satz 1.1.3.2, p. 24] and SACKER &

SELL [151, p. 430(1)]). Therefore, the corresponding past, future and all-time dichotomy spectra satisfy

$$\Sigma_A^{\leftarrow} = \Sigma_A^{\rightarrow} = \Sigma_A^{\leftrightarrow} = \left\{\, \mathrm{Re}\,\lambda : \lambda \text{ is an eigenvalue of } A \,\right\}.$$

A relation of this kind does not hold for the $(\tau, T)$-dichotomy spectrum. Nevertheless, by letting $T$ tend to $\infty$, we obtain the following statement.

**Theorem 4.32 (Spectra of autonomous linear systems).** *Consider the linear system (4.4). Then the limit relation*

$$\lim_{T \to \infty} \Sigma_A^{(0,T)} = \left\{\, \mathrm{Re}\,\lambda : \lambda \text{ is an eigenvalue of } A \,\right\}$$

*holds with respect to the Hausdorff distance.*

*Proof.* There exist $n \in \{1, \dots, N\}$ and reals $\lambda_1 < \lambda_2 < \cdots < \lambda_n$ with

$$\left\{\, \mathrm{Re}\,\lambda : \lambda \text{ is an eigenvalue of } A \,\right\} = \{\lambda_1, \dots, \lambda_n\}.$$

It is sufficient to show that for all $\varepsilon > 0$, there exists a $\tau > 0$ with

$$\{\lambda_1, \dots, \lambda_n\} \subset U_\varepsilon\!\left(\Sigma_A^{(0,T)}\right) \text{ and } \Sigma_A^{(0,T)} \subset \bigcup_{i=1}^{n} U_\varepsilon(\lambda_i) \quad \text{for all } T \geq \tau. \quad (4.5)$$

Let $\varepsilon > 0$. It is an elementary result in the theory of linear differential equations (see, e.g., COPPEL [54, p. 56]) that there exist nontrivial linear subspaces $U_1, \dots, U_n \subset \mathbb{R}^N$ with $U_1 \oplus \cdots \oplus U_n = \mathbb{R}^N$ and a real constant $K \geq 1$ such that for all $i \in \{1, \dots, n\}$,

$$\left\| e^{At}\xi \right\| \leq K \exp\!\left(\left(\lambda_i + \frac{\varepsilon}{4}\right)t\right)\|\xi\| \quad \text{for all } \xi \in U_1 \oplus \cdots \oplus U_i \text{ and } t \geq 0,$$
$$(4.6)$$

$$\left\| e^{At}\xi \right\| \geq \frac{1}{K} \exp\!\left(\left(\lambda_i - \frac{\varepsilon}{4}\right)t\right)\|\xi\| \quad \text{for all } \xi \in U_i \oplus \cdots \oplus U_n \text{ and } t \geq 0$$
$$(4.7)$$

is fulfilled. We choose $\tau > 0$ and $T \geq \tau$ with $K \exp(-\varepsilon\tau/4) < 1$. To prove the first condition of (4.5), we choose an $i \in \{1, \dots, n\}$ and assume to the contrary that $U_\varepsilon(\lambda_i) \cap \Sigma_A^{(0,T)} = \emptyset$. Thus, there exists an invariant projector $Q_{(\lambda_i - \varepsilon/2)}$ with

$$\left\| e^{AT}\xi \right\| < e^{(\lambda_i - \varepsilon/2)T}\|\xi\| \quad \text{for all } 0 \neq \xi \in \mathcal{R}\big(Q_{(\lambda_i - \varepsilon/2)}(0)\big) \quad (4.8)$$

and an invariant projector $Q_{(\lambda_i + \varepsilon/2)}$ with

$$\left\| e^{-AT}\xi \right\| < e^{-(\lambda_i + \varepsilon/2)T}\|\xi\| \quad \text{for all } 0 \neq \xi \in \mathcal{N}\big(Q_{(\lambda_i + \varepsilon/2)}(T)\big). \quad (4.9)$$

Because of Lemma 4.23, we have $\mathrm{rk}\, Q_{(\lambda_i - \varepsilon/2)} = \mathrm{rk}\, Q_{(\lambda_i + \varepsilon/2)} =: r$. In the case $r \geq \dim U_1 + \cdots + \dim U_i$, we have $\dim\big(\mathcal{R}(Q_{(\lambda_i - \varepsilon/2)}(0)) \cap (U_i \oplus \cdots \oplus U_n)\big) \geq 1$. Then there exists a nonzero element $\xi \in \mathcal{R}\big(Q_{(\lambda_i - \varepsilon/2)}(0)\big) \cap (U_i \oplus \cdots \oplus U_n)$, and this leads to the contradiction

$$
\|\xi\| = e^{-(\lambda_i - \varepsilon/2)T} e^{(\lambda_i - \varepsilon/2)T} \|\xi\| \overset{(4.8)}{>} e^{-(\lambda_i - \varepsilon/2)T} \big\| e^{AT} \xi \big\|
$$
$$
\overset{(4.7)}{\geq} e^{-(\lambda_i - \varepsilon/2)T} \frac{1}{K} e^{(\lambda_i - \varepsilon/4)T} \|\xi\| = \big( K e^{-\varepsilon T/4} \big)^{-1} \|\xi\| > \|\xi\| \, .
$$

If $r < \dim U_1 + \cdots + \dim U_i$, then $\dim\big(\mathcal{N}(Q_{(\lambda_i + \varepsilon/2)}(0)) \cap (U_1 \oplus \cdots \oplus U_i)\big) \geq 1$. Thus, there exists a nontrivial element $\xi \in \mathcal{N}\big(Q_{(\lambda_i + \varepsilon/2)}(0)\big) \cap (U_1 \oplus \cdots \oplus U_i)$, and this also yields the contradiction

$$
\|\xi\| = \big\| e^{-AT} e^{AT} \xi \big\| \overset{(4.9)}{<} e^{-(\lambda_i + \varepsilon/2)T} \big\| e^{AT} \xi \big\| \overset{(4.6)}{\leq} e^{-(\lambda_i + \varepsilon/2)T} K e^{(\lambda_i + \varepsilon/4)T} \|\xi\|
$$
$$
= K e^{-\varepsilon T/4} \|\xi\| < \|\xi\| \, .
$$

To prove the second condition of (4.5), let $\lambda \notin \cup_{i=1}^{n} U_\varepsilon(\lambda_i)$. We set $\lambda_0 := -\infty$ and $\lambda_{n+1} := \infty$. There exists a $i \in \{0, \ldots, n\}$ such that

$$
\lambda \geq \lambda_i + \varepsilon \quad \text{and} \quad \lambda \leq \lambda_{i+1} - \varepsilon \, .
$$

Now, we define the invariant projector $Q$ by

$$
\mathcal{R}(Q(0)) = U_1 \oplus \cdots \oplus U_i \quad \text{and} \quad \mathcal{N}(Q(0)) = U_{i+1} \oplus \cdots \oplus U_n \, .
$$

Thus, for all nonzero $\xi \in \mathcal{R}(Q(0))$, we have

$$
\big\| e^{AT} \xi \big\| \overset{(4.6)}{\leq} K e^{(\lambda_i + \varepsilon/4)T} \|\xi\| \leq K e^{(\lambda - 3\varepsilon/4)T} \|\xi\| < e^{\lambda T} \|\xi\| \, ,
$$

and for all nonzero $\xi \in \mathcal{N}(Q(T))$,

$$
\big\| e^{-AT} \xi \big\| \overset{(4.7)}{\leq} K e^{-(\lambda_{i+1} - \varepsilon/4)T} \|\xi\| \leq K e^{-(\lambda + 3\varepsilon/4)T} \|\xi\| < e^{-\lambda T} \|\xi\|
$$

is fulfilled. Hence, $\lambda \notin \Sigma_A^{(0,T)}$, and this finishes the proof of this theorem.   $\square$

*Remark 4.33.* Using Floquet Theory (see, e.g., CODDINGTON & LEVINSON [48, pp. 78–80] or CHICONE [45, Section 2.4, pp. 162–197]), one can extend the above theorem to periodic linear differential systems of the form

$$
\dot{x} = A(t)x \, , \tag{4.10}
$$

where $A : \mathbb{R} \to \mathbb{R}^{N \times N}$ fulfills $A(t) = A(t + \omega)$ for all $t \in \mathbb{R}$ with some $\omega > 0$. We denote the transition operator of (4.10) by $\Lambda$. For the past (future, all-time, respectively) dichotomy spectrum, we obtain

$$\Sigma_A^{\leftarrow} = \Sigma_A^{\rightarrow} = \Sigma_A^{\leftrightarrow} = \left\{ \ln |\lambda| : \lambda \text{ is an eigenvalue of } \Lambda(\omega,0) \right\}.$$

The matrix $\Lambda(\omega,0)$ is called *monodromy matrix* of (4.10). The $(0,T)$-dichotomy spectrum fulfills the limit relation

$$\lim_{T \to \infty} \Sigma_A^{(0,T)} = \left\{ \ln |\lambda| : \lambda \text{ is an eigenvalue of } \Lambda(\omega,0) \right\}$$

in the sense of Hausdorff distance.

*Example 4.34.* For fixed $T > 0$, we want to compute the $(0,T)$-dichotomy spectrum $\Sigma_A^{(0,T)}$ of system (4.4), where

$$A := \begin{pmatrix} 1 & 1 \\ 0 & 1 \end{pmatrix}.$$

Specifically, in this example, we use the norm $\| \cdot \|_1 : \mathbb{R}^2 \to \mathbb{R}_0^+$, defined by $\|(x_1,x_2)\|_1 := |x_1| + |x_2|$. Please note that, for $\gamma \in \mathbb{R}$, the relation

$$e^{(A-\gamma\mathbb{1})T} = \begin{pmatrix} e^{(1-\gamma)T} & Te^{(1-\gamma)T} \\ 0 & e^{(1-\gamma)T} \end{pmatrix}$$

is fulfilled (see, e.g., Aulbach [14]). Hence, for all $\xi = (\xi_1,\xi_2) \in \mathbb{R}^2$ with $\|\xi\|_1 = 1$, we have

$$\underbrace{\left\| e^{(A-\gamma\mathbb{1})T} \begin{pmatrix} 1 \\ 0 \end{pmatrix} \right\|_1}_{= e^{(1-\gamma)T}} \leq \left\| e^{(A-\gamma\mathbb{1})T} \begin{pmatrix} \xi_1 \\ \xi_2 \end{pmatrix} \right\|_1 \leq \underbrace{\left\| e^{(A-\gamma\mathbb{1})T} \begin{pmatrix} 0 \\ 1 \end{pmatrix} \right\|_1}_{= Te^{(1-\gamma)T} + e^{(1-\gamma)T}}.$$

The term $Te^{(1-\gamma)T} + e^{(1-\gamma)T}$ is strictly monotone decreasing in $\gamma \in \mathbb{R}$, and therefore, there exists a uniquely determined $\gamma_* = \gamma_*(T) > 1$ such that $Te^{(1-\gamma_*)T} + e^{(1-\gamma_*)T} = 1$.

Using these observations, it is easy to see that $\Sigma_A^{(0,T)} = \{1, \gamma_*\}$, since

- for $\gamma < 1$, the linear system (4.4) admits a nonhyperbolic $(0,T)$-dichotomy with growth rate $\gamma$ and invariant projector $Q_\gamma \equiv 0$,

- for $\gamma \in (1, \gamma_*)$, the linear system (4.4) admits a nonhyperbolic $(0,T)$-dichotomy with growth rate $\gamma$ and invariant projector $Q_\gamma$, determined by $\mathcal{R}(Q_\gamma(0)) = \{\beta(1,0) : \beta \in \mathbb{R}\}$ and $\mathcal{N}(Q_\gamma(0)) = \{\beta(0,1) : \beta \in \mathbb{R}\}$.

- for $\gamma > \gamma_*$, the linear system (4.4) admits a nonhyperbolic $(0,T)$-dichotomy with growth rate $\gamma$ and invariant projector $Q_\gamma \equiv \mathbb{1}$,

- for $\gamma \in \{1, \gamma_*\}$, the linear system (4.4) admits no nonhyperbolic $(0,T)$-dichotomy with growth rate $\gamma$.

Please note that Theorem 4.32 implies that $\lim_{T \to \infty} \gamma_*(T) = 1$.

## 4.5 Roughness

We consider the nonautonomous linear differential equation

$$\dot{x} = A(t)x \tag{4.11}$$

with a continuous function $A : \mathbb{R} \to \mathbb{R}^{N \times N}$ and a perturbed system

$$\dot{x} = (A(t) + B(t))x \tag{4.12}$$

with a continuous function $B : \mathbb{R} \to \mathbb{R}^{N \times N}$. The transition operators of (4.11) and (4.12) are denoted by $\Lambda$ and $\Lambda^*$, respectively.

**Theorem 4.35 (Roughness Theorem for nonhyperbolic exponential dichotomies on $\mathbb{I}$).** *Let $\mathbb{I}$ be an unbounded and closed interval, and suppose, (4.11) admits a nonhyperbolic exponential dichotomy on $\mathbb{I}$ with growth rate $\gamma$, constants $\alpha$, $K$ and invariant projector $Q$. If the relation*

$$\delta := \sup_{t \in \mathbb{i}} \|B(t)\| < \frac{\alpha}{4K^2}$$

*is fulfilled, then also the perturbed system (4.12) admits a nonhyperbolic exponential dichotomy on $\mathbb{I}$ with growth rate $\gamma$, constants $(\alpha - 2K\delta)$, $5K^2/2$ and an invariant projector $\hat{Q}$, more precisely, we have*

$$\left\|\Lambda^*(t,s)\hat{Q}(s)\right\| \leq \frac{5K^2}{2} e^{(\gamma - (\alpha - 2K\delta))(t-s)} \quad \textit{for all } t, s \in \mathbb{I} \textit{ with } t \geq s ,$$

$$\left\|\Lambda^*(t,s)(\mathbb{1} - \hat{Q}(s))\right\| \leq \frac{5K^2}{2} e^{(\gamma + \alpha - 2K\delta)(t-s)} \quad \textit{for all } t, s \in \mathbb{I} \textit{ with } t \leq s .$$

*In case $\mathbb{I} = \mathbb{R}_\kappa^-$, the invariant projector $\hat{Q}$ has the same range as $Q$, and if $\mathbb{I} = \mathbb{R}_\kappa^+$ holds, then $\hat{Q}$ has the same null space as $Q$. Finally, if $\mathbb{I} = \mathbb{R}$ is fulfilled, we get $\operatorname{rk} \hat{Q} = \operatorname{rk} Q$.*

*Proof.* See COPPEL [55, Proposition 1, p. 34] or COPPEL [54]. $\qquad\square$

*Remark 4.36.* The perturbations considered in this theorem are perturbations with respect to the uniform topology, generated by the norm

$$\|A\|_\infty := \sup_{t \in \mathbb{I}} \|A(t)\| \quad \text{for all } A \in C(\mathbb{I}, \mathbb{R}^{N \times N}) ,$$

where $C(\mathbb{I}, \mathbb{R}^{N \times N}) := \{X : \mathbb{I} \to \mathbb{R}^{N \times N} : X \text{ is continuous}\}$. It is possible to weaken this condition on the perturbation (see, e.g., PÖTZSCHE [139] or PLISS & SELL [133]). For instance, considering the topology of uniform convergence on compact sets, i.e., $\lim_{n \to \infty} A_n = A_0$ if and only if

$$\lim_{n \to \infty} \sup_{t \in \mathbb{J}} \|A_n(t) - A_0(t)\| = 0 \quad \text{for all compact sets } \mathbb{J} \subset \mathbb{I},$$

one can derive a similar but more stronger perturbation result as Theorem 4.35 (see also SACKER & SELL [154, Section 5, Remark on p. 346]).

**Theorem 4.37 (Roughness  Theorem  for  nonhyperbolic  $(\tau, T)$-dichotomies).** *Suppose, (4.11) admits a nonhyperbolic $(\tau, T)$-dichotomy with growth rate $\gamma$ and projector $Q$. Then there exists an $\varepsilon > 0$ with the following property: If*

$$\sup_{t \in [\tau, \tau+T]} \|B(t)\| < \varepsilon,$$

*then also the perturbed system (4.12) admits a nonhyperbolic $(\tau, T)$-dichotomy with growth rate $\gamma$ and projector $Q$.*

*Proof.* This statement follows directly from the continuity of the general solution (cf. Proposition A.3). $\qquad\qquad\Box$

# 5

# Nonlinear Systems

In the study of nonlinear systems, invariant manifolds play a central role, since they help to understand the often complicated dynamical behavior near an equilibrium, a periodic solution or—in the nonautonomous context—an arbitrary solution. The construction of stable and unstable invariant manifolds goes back to POINCARÉ [136] and HADAMARD [73]. In the sequel, the theory was extended from hyperbolic to nonhyperbolic systems, from finite to infinite dimension and from time-independent to time-dependent equations.

To mention only few references of the comprehensive amount of literature for autonomous differential equations, we refer to CARR [38], CHOW & LI & WANG [47], HIRSCH & PUGH & SHUB [80], KELLEY [91, 92], KIRCHGRABER & PALMER [93], PLISS [132], SHUB [170], VANDERBAUWHEDE [180] and WIGGINS [183]. In the nonautonomous context, see AULBACH [13], AULBACH & WANNER [21], SELL [168], WANNER [181] and YI [184].

In the first section of this chapter, invariant manifolds are constructed which apply to different time domains. It suffices to extend the results of AULBACH & WANNER [21] and SIEGMUND [171] slightly. Also, the relationship to the notions of attractivity and repulsivity is discussed. In Section 5.2, these results are applied in the context of nonautonomous bifurcation theory. It is shown that under special assumptions, zero is contained in the dichotomy spectrum of the linearization of a bifurcating solution. In Section 5.3, properties of attraction and repulsion for nonlinear systems are derived from the study of the linearization, and finally, Section 5.4 is devoted to the relationship between the bifurcation theory of adiabatic systems and the concept of finite-time bifurcation.

## 5.1 Invariant Manifolds

Let $\mathbb{I}$ be an unbounded interval of the form $\mathbb{R}$, $\mathbb{R}_\kappa^-$ or $\mathbb{R}_\kappa^+$, respectively. In this section, we consider a nonlinear differential equation

$$\dot{x} = A(t)x + F(t,x) \tag{5.1}$$

with a continuous function $A : \mathbb{I} \to \mathbb{R}^{N \times N}$ and a $C^1$-function $F : \mathbb{I} \times U \to \mathbb{R}^N$, where $U$ is an open neighborhood of 0 and $F(t,0) = 0$ for all $t \in \mathbb{I}$. The general solution of (5.1) will be denoted by $\lambda$. In addition to (5.1), we consider the corresponding linear differential equation

$$\dot{x} = A(t)x \tag{5.2}$$

with transition operator $\varLambda : \mathbb{I} \times \mathbb{I} \to \mathbb{R}^{N \times N}$. Let $Q_+ : \mathbb{I} \to \mathbb{R}^{N \times N}$ be an invariant projector of (5.2). Then $Q_- : \mathbb{I} \to \mathbb{R}^{N \times N}$, $Q_-(t) := 1 - Q_+(t)$ for all $t \in \mathbb{I}$, is also an invariant projector.

Please note that in the following, the symbol $Q_\pm$ simultaneously stands for $Q_+$ and $Q_-$, respectively. We proceed similarly with our further notation in this section.

Next, we introduce a nonautonomous counterpart of an invariant manifold for (5.1).

**Definition 5.1 (Nonautonomous invariant manifolds).** *Assume that for an interval $\mathbb{I} \subset \mathbb{R}$ and a neighborhood $V$ of 0, $C^1$-functions $s^\pm : \mathbb{I} \times V \to \mathbb{R}^N$ satisfy*

*(i)* $s^\pm(t,0) = 0$ *for all* $t \in \mathbb{I}$,

*(ii)* $\lim_{x \to 0} \frac{s^\pm(t,x)}{\|x\|} = 0$ *uniformly in* $t \in \mathbb{I}$,

*(iii)* $s^\pm(t,x) = s^\pm(t, Q_\pm(t)x) \in \mathcal{R}(Q_\mp(t))$ *for all* $t \in \mathbb{I}$ *and* $x \in V$.

*Then the graphs*

$$\mathcal{S}^\pm := \left\{ (\tau, \xi + s^\pm(\tau,\xi)) \in \mathbb{I} \times \mathbb{R}^N : \xi \in \mathcal{R}(Q_\pm(\tau)) \cap V \right\}$$

*are called* (local) nonautonomous invariant manifolds *of (5.1) if*

$$\left(t, \lambda(t,\tau,\xi)\right) \in \mathcal{S}^\pm \quad \begin{array}{l} \textit{for all } (\tau,\xi) \in \mathcal{S}^\pm \textit{ and } t \in \mathbb{I} \textit{ such that} \\ \lambda(\tau + c(t-\tau), \tau, \xi) \in V \textit{ for all } c \in [0,1]. \end{array}$$

*We call* $\mathcal{S}^\pm$ global nonautonomous invariant manifolds *if* $V = \mathbb{R}^N$.

Now, existence results for nonautonomous manifolds of (5.1) are proved and applications are discussed with respect to the notions of attractivity and repulsivity introduced in Chapter 2. Before doing so, some hypotheses on the linear part and the nonlinearity are needed.

We assume, the following hypotheses hold:

- *Hypothesis on linear part.* The linear system (5.2) admits a nonhyperbolic all-time (past, future, respectively) exponential dichotomy, more precisely, there exists an invariant projector $Q_+ : \mathbb{I} \to \mathbb{R}^{N \times N}$ such that the inequalities

$$\|A(t,s)Q_+(s)\| \leq Ke^{\alpha(t-s)} \quad \text{for all } t \geq s \,,$$
$$\|A(t,s)Q_-(s)\| \leq Ke^{\beta(t-s)} \quad \text{for all } t \leq s$$

hold with real constants $K \geq 1$ and $\alpha < \beta$.

- *Hypothesis on nonlinearity.* There exists a monotone increasing function $\Gamma : (0,1) \to \overline{\mathbb{R}}^+$ with $\lim_{s \searrow 0} \Gamma(s) = 0$ and

$$\sup_{x \in U, \|x\| \leq s} \sup_{t \in \mathbb{I}} \|D_2 F(t,x)\| \leq \Gamma(s) \quad \text{for all } s \in (0,1) \,.$$

*Remark 5.2.* The hypothesis on the nonlinear part of (5.1) is equivalent to $\lim_{x \to 0} \sup_{t \in \mathbb{I}} \|D_2 F(t,x)\| = 0$. In the above description, the function $\Gamma$ is needed to explain the dependence of some constants in the next theorems concerning the rate of this limit process.

First, the case of all-time invariant manifolds of (5.1) is treated.

**Theorem 5.3 (All-time invariant manifolds).** *In case $\mathbb{I} = \mathbb{R}$, there exist $\rho > 0$ and $C^1$-functions $s^\pm : \mathbb{R} \times U_\rho(0) \to \mathbb{R}^N$ such that the graphs*

$$\mathcal{S}^\pm := \left\{ \left(\tau, \xi + s^\pm(\tau, \xi)\right) \in \mathbb{R} \times \mathbb{R}^N : \xi \in \mathcal{R}(Q_\pm(\tau)) \cap U_\rho(0) \right\}$$

*are local nonautonomous invariant manifolds. Furthermore, the following statements are fulfilled:*

(i) *Case $\alpha < 0$ and $\operatorname{rk} Q_+ \geq 1$ (Trivial solution is not all-time repulsive). For all $\varepsilon > 0$, there exists an $r > 0$ such that for all $(\tau, \xi) \in \mathcal{S}^+$ with $\|\xi\| < r$, the solution $\lambda(\cdot, \tau, \xi)$ is $(\alpha + \varepsilon)^+$-quasibounded and we have $\lambda(t, \tau, \xi) \in U_\rho(0)$ for all $t \geq \tau$.*

(ii) *Case $\beta > 0$ and $\operatorname{rk} Q_- \geq 1$ (Trivial solution is not all-time attractive). For all $\varepsilon > 0$, there exists an $r > 0$ such that for all $(\tau, \xi) \in \mathcal{S}^-$ with $\|\xi\| < r$, the solution $\lambda(\cdot, \tau, \xi)$ is $(\beta - \varepsilon)^-$-quasibounded and we have $\lambda(t, \tau, \xi) \in U_\rho(0)$ for all $t \leq \tau$.*

(iii) *Case $\alpha < 0$ and $Q_+ \equiv \mathbb{1}$ (Trivial solution is all-time attractive). There exists an $r = r(\alpha, K, \Gamma) > 0$ with*

$$\lim_{t \to \infty} \sup_{\tau \in \mathbb{R}} d\big(\lambda\big(\tau + t, \tau, U_r(0)\big)\big|\{0\}\big) = 0 \,.$$

(iv) *Case $\beta > 0$ and $Q_- \equiv \mathbb{1}$ (Trivial solution is all-time repulsive). There exists an $r = r(\beta, K, \Gamma) > 0$ with*

$$\lim_{t \to \infty} \sup_{\tau \in \mathbb{R}} d\big(\lambda\big(\tau - t, \tau, U_r(0)\big)\big|\{0\}\big) = 0 \,.$$

*Proof.* The proof is divided into two steps.

*Step 1. Existence of $\mathcal{S}^\pm$.*

In case $Q_+ \equiv 0$ or $Q_+ \equiv 1$, the manifolds $\mathcal{S}^\pm$ are trivial, and nothing has to be shown. Therefore, we assume $Q_+ \not\equiv 0$ and $Q_+ \not\equiv 1$. In AULBACH & RASMUSSEN & SIEGMUND [18, Lemma 6.1] (see also RASMUSSEN [144, Lemma 6.3.7], COPPEL [55, Chapter 5] and SIEGMUND [173, Lemma 2.3]), it is shown that there exists a function $T : \mathbb{R} \to \mathbb{R}^{N \times N}$ of invertible matrices such that the so-called *Lyapunov transformation* $y = T(t)x$ of system (5.1) leads to the following system with decoupled linearization:

$$\dot{y} = \underbrace{\begin{pmatrix} B^+(t) & 0 \\ 0 & B^-(t) \end{pmatrix}}_{=:B(t)} y + \underbrace{T(t)F\big(t, T(t)^{-1}y\big)}_{=:G(t,y)},$$

where $B^+ : \mathbb{R} \to \mathbb{R}^{N^+ \times N^+}$ and $B^- : \mathbb{R} \to \mathbb{R}^{N^- \times N^-}$ with $N^+ := \operatorname{rk} Q_+$ and $N^- := \operatorname{rk} Q_-$. The transition operators $\Psi^+$ and $\Psi^-$ of the linear differential equations $\dot{y}_+ = B^+(t)y_+$ and $\dot{y}_- = B^-(t)y_-$, respectively, fulfill

$$\|\Psi^+(t,s)\| \le 2K^2 e^{\alpha(t-s)} \quad \text{for all } t \ge s \quad \text{and}$$
$$\|\Psi^-(t,s)\| \le 2K^2 e^{\beta(t-s)} \quad \text{for all } t \le s \,.$$

It is also shown that

$$\|T(t)\| \le \sqrt{2}\,K \quad \text{and} \quad \|T^{-1}(t)\| \le \sqrt{2} \quad \text{for all } t \in \mathbb{R}\,. \tag{5.3}$$

Thus, the hypothesis on nonlinearity implies the limit relation

$$\lim_{y \to 0} \sup_{t \in \mathbb{R}} \|D_2 G(t,y)\| = 0\,. \tag{5.4}$$

We fix a smooth cut-off function $\chi : \mathbb{R}^N \to [0,1]$ (see, e.g., ABRAHAM & MARSDEN & RATIU [1, Lemma 4.2.13]) such that

$$\chi(x) = 1 \text{ for all } x \text{ with } \|x\| \le 1 \quad \text{and} \quad \chi(x) = 0 \text{ for all } x \text{ with } \|x\| \ge 2\,.$$

For any $\sigma > 0$ with $U_{2\sigma}(0) \subset U$, we define the function $G_\sigma : \mathbb{R} \times \mathbb{R}^N \to \mathbb{R}^N$ by

$$G_\sigma(t,x) := \begin{cases} \chi\!\left(\frac{x}{\sigma}\right)G(t,x) & \text{for all } t \in \mathbb{R} \text{ and } x \in U \\ 0 & \text{for all } t \in \mathbb{R} \text{ and } x \notin U \end{cases}.$$

Due to the mean value inequality (see, e.g., LANG [102, Corollary 4.3, p. 342]), the relation $G(\cdot, 0) \equiv 0$ leads to

$$\|G(t,x)\| \le \|x\| \sup_{s \in [0,1]} \|D_2 G(t, sx)\| \quad \text{for all } x \in \mathbb{R}^N \text{ and } t \in \mathbb{R}\,.$$

Since $D_2 G_\sigma(t,x) = \chi\!\left(\frac{x}{\sigma}\right)D_2 G(t,x) + \frac{1}{\sigma}D\chi\!\left(\frac{x}{\sigma}\right)G(t,x)$, for all $t \in \mathbb{R}$, we have

$$\sup_{x \in \mathbb{R}^N} \|D_2 G_\sigma(t, x)\|$$

$$\leq \sup_{\|x\| \leq 2\sigma} \|D_2 G(t, x)\| + \frac{1}{\sigma} \sup_{\|x\| \leq 2} D\chi(x) \sup_{\|x\| \leq 2\sigma} \left( \|x\| \sup_{s \in [0,1]} \|D_2 G(t, sx)\| \right).$$

Hence,

$$\sup_{(t,x) \in \mathbb{R} \times \mathbb{R}^N} \|D_2 G_\sigma(t, x)\| \leq \left( 1 + 2 \sup_{x \in \mathbb{R}^N} D\chi(x) \right) \sup_{\|x\| \leq 2\sigma, t \in \mathbb{R}} \|D_2 G(t, x)\|.$$

Due to (5.4), this implies

$$\lim_{\sigma \to 0} \sup_{(t,x) \in \mathbb{R} \times \mathbb{R}^N} \|D_2 G_\sigma(t, x)\| = 0,$$

and thus, there exists a $\tilde{\rho} > 0$ such that

$$\dot{y} = B(t)y + G_{\tilde{\rho}}(t, y) \tag{5.5}$$

fulfills the (global) hypotheses of SIEGMUND [171, Satz 4.16 and Satz 4.30]. Denoting the general solution of (5.5) by $\tilde{\lambda}$, this means that there exist $C^1$-functions $\tilde{s}^\pm : \mathbb{R} \times \mathbb{R}^N \to \mathbb{R}^N$ fulfilling

$$\tilde{s}^+(t, \xi) = \tilde{s}^+ \left( t, (\xi_1, \ldots, \xi_{N^+}, 0, \ldots, 0) \right) \in \{(0, \ldots, 0)\} \times \mathbb{R}^{N^-} \subset \mathbb{R}^N$$

and

$$\tilde{s}^-(t, \xi) = \tilde{s}^- \left( t, (0, \ldots, 0, \xi_{N^+ + 1}, \ldots, \xi_N) \right) \in \mathbb{R}^{N^+} \times \{(0, \ldots, 0)\} \subset \mathbb{R}^N$$

for all $t \in \mathbb{R}$ and $\xi \in \mathbb{R}^N$ such that the graphs

$$\tilde{\mathcal{S}}^+ := \left\{ (\tau, \xi + \tilde{s}^+(\tau, \xi)) \in \mathbb{R} \times \mathbb{R}^N : \xi_i = 0 \text{ for } i > N^+ \right\}$$

and

$$\tilde{\mathcal{S}}^- := \left\{ (\tau, \xi + \tilde{s}^-(\tau, \xi)) \in \mathbb{R} \times \mathbb{R}^N : \xi_i = 0 \text{ for } i \leq N^+ \right\}$$

are global nonautonomous invariant manifolds with

$$\tilde{\mathcal{S}}^\pm := \left\{ (\tau, \xi) \in \mathbb{R} \times \mathbb{R}^N : \tilde{\lambda}(\cdot, \tau, \xi) \text{ is } \left( \frac{\alpha + \beta}{2} \right)^\pm \text{-quasibounded} \right\}.$$

We define $\hat{s}^\pm : \mathbb{R} \times \mathbb{R}^N \to \mathbb{R}^N$ by

$$\hat{s}^\pm(t, x) := T(t)^{-1} \tilde{s}^\pm(t, T(t)x) \quad \text{for all } t \in \mathbb{R} \text{ and } x \in \mathbb{R}^N.$$

Then $\hat{s}^\pm$ leads to the nonautonomous sets $\hat{\mathcal{S}}^\pm$, which also can be defined by

$$\hat{\mathcal{S}}^\pm(t) := T(t)^{-1} \tilde{\mathcal{S}}^\pm(t) \quad \text{for all } t \in \mathbb{R}. \tag{5.6}$$

Let $\hat{\lambda}$ denote the general solution of the system

$$\dot{x} = A(t)x + T(t)^{-1}G_\sigma(t, T(t)x)\,, \tag{5.7}$$

which is obtained via the transformation $x = T(t)^{-1}y$ from system (5.5). Then the representation

$$\hat{\mathcal{S}}^\pm := \left\{ (\tau, \xi) \in \mathbb{R} \times \mathbb{R}^N : \hat{\lambda}(\cdot, \tau, \xi) \text{ is } \left( \frac{\alpha + \beta}{2} \right)^\pm \text{-quasibounded} \right\}$$

is fulfilled (see RASMUSSEN [144, Satz 6.3.8] or AULBACH & RASMUSSEN & SIEGMUND [18]), and due to AULBACH & WANNER [21, p. 83–84, formulae (69), (70)] and (5.3), there exists an $M_1 \geq 1$ with

$$\left\|\hat{\lambda}(t, \tau, \xi)\right\| \leq M_1 \|\xi\| e^{\frac{\alpha+\beta}{2}(t-\tau)} \quad \text{for all } t \geq \tau \text{ and } \xi \in \hat{\mathcal{S}}^+(\tau) \tag{5.8}$$

and

$$\left\|\hat{\lambda}(t, \tau, \xi)\right\| \leq M_1 \|\xi\| e^{\frac{\alpha+\beta}{2}(t-\tau)} \quad \text{for all } t \leq \tau \text{ and } \xi \in \hat{\mathcal{S}}^-(\tau)\,.$$

Because of (5.3),

$$\begin{array}{l} \text{there exists a } \hat{\rho} > 0 \text{ such that the systems (5.7) and (5.1)} \\ \text{coincide on } t \in \mathbb{R} \text{ and } x \in U_{\hat{\rho}}(0)\,. \end{array} \tag{5.9}$$

Moreover, there exists an $M_2 > 0$ such that $\left\|\hat{s}^\pm(t, x)\right\| \leq M_2 \|x\|$ for all $t \in \mathbb{R}$ and $x \in \mathbb{R}^N$ (see the definition of $\hat{s}^\pm$, (5.3) and SIEGMUND [171, Satz 4.16 (c)]). This implies the existence of a $\rho > 0$ such that with the functions

$$s^\pm : \mathbb{R} \times U_\rho(0) \to \mathbb{R}^N\,, \quad s^\pm(t, x) := \hat{s}^\pm(t, x) \quad \text{for all } t \in \mathbb{R} \text{ and } x \in U_\rho(0)\,,$$

the sets

$$\mathcal{S}^\pm := \left\{ \left(\tau, \xi + s^\pm(\tau, \xi)\right) \in \mathbb{R} \times \mathbb{R}^N : \xi \in \mathcal{R}(Q_\pm(\tau)) \cap U_\rho(0) \right\}$$

are subsets of $\mathbb{R} \times U_{\hat{\rho}}(0)$. Furthermore, $\mathcal{S}^\pm$ are local nonautonomous invariant manifolds of (5.1), since the conditions of Definition 5.1 are easily verified ($\tilde{\mathcal{S}}^\pm$ are global nonautonomous invariant manifolds, and (5.9), (5.6) and (5.3) are fulfilled). For further reference, please note that (5.8) and (5.9) imply

$$\left\|\lambda(t, \tau, \xi)\right\| \leq M_1 \|\xi\| e^{\frac{\alpha+\beta}{2}(t-\tau)} \quad \begin{array}{l} \text{for all } t \geq \tau \text{ and } \xi \in \mathcal{S}^+(\tau) \text{ with} \\ \lambda\big(\tau + c(t - \tau), \tau, \xi\big) \in U_{\hat{\rho}}(0) \\ \text{for all } c \in [0, 1]\,. \end{array} \tag{5.10}$$

*Step 2. The statements (i)–(iv) are fulfilled.*
(i) Suppose that $\alpha < 0$, and choose $\varepsilon > 0$ arbitrarily. W.l.o.g., assume that $\alpha + \varepsilon < 0$. By applying Step 1 with the constants $\alpha$ and $\min\{\beta, \alpha + 2\varepsilon\}$ instead of $\alpha$ and $\beta$, we get another local nonautonomous invariant manifold

$\bar{\mathcal{S}}^+$, obtained as graph of a function $\bar{s}^+ : \mathbb{R} \times U_{\bar{\rho}}(0) \to \mathbb{R}^N$ with $0 < \bar{\rho} < \rho$. Then, because of $(\alpha + \alpha + 2\varepsilon)/2 = \alpha + \varepsilon$, (5.10) reads as

$$\|\lambda(t,\tau,\xi)\| \leq M_1 \|\xi\| e^{\min\{\frac{\alpha+\beta}{2}, \alpha+\varepsilon\}(t-\tau)} \qquad \begin{array}{l} \text{for all } t \geq \tau \text{ and } \xi \in \bar{\mathcal{S}}^+(\tau) \text{ with} \\ \lambda(\tau + c(t-\tau), \tau, \xi) \in U_{\bar{\rho}}(0) \\ \text{for all } c \in [0,1]\,. \end{array}$$

Due to $\alpha + \varepsilon < 0$, this means that there exists an $r > 0$ such that

$$\lambda(t,\tau,\xi) \in U_\rho(0) \quad \text{and} \quad \|\lambda(t,\tau,\xi)\| \leq M_1 \|\xi\| e^{\min\{\frac{\alpha+\beta}{2}, \alpha+\varepsilon\}(t-\tau)}$$
$$\text{for all } (\tau,\xi) \in \bar{\mathcal{S}}^+ \cap (\mathbb{R} \times U_r(0)) \text{ and } t \geq \tau\,.$$

Thus, for $(\tau,\xi) \in \bar{\mathcal{S}}^+ \cap (\mathbb{R} \times U_r(0))$, $\lambda(\cdot,\tau,\xi)$ is $((\alpha+\beta)/2)^+$-quasibounded and $(\alpha+\varepsilon)^+$-quasibounded. From the $((\alpha+\beta)/2)^+$-quasiboundedness, we get

$$\bar{\mathcal{S}}^+ \cap (\mathbb{R} \times U_r(0)) = \mathcal{S}^+ \cap (\mathbb{R} \times U_r(0))$$

from the dynamic characterization (of the global manifolds) in Step 1. Thus, the proof of (i) is finished.

(ii) can be shown analogously to (i).

(iii) We choose an $L > 0$ such that $\alpha + KL < 0$. Let $\chi$ denote the cut-off function from Step 1. Then we define for any $\sigma > 0$ with $U_{2\sigma}(0) \subset U$ the function $F_\sigma : \mathbb{R} \times \mathbb{R}^N \to \mathbb{R}^N$ by

$$F_\sigma(t,x) := \begin{cases} \chi\left(\frac{x}{\sigma}\right) F(t,x) & \text{for all } t \in \mathbb{R} \text{ and } x \in U \\ 0 & \text{for all } t \in \mathbb{R} \text{ and } x \notin U \end{cases}.$$

Analogously to Step 1, the relation

$$\lim_{\sigma \to 0} \sup_{(t,x) \in \mathbb{R} \times \mathbb{R}^N} \|D_2 F_\sigma(t,x)\| = 0$$

follows, and the limit behavior only depends on $\Gamma$ and $\chi$. This means that there exists an $\tilde{r} = \tilde{r}(\alpha, K, \Gamma, \chi) > 0$ such that

$$\dot{y} = A(t)y + F_{\tilde{r}}(t,y) \tag{5.11}$$

fulfills the hypotheses of AULBACH & WANNER [21, Lemma 3.4, p. 70] with the constants $\alpha$, $K$ and $L$. We denote the general solution of (5.11) by $\tilde{\lambda}$. Then, due to [21, Lemma 3.4, p. 70], we obtain

$$\|\tilde{\lambda}(t,\tau,\xi)\| \leq K\|\xi\| e^{\frac{\alpha+KL}{2}(t-\tau)} \quad \text{for all } t \geq \tau \text{ and } \xi \in \mathbb{R}^N\,.$$

We define $r := \frac{\tilde{r}}{K}$. Since (5.11) coincides with (5.1) on $\mathbb{R} \times U_{\tilde{r}}(0)$, we get

$$\|\lambda(t,\tau,\xi)\| \leq K\|\xi\| e^{\frac{\alpha+KL}{2}(t-\tau)} \quad \text{for all } t \geq \tau \text{ and } \xi \in U_r(0)\,.$$

This implies the assertion.

(iv) can be proved similarly to (iii) using Lemma 3.7 of AULBACH & WANNER [21] instead of Lemma 3.4. $\qquad\qquad\square$

*Remark 5.4.* An alternative way for the construction of nonautonomous invariant manifolds for the ODE (5.1) without applying the Lyapunov transformation as in Step 1 of the preceding proof can be found in PÖTZSCHE [138].

By applying the preceding theorem, in the next two theorems, the existence of past and future invariant manifolds is proved.

**Theorem 5.5 (Past invariant manifolds).** *In case* $\mathbb{I} = \mathbb{R}_\kappa^-$, *there exist* $\rho > 0$ *and* $C^1$-*functions* $s^\pm : \mathbb{I} \times U_\rho(0) \to \mathbb{R}^N$ *such that the graphs*

$$\mathcal{S}^\pm := \left\{ \left(\tau, \xi + s^\pm(\tau, \xi)\right) \in \mathbb{I} \times \mathbb{R}^N : \xi \in \mathcal{R}(Q_\pm(\tau)) \cap U_\rho(0) \right\}$$

*are local nonautonomous invariant manifolds. Furthermore, the following statements are fulfilled:*

(i) *Case* $\alpha < 0$ *and* $\mathrm{rk}\, Q_+ \geq 1$ *(Trivial solution is not past repulsive).*
*For all* $\varepsilon > 0$, *there exist* $r > 0$ *and* $M \geq 1$ *such that for all* $(\tau, \xi) \in \mathcal{S}^+$
*with* $\|\xi\| < r$ *and for all* $\kappa \geq t \geq \tau$, *we have* $\lambda(t, \tau, \xi) \in U_\rho(0)$ *and*

$$\|\lambda(t, \tau, \xi)\| \leq M e^{(\alpha + \varepsilon)(t - \tau)} \|\xi\|\,.$$

(ii) *Case* $\beta > 0$ *and* $\mathrm{rk}\, Q_- \geq 1$ *(Trivial solution is not past attractive).*
*For all* $\varepsilon > 0$, *there exists an* $r > 0$ *such that for all* $(\tau, \xi) \in \mathcal{S}^-$ *with* $\|\xi\| < r$, *the solution* $\lambda(\cdot, \tau, \xi)$ *is* $(\beta - \varepsilon)^-$-*quasibounded and we have* $\lambda(t, \tau, \xi) \in U_\rho(0)$ *for all* $t \leq \tau \leq \kappa$.

(iii) *Case* $\alpha < 0$ *and* $Q_+ \equiv \mathbb{1}$ *(Trivial solution is past attractive).*
*There exists an* $r = r(\alpha, K, \Gamma) > 0$ *with*

$$\lim_{t \to \infty} \sup_{\tau \in \mathbb{I}} d\left(\lambda\left(\tau, \tau - t, U_r(0)\right) \big| \{0\}\right) = 0\,.$$

(iv) *Case* $\beta > 0$ *and* $Q_- \equiv \mathbb{1}$ *(Trivial solution is past repulsive).*
*There exists an* $r = r(\beta, K, \Gamma) > 0$ *with*

$$\lim_{t \to \infty} \sup_{\tau \in \mathbb{I}} d\left(\lambda\left(\tau - t, \tau, U_r(0)\right) \big| \{0\}\right) = 0\,.$$

*Proof.* We first observe that all assertions of Theorem 5.3 also hold in case (5.1) is a differential equation of Carathéodory type, since equations of this form are treated in SIEGMUND [171] and AULBACH & WANNER [22]. This is important, because we want to apply this theorem to the Carathéodory differential equation

$$\dot{x} = B(t)x + G(t, x) \tag{5.12}$$

with functions $B : \mathbb{R} \to \mathbb{R}^{N \times N}$ and $G : \mathbb{R} \times U \to \mathbb{R}^N$ defined as follows. Let $C \in \mathbb{R}^{N \times N}$ be the matrix fulfilling

$$Cx = \alpha x \quad \text{for all } x \in \mathcal{R}(Q_+(\kappa)) \quad \text{and} \quad Cx = \beta x \quad \text{for all } x \in \mathcal{R}(Q_-(\kappa))\,.$$

Then we define the functions $B$ and $G$ by

$$B(t) := \begin{cases} A(t) : t \leq \kappa \\ C \quad : t > \kappa \end{cases} \quad \text{and} \quad G(t, x) := \begin{cases} F(t, x) : t \leq \kappa, \ x \in U \\ 0 \quad : t > \kappa, \ x \in U \end{cases}.$$

It is easy to see that equation (5.12) fulfills the hypotheses of Theorem 5.3 with the invariant projector $\hat{Q}_+ : \mathbb{R} \to \mathbb{R}^{N \times N}$ defined by

$$\hat{Q}_+ := \begin{cases} Q_+(t) : t \leq \kappa \\ Q_+(\kappa) : t > \kappa \end{cases}.$$

Then there exist nonautonomous invariant manifolds of (5.12) which, by restriction to $\mathbb{R}_\kappa^- \times \mathbb{R}^N$, are nonautonomous invariant manifolds of (5.1). The statements (i)–(iv) follow directly. □

An analogous statement is fulfilled by considering $\mathbb{R}_\kappa^+$ instead of $\mathbb{R}_\kappa^-$.

**Theorem 5.6 (Future invariant manifolds).** *In case $\mathbb{I} = \mathbb{R}_\kappa^+$, there exist $\rho > 0$ and $C^1$-functions $s^\pm : \mathbb{I} \times U_\rho(0) \to \mathbb{R}^N$ such that the graphs*

$$\mathcal{S}^\pm := \left\{ \left( \tau, \xi + s^\pm(\tau, \xi) \right) \in \mathbb{I} \times \mathbb{R}^N : \xi \in \mathcal{R}(Q_\pm(\tau)) \cap U_\rho(0) \right\}$$

*are local nonautonomous invariant manifolds. Furthermore, the following statements are fulfilled:*

(i) *Case $\alpha < 0$ and $\mathrm{rk}\, Q_+ \geq 1$ (Trivial solution is not future repulsive). For all $\varepsilon > 0$, there exists an $r > 0$ such that for all $(\tau, \xi) \in \mathcal{S}^+$ with $\|\xi\| < r$, the solution $\lambda(\cdot, \tau, \xi)$ is $(\alpha + \varepsilon)^+$-quasibounded and we have $\lambda(t, \tau, \xi) \in U_\rho(0)$ for all $t \geq \tau \geq \kappa$.*

(ii) *Case $\beta > 0$ and $\mathrm{rk}\, Q_- \geq 1$ (Trivial solution is not future attractive). For all $\varepsilon > 0$, there exist $r > 0$ and $M \geq 1$ such that for all $(\tau, \xi) \in \mathcal{S}^-$ with $\|\xi\| < r$ and for all $\kappa \leq t \leq \tau$, we have $\lambda(t, \tau, \xi) \in U_\rho(0)$ and*

$$\|\lambda(t, \tau, \xi)\| \leq M e^{(\beta - \varepsilon)(t - \tau)} \|\xi\|.$$

(iii) *Case $\alpha < 0$ and $Q_+ \equiv \mathbb{1}$ (Trivial solution is future attractive). There exists an $r = r(\alpha, K, \Gamma) > 0$ with*

$$\lim_{t \to \infty} \sup_{\tau \in \mathbb{I}} d\left( \lambda(\tau + t, \tau, U_r(0)) \big| \{0\} \right) = 0.$$

(iv) *Case $\beta > 0$ and $Q_- \equiv \mathbb{1}$ (Trivial solution is future repulsive). There exists an $r = r(\beta, K, \Gamma) > 0$ with*

$$\lim_{t \to \infty} \sup_{\tau \in \mathbb{I}} d\left( \lambda(\tau, \tau + t, U_r(0)) \big| \{0\} \right) = 0.$$

*Proof.* See proof of Theorem 5.5. □

*Remark 5.7.*

(i) The sets $\mathcal{S}^+$ and $\mathcal{S}^-$ of the above theorems are also denoted as *all-time (past, future, respectively) pseudo-stable* and *pseudo-unstable invariant manifolds* of (5.1), respectively. To be more specific, $\mathcal{S}^+$ describes an *all-time (past, future, respectively)*

$$\left.\begin{array}{r}\text{center-stable}\\\text{stable}\\\text{strongly stable}\end{array}\right\}\text{ invariant manifold in case }\left\{\begin{array}{c}\beta>0\\\alpha<0<\beta\\\beta<0\end{array}\right..$$

Accordingly, $\mathcal{S}^-$ describes an *all-time (past, future, respectively)*

$$\left.\begin{array}{r}\text{center-unstable}\\\text{unstable}\\\text{strongly unstable}\end{array}\right\}\text{ invariant manifold in case }\left\{\begin{array}{c}\alpha<0\\\alpha<0<\beta\\\alpha>0\end{array}\right..$$

This terminology corresponds to the autonomous situation of invariant manifolds considered, e.g., in CHOW & LI & WANG [47]. Center manifolds are obtained as intersections of center-stable and center-unstable invariant manifolds.

(ii) In the hyperbolic situation ($\alpha < 0 < \beta$), the all-time invariant manifolds $\mathcal{S}^\pm$ of Theorem 5.3 are uniquely determined. Easy examples (see, e.g., HALE & KOÇAK [78, Example 10.13, p. 322]), however, show that center-stable, center-unstable or center manifolds are nonunique in general. Since global invariant manifolds are uniquely determined, different cut-off-techniques (as used in the proof of Theorem 5.3) lead to different manifolds. In the situation of Theorem 5.5 and Theorem 5.6, the question of nonuniqueness is more subtle. In the hyperbolic case, only the pseudo-unstable manifold $\mathcal{S}^-$ of Theorem 5.5 and the pseudo-stable manifold $\mathcal{S}^+$ of Theorem 5.6 are uniquely determined. This corresponds to Remark 4.12 in the linear situation.

## 5.2 An Application to Bifurcation Theory

In autonomous bifurcation theory, it is necessary that at least one eigenvalue of the linearization in a bifurcating equilibrium crosses the imaginary axis. In this section, this fact is generalized with respect to the notions of past, future and all-time bifurcation. For a similar result in the context of random dynamical systems (concerning the Lyapunov exponents of ergodic invariant measures), we refer to ARNOLD & XU [9] and ARNOLD [5, Theorem 9.2.3, p. 471].

Let $\mathbb{I}$ be an unbounded interval of the form $\mathbb{R}_\kappa^-$, $\mathbb{R}_\kappa^+$ or $\mathbb{R}$, respectively. In this section, we consider a nonlinear differential equation

$$\dot{x} = A(t, \alpha)x + F(t, x, \alpha) \tag{5.13}_\alpha$$

with a continuous matrix-valued function $A : \mathbb{I} \times (\alpha_-, \alpha_+) \to \mathbb{R}^{N \times N}$ and a $C^1$-function $F : \mathbb{I} \times U \times (\alpha_-, \alpha_+) \to \mathbb{R}^N$, where $U$ is supposed to be a neighborhood of 0. Furthermore, we assume that $F(t, 0, \alpha) = 0$ for all $\alpha \in (\alpha_-, \alpha_+)$ and $t \in \mathbb{I}$.

**Theorem 5.8 (Linearization and bifurcation).** *We suppose that the trivial solution of* $(5.13)_\alpha$ *admits a past (future, all-time, respectively) supercritical bifurcation at the parameter value* $\alpha_0 \in (\alpha_-, \alpha_+)$ *and there exists an* $\hat{\alpha} > \alpha_0$ *with*

$$\lim_{x \to 0} \sup_{t \in \mathbb{I}, \, \alpha \in [\alpha_0, \hat{\alpha}]} \| D_2 F(t, x, \alpha) \| = 0 \tag{5.14}$$

*and*

$$\lim_{\alpha \searrow \alpha_0} \sup_{t \in \mathbb{I}} \| A(t, \alpha) - A(t, \alpha_0) \| = 0. \tag{5.15}$$

*Then we have*

$$0 \in \Sigma_{A(\cdot, \alpha_0)}^{\leftarrow}, \; \Sigma_{A(\cdot, \alpha_0)}^{\rightarrow}, \; \Sigma_{A(\cdot, \alpha_0)}^{\leftrightarrow}, \; respectively.$$

*An analogous statement is fulfilled in case of a subcritical bifurcation.*

*Proof.* We only treat the case of an all-time bifurcation, since the other proofs are similar. Arguing negatively, we suppose that zero is not contained in $\Sigma_{A(\cdot, \alpha_0)}^{\leftrightarrow}$. We distinguish the following two cases.
*Case 1.* There exists an $\tilde{\alpha} \in (\alpha_0, \hat{\alpha})$ such that the trivial solution of $(5.13)_\alpha$ is all-time attractive for all $\alpha \in (\alpha_0, \tilde{\alpha})$.
First, assume that $\Sigma_{A(\cdot, \alpha_0)}^{\leftrightarrow} \cap (0, \infty] \neq \emptyset$. Since $0 \notin \Sigma_{A(\cdot, \alpha_0)}^{\leftrightarrow}$, this means that the linear differential equation $\dot{x} = A(t, \alpha_0)x$ admits a nonhyperbolic all-time exponential dichotomy with growth rate $\gamma > 0$ and an invariant projector $Q_{\alpha_0}$ such that $\mathrm{rk}\, Q_{\alpha_0} < N$ (please note that due to Theorem 4.24, the all-time spectrum is closed). Due to Theorem 4.35 and (5.15), there exists an $\alpha_1 > \alpha_0$ such that for all $\alpha \in (\alpha_0, \alpha_1)$, the linear differential equation

$$\dot{x} = A(t, \alpha)x$$

admits a nonhyperbolic all-time exponential dichotomy with growth rate $\gamma$ and an invariant projector $Q_\alpha$ such that $\mathrm{rk}\, Q_\alpha < N$. Hence,

$$\Sigma_{A(\cdot, \alpha)}^{\leftrightarrow} \cap (\gamma, \infty] \neq \emptyset \quad \text{for all } \alpha \in (\alpha_0, \alpha_1).$$

This means that at least one spectral interval of $\Sigma_{A(\cdot, \alpha)}^{\leftrightarrow}$, $\alpha \in (\alpha_0, \alpha_1)$, lies in $(\gamma, \infty]$, and hence, Theorem 5.3 (ii) is applicable with $\beta = \gamma > 0$ and $\mathrm{rk}\, Q_- = N - \mathrm{rk}\, Q_\alpha \geq 1$. This implies that the trivial solution of $(5.13)_\alpha$ is not all-time attractive, which is a contradiction to the hypothesis of Case 1, and thus, there exists a $\delta < 0$ with

$$\Sigma^{\leftrightarrow}_{A(\cdot,\alpha_0)} \subset [-\infty, \delta)$$

(again, we use the fact that all-time dichotomy spectra are closed). Because of Theorem 4.35 and (5.15) (cf. the argumentation above), there exists an $\alpha_2 \in (\alpha_0, \alpha_1)$ with

$$\Sigma^{\leftrightarrow}_{A(\cdot,\alpha)} \subset [-\infty, \delta] \quad \text{for all } \alpha \in [\alpha_0, \alpha_2].$$

We apply Theorem 5.3 (iii) and obtain that, since due to (5.14), the function $\Gamma : (0,1) \to \overline{\mathbb{R}}^+$ can be chosen independently of $\alpha$, the lower bound $r$ of this theorem for the radius of all-time attraction $\mathfrak{A}_0^{\leftrightarrow}$ is also independent of $\alpha$. Hence, $\mathfrak{A}_0^{\leftrightarrow}$ does not converge to zero in the limit $\alpha \searrow \alpha_0$. This contradiction finishes the proof of this case.

*Case 2. There exists an $\tilde{\alpha} \in (\alpha_0, \hat{\alpha})$ such that the trivial solution of (5.13)$_\alpha$ is all-time repulsive for all $\alpha \in (\alpha_0, \tilde{\alpha})$.*
This case is treated analogously to Case 1.                                   □

## 5.3 Linearized Attractivity and Repulsivity

In Section 5.1, properties of attractivity and repulsivity for a nonlinear system have been derived already by studying the linearization. In contrast to these considerations, in this section, more quantitative results are obtained, and furthermore, $C^1$-differentiability is not assumed but only continuity.

We first concentrate on the notions of past, future and all-time attractivity and repulsivity.

**Theorem 5.9 (Linearized attractivity and repulsivity, part I).** *Consider an unbounded interval $\mathbb{I}$ of the form $\mathbb{R}^-_\kappa$, $\mathbb{R}^+_\kappa$ or $\mathbb{R}$, respectively, and let*

$$\dot{x} = A(t)x + F(t,x) \tag{5.16}$$

*be a nonautonomous differential equation with continuous functions $A : \mathbb{I} \to \mathbb{R}^{N \times N}$ and $F : \mathbb{I} \times U \to \mathbb{R}^N$, $U \subset \mathbb{R}^N$ a neighborhood of $0$, such that $F(t,0) = 0$ for all $t \in \mathbb{I}$. Let $\lambda$ denote the general solution of (5.16) and $\Lambda : \mathbb{I} \times \mathbb{I} \to \mathbb{R}^{N \times N}$ denote the transition operator of the linearized equation $\dot{x} = A(t)x$. Then the following statements are fulfilled:*

(i) *In case there exist $\beta < 0$, $K \geq 1$ and $\delta > 0$ such that*

$$\|\Lambda(t,s)\| \leq K e^{\beta(t-s)} \quad \text{for all } t \geq s$$

*and*

$$\|F(t,x)\| \leq \frac{-\beta}{2K} \|x\| \quad \text{for all } t \in \mathbb{I} \text{ and } x \in U_\delta(0), \tag{5.17}$$

*we have*

$$d\big(\lambda\big(t,\tau,U_{\delta/K}(0)\big)\big|\{0\}\big) \leq \delta\,e^{(\beta/2)(t-\tau)} \quad \text{for all } \tau,t \in \mathbb{I} \text{ with } \tau \leq t\,,$$

*i.e., the trivial solution of (5.16) is past (future, all-time, respectively) attractive.*

(ii) *In case there exist* $\beta > 0$, $K \geq 1$ *and* $\delta > 0$ *such that*

$$\|A(t,s)\| \leq K e^{\beta(t-s)} \quad \text{for all } t \leq s$$

*and*

$$\|F(t,x)\| \leq \frac{\beta}{2K}\,\|x\| \quad \text{for all } t \in \mathbb{I} \text{ and } x \in U_\delta(0)\,, \tag{5.18}$$

*we have*

$$d\big(\lambda\big(t,\tau,U_{\delta/K}(0)\big)\big|\{0\}\big) \leq \delta\,e^{(\beta/2)(t-\tau)} \quad \text{for all } \tau,t \in \mathbb{I} \text{ with } t \leq \tau\,,$$

*i.e., the trivial solution of (5.16) is past (future, all-time, respectively) repulsive.*

*Proof.* We only prove (i), since (ii) can be shown analogously. Given $\tau \in \mathbb{I}$ and $\xi \in U_\delta(0)$, we now prove an estimate for the general solution under the additional assumption

$$\lambda(t,\tau,\xi) \in U_\delta(0) \quad \text{for all } t \geq \tau\,. \tag{5.19}$$

The solution $\lambda(\cdot,\tau,\xi)$ of (5.16) is also a solution of the inhomogeneous linear differential equation

$$\dot{x} = A(t)x + F(t,\lambda(t,\tau,\xi))\,.$$

Thus, the variation of constants formula (Proposition A.6) implies

$$\lambda(t,\tau,\xi) = \Lambda(t,\tau)\xi + \int_\tau^t \Lambda(t,s)F(s,\lambda(s,\tau,\xi))\,ds \quad \text{for all } t \geq \tau\,,$$

and hence,

$$
\begin{aligned}
\|\lambda(t,\tau,\xi)\| &\leq \|\Lambda(t,\tau)\|\,\|\xi\| + \int_\tau^t \|\Lambda(t,s)\|\,\|F(s,\lambda(s,\tau,\xi))\|\,ds \\
&\overset{(5.17)}{\leq} K e^{\beta(t-\tau)}\|\xi\| + \int_\tau^t K e^{\beta(t-s)}\frac{-\beta}{2K}\|\lambda(s,\tau,\xi)\|\,ds \quad \text{for all } t \geq \tau
\end{aligned}
$$

is fulfilled. This implies

$$e^{-\beta t}\|\lambda(t,\tau,\xi)\| \leq K e^{-\beta\tau}\|\xi\| + \frac{-\beta}{2}\int_\tau^t e^{-\beta s}\|\lambda(s,\tau,\xi)\|\,ds \quad \text{for all } t \geq \tau\,.$$

Hence, Gronwall's inequality (cf. Lemma A.8) yields the estimate

$$\|\lambda(t,\tau,\xi)\| \le K e^{(\beta/2)(t-\tau)}\|\xi\| \quad \text{for all } t \ge \tau. \tag{5.20}$$

We define $\eta := \delta/K$. Since $\beta/2 < 0$, for all $\tau \in \mathbb{I}$ and $\xi \in U_\eta(0)$, the assumption (5.19) is fulfilled, therefore, (5.20) holds for such $\tau$ and $\xi$. This implies

$$d\big(\lambda(t,\tau,U_\eta(0))\big|\{0\}\big) \le K\eta\, e^{(\beta/2)(t-\tau)} \quad \text{for all } \tau, t \in \mathbb{I} \text{ with } \tau \le t.$$

From this inequality, the required conditions for the past (future, all-time, respectively) attractivity are easily obtained.     □

In case of finite-time attractivity and repulsivity, the following result is obtained.

**Theorem 5.10 (Linearized attractivity and repulsivity, part II).** *Consider a compact interval* $\mathbb{I} := [\tau, \tau + T]$ *for some* $\tau \in \mathbb{R}$ *and* $T > 0$, *and let*

$$\dot{x} = A(t)x + F(t,x) \tag{5.21}$$

*be a nonautonomous differential equation with continuous functions* $A : \mathbb{I} \to \mathbb{R}^{N \times N}$ *and* $F : \mathbb{I} \times U \to \mathbb{R}^N$, $U \subset \mathbb{R}^N$ *a neighborhood of* $0$, *such that* $F(t,0) = 0$ *for all* $t \in \mathbb{I}$. *Let* $\lambda$ *denote the general solution of (5.21) and* $\Lambda : \mathbb{I} \times \mathbb{I} \to \mathbb{R}^{N \times N}$ *denote the transition operator of the linearized equation* $\dot{x} = A(t)x$, *and define*

$$K_+ := \sup\big\{\|\Lambda(t,s)\| : \tau \le s \le t \le \tau + T\big\}$$

*and*

$$K_- := \sup\big\{\|\Lambda(t,s)\| : \tau \le t \le s \le \tau + T\big\}.$$

*Then the following statements are fulfilled:*

(i) *In case*
$$\|\Lambda(\tau + T, \tau)\| < 1$$
*and there exist* $\delta > 0$ *and* $\beta > 1$ *with*

$$\|F(t,x)\| \le -\frac{\ln\big(\beta\,\|\Lambda(\tau+T,\tau)\|\big)}{TK_+}\,\|x\| \quad \text{for all } t \in \mathbb{I} \text{ and } x \in U_\delta(0),$$
$$\tag{5.22}$$

*there exists an* $\eta > 0$ *such that*

$$\|\lambda(\tau + T, \tau, \xi)\| \le \beta^{-1}\|\xi\| \quad \text{for all } \xi \in U_\eta(0),$$

*i.e., the trivial solution of (5.21) is* $(\tau, T)$-*attractive.*

(ii) *In case*
$$\|\Lambda(\tau, \tau + T)\| < 1$$
*and there exist* $\delta > 0$ *and* $\beta > 1$ *with*

$$\|F(t,x)\| \leq -\frac{\ln\left(\beta\,\|\Lambda(\tau,\tau+T)\|\right)}{TK_-}\,\|x\| \quad \text{for all } t \in \mathbb{I} \text{ and } x \in U_\delta(0),$$

$$(5.23)$$

*there exists an $\eta > 0$ such that*

$$\|\lambda(\tau,\tau+T,\xi)\| \leq \beta^{-1}\,\|\xi\| \quad \text{for all } \xi \in U_\eta(0),$$

*i.e., the trivial solution of* (5.21) *is* $(\tau,T)$*-repulsive.*

*Proof.* We only prove (i), since (ii) can be shown analogously. Due to the continuity of the general solution (cf. Proposition A.3), there exists an $\eta < \delta$ with

$$\|\lambda(t,\tau,\xi)\| < \delta \quad \text{for all } t \in \mathbb{I} \text{ and } \xi \in U_\eta(0).$$

We choose $\xi \in U_\eta(0)$ arbitrarily. Then the solution $\lambda(\cdot,\tau,\xi)$ of (5.21) is also a solution of the linear differential equation

$$\dot{x} = A(t)x + F(t,\lambda(t,\tau,\xi)).$$

Thus, the variation of the constants formula (cf. Proposition A.6) implies

$$\lambda(t,\tau,\xi) = \Lambda(t,\tau)\xi + \int_\tau^t \Lambda(t,s)F(s,\lambda(s,\tau,\xi))\,ds \quad \text{for all } t \in \mathbb{I}.$$

Hence, for all $t \in \mathbb{I}$, the relation

$$\|\lambda(t,\tau,\xi)\| \leq \|\Lambda(t,\tau)\xi\| + \int_\tau^t \|\Lambda(t,s)\|\,\big\|F(s,\lambda(s,\tau,\xi))\big\|\,ds$$

$$\leq \|\Lambda(t,\tau)\|\,\|\xi\| - K_+ \frac{\ln\left(\beta\,\|\Lambda(\tau+T,\tau)\|\right)}{TK_+} \int_\tau^t \|\lambda(s,\tau,\xi)\|\,ds$$

$$= \|\Lambda(t,\tau)\|\,\|\xi\| - \frac{\ln\left(\beta\,\|\Lambda(\tau+T,\tau)\|\right)}{T} \int_\tau^t \|\lambda(s,\tau,\xi)\|\,ds\,.$$

We apply Gronwall's inequality (cf. Lemma A.8) and obtain for all $\xi \in U_\eta(0)$,

$$\|\lambda(\tau+T,\tau,\xi)\| \leq \|\Lambda(\tau+T,\tau)\|\,\|\xi\|\,\exp\left(-\ln\left(\beta\,\|\Lambda(\tau+T,\tau)\|\right)\right)$$

$$= \beta^{-1}\,\|\xi\|\,.$$

This finishes the proof of this theorem.    □

*Remark 5.11.*

(i) Concerning Theorem 5.9 and Theorem 5.10, the past (future, all-time, $(\tau,T)$-, respectively) dichotomy spectrum of the linearization $\dot{x} = A(t)x$ is a subset of $\mathbb{R}^-$.

(ii) The conditions (5.17), (5.18) of Theorem 5.9 and (5.22), (5.23) of Theorem 5.10 are fulfilled if we have

$$\lim_{x \to 0} \sup_{t \in \mathbb{I}} \frac{\|F(t,x)\|}{\|x\|} = 0 \, .$$

This limit relation is only sufficient but not necessary for the above mentioned conditions.

## 5.4 Bifurcation Theory of Adiabatic Systems

In this section, a relationship between the bifurcation theory of adiabatic systems (see, e.g., BENOÎT [27], BERGLUND [29] or LEBOVITZ & SCHAAR [106, 107]) and the concept of finite-time bifurcation is pointed out.

The bifurcation theory of adiabatic systems is usually called *dynamic bifurcation theory* (see title of BENOÎT [27]). We will not employ this term here, since it is unfortunately used in a different sense both in autonomous bifurcation theory (as opposed to static bifurcation theory, cf. Subsection 2.6.1) and random bifurcation theory (as opposed to phenomenological bifurcation theory, cf. Subsection 2.6.3).

Let $I$ be an open interval and $D \subset \mathbb{R}^N$ be an open set, and consider an autonomous differential equation

$$\dot{x} = f(\alpha, x) \, , \qquad\qquad (5.24)_\alpha$$

depending on a parameter $\alpha$ with a $C^1$-function $f : I \times D \to \mathbb{R}^N$. To mimic the situation of a slowly varying parameter, for $\varepsilon > 0$, we also look at the system

$$\dot{x} = f(\varepsilon t, x) \, ,$$

which can be transformed via the *slow time* $t \mapsto \varepsilon t$ into the so-called *adiabatic* or *singularly-perturbed system*

$$\dot{x} = \frac{1}{\varepsilon} f(t, x) \, . \qquad\qquad (5.25)_\varepsilon$$

The central question of the bifurcation theory of adiabatic systems is: How do solutions of $(5.25)_\varepsilon$ behave in the limit $\varepsilon \searrow 0$ in case $(5.24)_\alpha$ admits an autonomous bifurcation?

We assume that $(5.24)_\alpha$ admits a bifurcation of the following type.

**Standing Hypothesis.** For fixed $\alpha_- < \alpha_+ \in I$, we consider two different continuous functions $s_1, s_2 : [\alpha_-, \alpha_+] \to D$ such that

$$s_1(\alpha_-) = s_2(\alpha_-) \quad \text{and} \quad f(\alpha, s_1(\alpha)) = f(\alpha, s_2(\alpha)) = 0 \text{ for all } \alpha \in [\alpha_-, \alpha_+] \, .$$

We suppose that

- $s_1$ describes attractive equilibria of $(5.24)_\alpha$, i.e., for all $\alpha \in (\alpha_-, \alpha_+]$, the eigenvalues of $D_2 f(\alpha, s_1(\alpha))$ have a negative real part,

- $s_2$ describes hyperbolic equilibria of $(5.24)_\alpha$, i.e., for all $\alpha \in (\alpha_-, \alpha_+]$, the eigenvalues of $D_2 f(\alpha, s_2(\alpha))$ have a non-vanishing real part.

The existence of such a bifurcation implies that $D_2 f(\alpha_-, s_1(\alpha_-))$ has an eigenvalue with vanishing real part.

In the bifurcation theory of adiabatic systems, the occurrence of the following two possibilities is discussed:

(a) There exists a family of solutions $\nu_\varepsilon : [\alpha_-, \alpha_+] \to D$ of $(5.25)_\varepsilon$, $\varepsilon > 0$ small, which converge to the attractive equilibrium branch in the limit $\varepsilon \to 0$.

(b) There exists a family of solutions $\nu_\varepsilon : [\alpha_-, \alpha_+] \to D$ of $(5.25)_\varepsilon$, $\varepsilon > 0$ small, which follow for some time interval (which does not depend on $\varepsilon$) the equilibrium branch $s_2$ and then jump to the stable branch $s_1$.

The phenomenon (b) is called *bifurcation delay* or *delayed exchange of stabilities*. The corresponding solutions are said to be *canard solutions*. The property (a), which we will discuss in this section, is generalized by the following definition.

**Definition 5.12 (Adiabatic solutions).** *Let $\alpha_0 < \alpha_1 \in I$. A continuous function $s : [\alpha_0, \alpha_1] \to D$ with*

$$f(\alpha, s(\alpha)) = 0 \quad \text{for all } \alpha \in [\alpha_0, \alpha_1]$$

*is called equilibrium branch which admits* adiabatic solutions *if there exist $\tilde{\varepsilon} > 0$ and a function $\nu : [\alpha_0, \alpha_1] \times (0, \tilde{\varepsilon}) \to D$ such that $\nu(\cdot, \varepsilon)$ is a solution of $(5.25)_\varepsilon$ and*

$$\lim_{\varepsilon \searrow 0} \sup_{\alpha \in [\alpha_0, \alpha_1]} \|\nu(\alpha, \varepsilon) - s(\alpha)\| = 0 .$$

In case the equilibria of $(5.24)_\alpha$ described by the function $s$ are hyperbolic, the existence of adiabatic solutions follows from the following theorem. A proof can be found, e.g., in BERGLUND [29, Theorem 5.1, p. 140].

**Theorem 5.13 (Existence of adiabatic solutions).** *Let $\alpha_0 < \alpha_1 \in I$, and consider a continuous function $s : [\alpha_0, \alpha_1] \to D$ such that*

$$f(\alpha, s(\alpha)) = 0 \quad \text{and} \quad D_2 f(\alpha, s(\alpha)) \text{ is hyperbolic} \quad \text{for all } \alpha \in [\alpha_0, \alpha_1].$$

*Then the equilibrium branch $s$ admits adiabatic solutions.*

In the next lemma, linearizations near a branch of stable equilibria are examined.

**Lemma 5.14.** *Given $\alpha_0 < \alpha_1 \in I$, let $s : [\alpha_0, \alpha_1] \to D$ be a continuous function such that $f(\alpha, s(\alpha)) = 0$ and $D_2 f(\alpha, s(\alpha))$ has only eigenvalues with negative real part for all $\alpha \in [\alpha_0, \alpha_1]$. Then there exist constants $\delta > 0$, $K \geq 1$ and $\gamma < 0$ such that for all continuous functions $h : [\alpha_0, \alpha_1] \to \mathbb{R}^N$ with*

$$\|h(\alpha) - s(\alpha)\| < \delta \quad \text{for all } \alpha \in [\alpha_0, \alpha_1],$$

*the transition operator $\Lambda_\varepsilon$ of the linear system*

$$\dot{x} = \frac{1}{\varepsilon} D_2 f(t, h(t)) \, x$$

*fulfills*

$$\|\Lambda_\varepsilon(\alpha_1, \alpha_0)\| \leq K \exp\left(\frac{\gamma}{\varepsilon}(\alpha_1 - \alpha_0)\right).$$

*Proof.* We define $A(\alpha) := D_2 f(\alpha, s(\alpha))$ for all $\alpha \in [\alpha_0, \alpha_1]$. Since $A(\cdot)$ is continuous on the compact interval $[\alpha_0, \alpha_1]$ and all eigenvalues of $A(\alpha)$ for $\alpha \in [\alpha_0, \alpha_1]$ have negative real part, there exist $\tilde{K} \geq 1$ and $\tilde{\gamma} < 0$ with

$$\left\| e^{A(\alpha)t} \right\| \leq \tilde{K} \, e^{\tilde{\gamma}t} \quad \text{for all } t \geq 0 \text{ and } \alpha \in [\alpha_0, \alpha_1].$$

Due to the uniform continuity of $D_2 f(\cdot, \cdot)$ on compact sets, there exists a $\delta > 0$ such that for all $\alpha \in [\alpha_0, \alpha_1]$ and $x \in U_\delta(s(\alpha))$, we have

$$\|D_2 f(\alpha, x) - \underbrace{D_2 f(\alpha, s(\alpha))}_{A(\alpha)}\| < -\frac{\tilde{\gamma}}{16 \tilde{K}^2}. \tag{5.26}$$

Since $A(\cdot)$ is uniform continuous on $[\alpha_0, \alpha_1]$, there exist $n \in \mathbb{N}$ and constants $\beta_i \in [\alpha_0, \alpha_1]$, $i \in \{0, \ldots, n\}$, with $\alpha_0 = \beta_0 < \beta_1 < \cdots < \beta_n = \alpha_1$ such that for all $i \in \{1, \ldots, n\}$, we have

$$\|A(\beta_{i-1}) - A(\alpha)\| < -\frac{\tilde{\gamma}}{16 \tilde{K}^2} \quad \text{for all } \alpha \in [\beta_{i-1}, \beta_i]. \tag{5.27}$$

Let $h : [\alpha_0, \alpha_1] \to \mathbb{R}^N$ be a continuous function fulfilling

$$\|h(\alpha) - s(\alpha)\| < \delta \quad \text{for all } \alpha \in [\alpha_0, \alpha_1],$$

and consider the linear system

$$\dot{x} = D_2 f(\varepsilon t, h(\varepsilon t)) \, x \tag{5.28}$$

for fixed $\varepsilon > 0$. The transition operator of (5.28) is denoted by $\Psi_\varepsilon$. Due to (5.26) and (5.27), for fixed $i \in \{1, \ldots, n\}$,

$$\left\| D_2 f(\varepsilon t, h(\varepsilon t)) - A(\beta_{i-1}) \right\| < -\frac{\tilde{\gamma}}{8 \tilde{K}^2} \quad \text{for all } t \in \left[\frac{\beta_{i-1}}{\varepsilon}, \frac{\beta_i}{\varepsilon}\right]$$

is fulfilled. Therefore, Theorem 4.35 implies

$$\left\| \Psi_\varepsilon \left( \frac{\beta_i}{\varepsilon}, \frac{\beta_{i-1}}{\varepsilon} \right) \right\| \leq \frac{5\tilde{K}^2}{2} \exp\left( \left( \tilde{\gamma} - \frac{\tilde{\gamma}}{2K} \right) \frac{\beta_i - \beta_{i-1}}{\varepsilon} \right).$$

Hence, the relation

$$\left\| \Psi_\varepsilon \left( \frac{\alpha_1}{\varepsilon}, \frac{\alpha_0}{\varepsilon} \right) \right\| \leq \left( \frac{5\tilde{K}^2}{2} \right)^n \exp\left( \left( \tilde{\gamma} - \frac{\tilde{\gamma}}{2K} \right) \frac{\alpha_1 - \alpha_0}{\varepsilon} \right)$$

$$=: K \exp\left( \frac{\gamma}{\varepsilon}(\alpha_1 - \alpha_0) \right)$$

holds. This implies the assertion, since $\Lambda_\varepsilon(\alpha_1, \alpha_0) = \Psi_\varepsilon(\alpha_1/\varepsilon, \alpha_0/\varepsilon)$. $\qquad\square$

Using the preceding lemma, we are able to prove the following relationship between adiabatic solutions of attractive equilibrium branches and the concept of finite-time attractivity.

**Corollary 5.15.** *Let $\alpha_0 < \alpha_1 \in I$, and consider an equilibrium branch $s : [\alpha_0, \alpha_1] \to D$ which admits adiabatic solutions such that all eigenvalues of $D_2 f(\alpha, s(\alpha))$ have negative real part for $\alpha \in [\alpha_0, \alpha_1]$. Moreover, let $\mu : [\alpha_0, \alpha_1] \times (0, \tilde{\varepsilon}) \to \mathbb{R}^N$ be a function describing corresponding adiabatic solutions. Then there exists an $\hat{\varepsilon} > 0$ such that for all $\varepsilon \in (0, \hat{\varepsilon})$, the solution $\mu(\cdot, \varepsilon)$ is $(\alpha_0, \alpha_1 - \alpha_0)$-attractive.*

*Proof.* Lemma 5.14 implies the existence of $\delta > 0$, $K \geq 1$ and $\gamma < 0$ with the properties mentioned in the lemma. We choose $\varepsilon^* > 0$ such that

$$\sup_{\alpha \in [\alpha_0, \alpha_1]} \|\mu(\alpha, \varepsilon) - s(\alpha)\| < \delta \quad \text{for all } \varepsilon \in (0, \varepsilon^*)$$

and

$$K \exp\left( \frac{\gamma}{\hat{\varepsilon}}(\alpha_1 - \alpha_0) \right) < 1.$$

By applying Lemma 5.14, we obtain that the transition operator $\Lambda_\varepsilon$ of the variational equation

$$\dot{x} = \frac{1}{\varepsilon} D_2 f(t, \mu(t, \varepsilon)) \, x$$

satisfies $\|\Lambda_\varepsilon(\alpha_1, \alpha_0)\| < 1$. Hence, Theorem 5.10 implies that there exists an $\hat{\varepsilon} \in (0, \varepsilon^*)$ such that for all $\varepsilon \in (0, \hat{\varepsilon})$, the solution $\mu(\cdot, \varepsilon)$ is $(\alpha_0, \alpha_1 - \alpha_0)$-attractive (due to Remark 5.11 (ii), the condition on the nonlinearity is fulfilled for small $\varepsilon$). $\qquad\square$

For the main result of this section, recall the Standing Hypothesis from the beginning of this section.

**Theorem 5.16 (Relationship to the concept of finite-time bifurcation).** *We assume that the equilibrium branch $s_1$ admits adiabatic solutions, i.e., there exists a function $\mu : [\alpha_-, \alpha_+] \times (0, \tilde{\varepsilon}) \to \mathbb{R}^N$ such that $\mu(\cdot, \varepsilon)$ is a solution of $(5.25)_\varepsilon$ and we have*

$$\lim_{\varepsilon \searrow 0} \sup_{\alpha \in [\alpha_-, \alpha_+]} \|\mu(\alpha, \varepsilon) - s_1(\alpha)\| = 0\,.$$

*Then, for sufficiently small $\alpha > \alpha_-$ and $\varepsilon > 0$, $\mu(\cdot, \varepsilon)$ is $(\alpha, \alpha_+ - \alpha)$-attractive, and the limit relation*

$$\lim_{\alpha \searrow \alpha_-} \limsup_{\varepsilon \searrow 0} \mathfrak{A}^{(\alpha, \alpha_+ - \alpha)}_{\mu(\cdot, \varepsilon)} = 0$$

*is satisfied.*

*Proof.* Since $\lim_{\alpha \to \alpha_-} s_1(\alpha) = \lim_{\alpha \to \alpha_-} s_2(\alpha)$, there exists an $\hat{\alpha} \in (\alpha_-, \alpha_+)$ with

$$\|s_1(\alpha) - s_2(\alpha)\| < \frac{1}{3} \|s_1(\alpha_+) - s_2(\alpha_+)\| \quad \text{for all } \alpha \in [\alpha_-, \hat{\alpha}]\,. \qquad (5.29)$$

Now, we prove the following statement which is obviously sufficient for the assertion: For all $\alpha \in (\alpha_-, \hat{\alpha}]$, there exists an $\hat{\varepsilon} > 0$ such that for all $\varepsilon \in (0, \hat{\varepsilon})$, the solution $\mu(\cdot, \varepsilon)$ is $(\alpha, \alpha_+ - \alpha)$-attractive and

$$\mathfrak{A}^{(\alpha, \alpha_+ - \alpha)}_{\mu(\cdot, \varepsilon)} \leq \frac{3}{2} \|s_1(\alpha) - s_2(\alpha)\|$$

is fulfilled. We choose $\alpha \in (\alpha_-, \hat{\alpha}]$ arbitrarily. It follows from Theorem 5.13 that there exists a function $\nu : [\alpha, \alpha_+] \times (0, \tilde{\varepsilon}) \to \mathbb{R}^N$ such that $\nu(\cdot, \varepsilon)$ is a solution of $(5.25)_\varepsilon$ and

$$\lim_{\varepsilon \searrow 0} \sup_{\tilde{\alpha} \in [\alpha, \alpha_+]} \|\nu(\tilde{\alpha}, \varepsilon) - s_2(\tilde{\alpha})\| = 0\,.$$

Thus, there exists an $\hat{\varepsilon} > 0$ such that for all $\varepsilon \in (0, \hat{\varepsilon})$, we have

$$\sup_{\tilde{\alpha} \in [\alpha, \alpha_+]} \|\nu(\tilde{\alpha}, \varepsilon) - s_2(\tilde{\alpha})\| < \frac{1}{4} \|s_1(\alpha) - s_2(\alpha)\|\,,$$

$$\sup_{\tilde{\alpha} \in [\alpha, \alpha_+]} \|\mu(\tilde{\alpha}, \varepsilon) - s_1(\tilde{\alpha})\| < \frac{1}{4} \|s_1(\alpha) - s_2(\alpha)\|$$

and $\mu(\cdot, \varepsilon)$ is $(\alpha, \alpha_+ - \alpha)$-attractive (cf. Corollary 5.15). For all $\varepsilon \in (0, \hat{\varepsilon})$, this implies the relations

$$\begin{aligned}
&\|\nu(\alpha, \varepsilon) - \mu(\alpha, \varepsilon)\| \\
&= \|\nu(\alpha, \varepsilon) - s_2(\alpha) + s_2(\alpha) - s_1(\alpha) + s_1(\alpha) - \mu(\alpha, \varepsilon)\| \\
&\leq \|\nu(\alpha, \varepsilon) - s_2(\alpha)\| + \|s_2(\alpha) - s_1(\alpha)\| + \|s_1(\alpha) - \mu(\alpha, \varepsilon)\| \\
&< \frac{3}{2} \|s_1(\alpha) - s_2(\alpha)\|
\end{aligned}$$

and

$$\Big\| \underbrace{\lambda_\varepsilon(\alpha_+, \alpha, \nu(\alpha, \varepsilon))}_{\nu(\alpha_+, \varepsilon)} - \mu(\alpha_+, \varepsilon) \Big\| \;>\; \frac{1}{2} \|s_1(\alpha_+) - s_2(\alpha_+)\|$$

$$\overset{(5.29)}{>} \frac{3}{2} \|s_1(\alpha) - s_2(\alpha)\|,$$

where $\lambda_\varepsilon$ denotes the general solution of $(5.25)_\varepsilon$. This finishes the proof of this theorem.   □

**6**

# Bifurcations in Dimension One

The aim of this chapter is to develop nonautonomous counterparts for the classical one-dimensional bifurcation patterns such as the transcritical and pitchfork bifurcation, both for nonautonomous bifurcations and transitions.

In this chapter, only the continuous case of ordinary differential equations is treated. For analogous results in the context of difference equations, see RASMUSSEN [145].

Recently, LANGA & ROBINSON & SUÁREZ [105] also studied the occurrence of one-dimensional nonautonomous bifurcations, which they understand as merging processes of two distinct solutions with different stability behavior. As in this chapter, their theorems are formulated in terms of Taylor coefficients for the right hand side of an ordinary differential equation. These conditions, however, are of a quite different form than the results obtained in this chapter. This difference is due to fact that, in [105], explicitly solvable models are used to formulate these conditions.

Stochastic versions (in the sense of a D-bifurcation, cf. Subsection 2.6.3) of the transcritical and pitchfork bifurcation are examined in the thesis of STEINKAMP [177] (see also CRAUEL & IMKELLER & STEINKAMP [59]).

## 6.1 Nonautonomous Transcritical Bifurcation

In this section, nonautonomous generalizations of the classical transcritical bifurcation are derived. First, the case of unbounded time domains is treated.

**Theorem 6.1 (Nonautonomous transcritical bifurcation, part I).** *Let* $x_- < 0 < x_+$ *and* $\alpha_- < \alpha_+$ *be in* $\overline{\mathbb{R}}$ *and* $\mathbb{I}$ *be an unbounded interval of the form* $\mathbb{R}_\kappa^-$, $\mathbb{R}_\kappa^+$ *or* $\mathbb{R}$, *respectively, and consider the nonautonomous differential equation*

$$\dot{x} = a(t, \alpha)x + b(t, \alpha)x^2 + r(t, x, \alpha) \qquad (6.1)_\alpha$$

*with continuous functions* $a : \mathbb{I} \times (\alpha_-, \alpha_+) \to \mathbb{R}$, $b : \mathbb{I} \times (\alpha_-, \alpha_+) \to \mathbb{R}$ *and* $r : \mathbb{I} \times (x_-, x_+) \times (\alpha_-, \alpha_+) \to \mathbb{R}$ *fulfilling* $r(\cdot, 0, \cdot) \equiv 0$. *Let* $\Lambda_\alpha : \mathbb{I} \times \mathbb{I} \to \mathbb{R}$ *denote the transition operator of the linearized equation* $\dot{x} = a(t, \alpha)x$, *and assume, there exists an* $\alpha_0 \in (\alpha_-, \alpha_+)$ *such that the following hypotheses hold:*

- **Hypothesis on linear part.** *There exist two functions* $\beta_1, \beta_2 : (\alpha_-, \alpha_+) \to \mathbb{R}$ *which are either both monotone increasing or both monotone decreasing and* $K \geq 1$ *such that* $\lim_{\alpha \to \alpha_0} \beta_1(\alpha) = \lim_{\alpha \to \alpha_0} \beta_2(\alpha) = 0$ *and*

$$\Lambda_\alpha(t, s) \leq K e^{\beta_1(\alpha)(t-s)} \quad \text{for all } \alpha \in (\alpha_-, \alpha_+) \text{ and } t, s \in \mathbb{I} \text{ with } t \geq s,$$
$$\Lambda_\alpha(t, s) \leq K e^{\beta_2(\alpha)(t-s)} \quad \text{for all } \alpha \in (\alpha_-, \alpha_+) \text{ and } t, s \in \mathbb{I} \text{ with } t \leq s.$$

- **Hypothesis on nonlinearity.** *The quadratic term either fulfills*

$$0 < \liminf_{\alpha \to \alpha_0} \inf_{t \in \mathbb{I}} b(t, \alpha) \leq \limsup_{\alpha \to \alpha_0} \sup_{t \in \mathbb{I}} b(t, \alpha) < \infty \qquad (6.2)$$

*or*

$$-\infty < \liminf_{\alpha \to \alpha_0} \inf_{t \in \mathbb{I}} b(t, \alpha) \leq \limsup_{\alpha \to \alpha_0} \sup_{t \in \mathbb{I}} b(t, \alpha) < 0, \qquad (6.3)$$

*and the remainder satisfies*

$$\lim_{x \to 0} \sup_{\alpha \in (\alpha_0 - |x|, \alpha_0 + |x|)} \sup_{t \in \mathbb{I}} \frac{|r(t, x, \alpha)|}{|x|^2} = 0 \qquad (6.4)$$

*and*

$$\limsup_{\alpha \to \alpha_0} \limsup_{x \to 0} \sup_{t \in \mathbb{I}} \frac{2K|r(t, x, \alpha)|}{|x| \max\{-\beta_1(\alpha), \beta_2(\alpha)\}} < 1. \qquad (6.5)$$

*Then there exist* $\hat{\alpha}_- < 0 < \hat{\alpha}_+$ *such that the following statements are fulfilled:*

(i) *In case the functions* $\beta_1$ *and* $\beta_2$ *are monotone increasing, the trivial solution is past (future, all-time, respectively) attractive for* $\alpha \in (\hat{\alpha}_-, \alpha_0)$ *and past (future, all-time, respectively) repulsive for* $\alpha \in (\alpha_0, \hat{\alpha}_+)$. *The differential equation* $(6.1)_\alpha$ *admits a past (future, all-time, respectively) bifurcation, since the corresponding radii of past (future, all-time, respectively) attraction and repulsion satisfy*

$$\lim_{\alpha \nearrow \alpha_0} \mathfrak{A}_0^\alpha = 0 \quad \text{and} \quad \lim_{\alpha \searrow \alpha_0} \mathfrak{R}_0^\alpha = 0.$$

(ii) *In case the functions* $\beta_1$ *and* $\beta_2$ *are monotone decreasing, the trivial solution is past (future, all-time, respectively) repulsive for* $\alpha \in (\hat{\alpha}_-, \alpha_0)$ *and past (future, all-time, respectively) attractive for* $\alpha \in (\alpha_0, \hat{\alpha}_+)$. *The differential equation* $(6.1)_\alpha$ *admits a past (future, all-time, respectively) bifurcation, since the corresponding radii of past (future, all-time, respectively) repulsion and attraction satisfy*

$$\lim_{\alpha \nearrow \alpha_0} \mathfrak{R}_0^\alpha = 0 \quad \text{and} \quad \lim_{\alpha \searrow \alpha_0} \mathfrak{A}_0^\alpha = 0.$$

*Proof.* First of all, we assume w.l.o.g. that $K > 1$. Let $\lambda_\alpha$ denote the general solution of $(6.1)_\alpha$. We will only prove assertion (i), since the proof of (ii) is similar. The functions $\beta_1$ and $\beta_2$ are therefore monotone increasing. W.l.o.g., we only treat the case (6.2). We choose $\hat{\alpha}_- < \alpha_0 < \hat{\alpha}_+$ such that

$$0 < \inf_{\alpha \in (\hat{\alpha}_-,\hat{\alpha}_+), t \in \mathbb{I}} b(t,\alpha) \leq \sup_{\alpha \in (\hat{\alpha}_-,\hat{\alpha}_+), t \in \mathbb{I}} b(t,\alpha) < \infty \qquad (6.6)$$

(cf. (6.2)) and

$$\limsup_{x \to 0} \sup_{t \in \mathbb{I}} \frac{|r(t,x,\alpha)|}{|x|} < \frac{-\min\{\beta_1(\alpha), -\beta_2(\alpha)\}}{2K} \quad \text{for all } \alpha \in (\hat{\alpha}_-,\hat{\alpha}_+)$$

(cf. (6.5)). Because of these two relations, Theorem 5.9 can be applied, and the attractivity and repulsivity of the trivial solutions as stated in the theorem follows. Assume to the contrary that

$$\eta := \limsup_{\alpha \nearrow \alpha_0} \mathfrak{A}_0^\alpha > 0$$

holds. Due to (6.6) and (6.4), there exist $\tilde{\alpha}_- \in (\hat{\alpha}_-, \alpha_0)$, $\xi \in (0,\eta)$ and $L \in \big(0, \xi/(4K)\big)$ with

$$b(t,\alpha)x^2 + r(t,x,\alpha) > L \quad \text{for all } t \in \mathbb{I}, \, \alpha \in (\tilde{\alpha}_-,\alpha_0) \text{ and } x \in \left[\frac{\xi}{2K^2}, \xi\right].$$

$$(6.7)$$

We fix $\hat{\alpha} \in (\tilde{\alpha}_-,\alpha_0)$ such that $\mathfrak{A}_0^{\hat{\alpha}} > \xi$ and $\beta_2(\hat{\alpha}) \geq \beta := -2KL/\xi > -1/2$. For arbitrary $\tau \in \mathbb{I}$, the solution $\mu_\tau(\cdot) := \lambda_{\hat{\alpha}}(\cdot,\tau,\xi)$ of $(6.1)_{\hat{\alpha}}$ is also a solution of the inhomogeneous linear differential equation

$$\dot{x} = a(t,\hat{\alpha})x + b(t,\hat{\alpha})(\mu_\tau(t))^2 + r(t,\mu_\tau(t),\hat{\alpha}). \qquad (6.8)$$

Since $\mathfrak{A}_0^{\hat{\alpha}} > \xi = \mu_t(t)$ for all $t \in \mathbb{I}$, there exist $\tau, \tau_2 \in \mathbb{I}, \tau < \tau_2$, such that $\mu_\tau(\tau_2) \leq \xi/(2K^2)$. We choose $\tau_2$ minimal with this property, i.e., we have $\mu_\tau(t) \geq \xi/(2K^2)$ for all $t \in [\tau,\tau_2]$. Furthermore, we choose $\tau_1 \in [\tau,\tau_2]$ such that

$$\mu_\tau(\tau_1) = \frac{\xi}{2K} \quad \text{and} \quad \mu_\tau(t) \in \left[\frac{\xi}{2K^2}, \xi\right] \quad \text{for all } t \in [\tau_1,\tau_2].$$

Therefore, and due to (6.7) and the variation of constants formula (cf. Proposition A.6), applied to (6.8), the relation

$$\mu_\tau(\tau_2) = \Lambda_{\hat{\alpha}}(\tau_2,\tau_1)\mu_\tau(\tau_1) + \int_{\tau_1}^{\tau_2} \Lambda_{\hat{\alpha}}(\tau_2,t)\big(b(t,\hat{\alpha})(\mu_\tau(t))^2 + r(t,\mu_\tau(t),\hat{\alpha})\big)\,dt$$

$$> \frac{\xi}{2K^2}e^{\beta(\tau_2-\tau_1)} + \frac{L}{K}\int_{\tau_1}^{\tau_2} e^{\beta(\tau_2-t)}\,dt$$

$$= e^{\beta(\tau_2-\tau_1)}\underbrace{\left(\frac{\xi}{2K^2} + \frac{L}{K\beta}\right)}_{=0} - \frac{L}{K\beta} = \frac{\xi}{2K^2}$$

holds ($K > 1$ implies $\tau_1 < \tau_2$). This is a contradiction and proves $\lim_{\alpha \nearrow \alpha_0} \mathfrak{A}_0^\alpha = 0$. Analogously, one can show $\lim_{\alpha \searrow \alpha_0} \mathfrak{R}_0^\alpha = 0$ and treat the case (6.3). □

*Remark 6.2.*

(i) In the limit $\alpha \to \alpha_0$, the attractivity or repulsivity of the trivial solution is only lost in one direction, i.e., nonautonomous transcritical bifurcations are partial bifurcations. For instance, in case the functions $\beta_1, \beta_2$ are monotone increasing and (6.2) is satisfied, there exists a $\gamma < 0$ such that $(\gamma, 0]$ is attracted by the trivial solution of $(6.4)_\alpha$ for $\alpha \in (\hat{\alpha}_-, \alpha_0)$ in the sense of past, future or all-time attractivity, respectively.

(ii) The hypothesis on the linear part implies that the past (future, all-time, respectively) dichotomy spectrum of the linearization $\dot{x} = a(t, \alpha)x$ converges to $\{0\}$ in Hausdorff distance in the limit $\alpha \to \alpha_0$.

(iii) Condition (6.5) is only used to obtain the attractivity or repulsivity of the trivial solution by applying Theorem 5.9. Alternatively, one can directly postulate that the trivial solution changes the stability at the parameter value $\alpha_0$ from, say, attractivity to repulsivity.

(iv) Please note that the above bifurcation result is essentially the combination of two scenarios which are independent of each other. This means that it is possible to consider $(6.1)_\alpha$ only for $\alpha > \alpha_0$ or $\alpha < \alpha_0$, respectively, in order to obtain the results which apply for these parameter values.

The following example shows that Theorem 6.1 is indeed a nonautonomous generalization of the well-known autonomous result.

*Example 6.3.* Let $x_- < 0 < x_+$ and $\alpha_- < 0 < \alpha_+$ be in $\mathbb{R}$, and consider the autonomous differential equation

$$\dot{x} = f(x, \alpha), \tag{6.9}$$

where the $C^4$-function $f : (x_-, x_+) \times (\alpha_-, \alpha_+) \to \mathbb{R}$ satisfies the following assumptions:

(i) $f(0, \alpha) = 0$ for all $\alpha \in (\alpha_-, \alpha_+)$,

(ii) $D_1 f(0, 0) = 0$,

(iii) $D_1 D_2 f(0, 0) \neq 0$,

(iv) $D_1^2 f(0, 0) \neq 0$.

Please note that (i) implies $D_2^n f(0, \alpha) = 0$ for all $\alpha \in (\alpha_-, \alpha_+)$ and $n \in \mathbb{N}$. Then (6.9) admits an autonomous transcritical bifurcation (see, e.g., WIGGINS [182, p. 265 f.] and AULBACH [14, Satz 7.10.6]), i.e., there exist a neighborhood $U \times V$ of $(0, 0)$ in $\mathbb{R}^2$ and a $C^1$-function $h : U \to V$ with $h(0) = 0$ and

$$f(x, h(x)) = 0 \quad \text{for all } x \in U.$$

Except the trivial equilibria and the equilibria described by $h$, there are no other equilibria in $U \times V$. Now, we will show that this example fulfills the hypotheses of Theorem 6.1. Thereto, we write the second order Taylor expansion of $f$ (see, e.g., LANG [102, p. 349]):

$$f(x, \alpha) = \underbrace{D_1 D_2 f(0,0)\alpha}_{=:\bar{a}(\alpha)} x + \underbrace{\frac{1}{2} D_1^2 f(0,0)}_{=:\bar{b}(\alpha)} x^2 + r(x, \alpha),$$

where

$$r(x, \alpha) = \int_0^1 \frac{(1-t)^2}{2} \left( D_1^3 f(tx, t\alpha)x^3 + 3D_1^2 D_2 f(tx, t\alpha)x^2\alpha + \right.$$
$$\left. 3D_1 D_2^2 f(tx, t\alpha)x\alpha^2 + D_2^3 f(tx, t\alpha)\alpha^3 \right) dt.$$

The hypothesis on the linear part is fulfilled (with $\beta_1(\alpha) := \beta_2(\alpha) := \bar{a}(\alpha)$ and $K := 1$), and (6.2) or (6.3) holds, since the above defined function $\bar{b}$ is constant. Furthermore, the representation for the remainder implies that

$$\lim_{x \to 0} \sup_{\alpha \in (-|x|, |x|)} \frac{|r(x, \alpha)|}{|x|^2} = 0$$

and

$$\limsup_{x \to 0} \frac{|r(x, \alpha)|}{|x|} \leq \alpha^2 \int_0^1 \frac{(1-t)^2}{2} \left( |3D_1 D_2^2 f(0, t\alpha)| + t|D_1 D_2^3 f(0, t\alpha)\alpha| \right) dt.$$

This means that (6.5) holds, since $\max\{-\beta_1(\alpha), \beta_2(\alpha)\}$ depends linearly in $\alpha$. Therefore, all hypotheses of Theorem 6.1 are fulfilled, and thus, this example shows that Theorem 6.1 is a proper generalization of the well-known autonomous transcritical bifurcation pattern.

In case of compact time domains, the following result is obtained.

**Theorem 6.4 (Nonautonomous transcritical bifurcation, part II).** *Let $x_- < 0 < x_+$ and $\alpha_- < \alpha_+$ be in $\mathbb{R}$ and $\mathbb{I} := [\tau, \tau + T]$, and consider the nonautonomous differential equation*

$$\dot{x} = a(t, \alpha)x + b(t, \alpha)x^2 + r(t, x, \alpha) \tag{6.10}_\alpha$$

*with continuous functions $a : \mathbb{I} \times (\alpha_-, \alpha_+) \to \mathbb{R}$, $b : \mathbb{I} \times (\alpha_-, \alpha_+) \to \mathbb{R}$ and $r : \mathbb{I} \times (x_-, x_+) \times (\alpha_-, \alpha_+) \to \mathbb{R}$ fulfilling $r(\cdot, 0, \cdot) \equiv 0$. Let $\Lambda_\alpha : \mathbb{I} \times \mathbb{I} \to \mathbb{R}$ denote the transition operator of the linearized equation $\dot{x} = a(t, \alpha)x$. We define*

$$K(\alpha) := \sup\{\Lambda_\alpha(t, s) : t, s \in \mathbb{I}\} \quad \text{for all } \alpha \in (\alpha_-, \alpha_+)$$

*and assume, there exists an $\alpha_0 \in (\alpha_-, \alpha_+)$ such that the following hypotheses hold:*

- Hypothesis on linear part. *We either have*

$$\Lambda_\alpha(\tau + T, \tau) < 1 \quad \text{for all } \alpha \in (\alpha_-, \alpha_0) \text{ and}$$
$$\Lambda_\alpha(\tau + T, \tau) > 1 \quad \text{for all } \alpha \in (\alpha_0, \alpha_+) \tag{6.11}$$

*or*

$$\Lambda_\alpha(\tau + T, \tau) > 1 \quad \text{for all } \alpha \in (\alpha_-, \alpha_0) \text{ and}$$
$$\Lambda_\alpha(\tau + T, \tau) < 1 \quad \text{for all } \alpha \in (\alpha_0, \alpha_+). \tag{6.12}$$

- Hypothesis on nonlinearity. *The quadratic term either fulfills*

$$\liminf_{\alpha \to \alpha_0} \inf_{t \in \mathbb{I}} b(t, \alpha) > 0 \tag{6.13}$$

*or*

$$\limsup_{\alpha \to \alpha_0} \sup_{t \in \mathbb{I}} b(t, \alpha) < 0, \tag{6.14}$$

*and the remainder satisfies*

$$\lim_{x \to 0} \sup_{\alpha \in (\alpha_0 - |x|, \alpha_0 + |x|)} \sup_{t \in \mathbb{I}} \frac{|r(t, x, \alpha)|}{|x|^2} = 0 \tag{6.15}$$

*and*

$$\limsup_{\alpha \to \alpha_0} \limsup_{x \to 0} \sup_{t \in \mathbb{I}} -\frac{TK(\alpha)|r(t, x, \alpha)|}{|x| \ln \left( \min \left\{ \Lambda_\alpha(\tau + T, \tau), \Lambda_\alpha(\tau, \tau + T) \right\} \right)} < 1. \tag{6.16}$$

*Then there exist $\hat{\alpha}_- < 0 < \hat{\alpha}_+$ such that the following statements are fulfilled:*

(i) *In case (6.11), the trivial solution is $(\tau, T)$-attractive for $\alpha \in (\hat{\alpha}_-, \alpha_0)$ and $(\tau, T)$-repulsive for $\alpha \in (\alpha_0, \hat{\alpha}_+)$. The differential equation $(6.10)_\alpha$ admits a $(\tau, T)$-bifurcation, since the corresponding radii of $(\tau, T)$-attraction and repulsion satisfy*

$$\lim_{\alpha \nearrow \alpha_0} \mathfrak{A}_0^\alpha = 0 \quad \text{and} \quad \lim_{\alpha \searrow \alpha_0} \mathfrak{R}_0^\alpha = 0.$$

(ii) *In case (6.11), the trivial solution is $(\tau, T)$-repulsive for $\alpha \in (\hat{\alpha}_-, \alpha_0)$ and $(\tau, T)$-attractive for $\alpha \in (\alpha_0, \hat{\alpha}_+)$. The differential equation $(6.10)_\alpha$ admits a $(\tau, T)$-bifurcation, since the corresponding radii of $(\tau, T)$-repulsion and attraction satisfy*

$$\lim_{\alpha \nearrow \alpha_0} \mathfrak{R}_0^\alpha = 0 \quad \text{and} \quad \lim_{\alpha \searrow \alpha_0} \mathfrak{A}_0^\alpha = 0.$$

*Proof.* Let $\lambda_\alpha$ denote the general solution of $(6.10)_\alpha$. We will only prove assertion (i), since the proof of (ii) is similar. Therefore, (6.11) is fulfilled. W.l.o.g., we only treat the case (6.13). We choose $\hat{\alpha}_- < 0 < \hat{\alpha}_+$ such that

$$\inf_{\alpha \in (\hat{\alpha}_-, \hat{\alpha}_+), t \in \mathbb{I}} b(t, \alpha) > 0 \tag{6.17}$$

and for all $\alpha \in (\hat{\alpha}_-, \hat{\alpha}_+)$, we have

$$\limsup_{x \to 0} \sup_{t \in \mathbb{I}} \frac{|r(t, x, \alpha)|}{|x|} \leq -\gamma \frac{\ln \left( \min \{ \Lambda_\alpha(\tau + T, \tau), \Lambda_\alpha(\tau, \tau + T) \} \right)}{TK(\alpha)}$$

for some $\gamma \in (0, 1)$. Because of these two relations, Theorem 5.10 can be applied, and the attractivity and repulsivity of the trivial solutions as stated in the theorem follows. We define

$$K_- := \inf \left\{ \Lambda_\alpha(t, s) : t, s \in \mathbb{I}, \alpha \in [\hat{\alpha}_-, \alpha_0] \right\} \in (0, 1).$$

Assume to the contrary that

$$\eta := \limsup_{\alpha \nearrow \alpha_0} \mathfrak{A}_0^\alpha > 0$$

holds. Due to (6.17) and (6.15), there exist $\tilde{\alpha}_- \in (\hat{\alpha}_-, \alpha_0)$, $\xi \in (0, K_- \eta)$ and $L > 0$ with

$$b(t, \alpha)x^2 + r(t, x, \alpha) > L \quad \text{for all } t \in \mathbb{I}, \ \alpha \in (\tilde{\alpha}_-, \alpha_0) \text{ and } x \in \left[ K_- \xi, \frac{\xi}{K_-} \right].$$
$$(6.18)$$

We fix $\hat{\alpha} \in (\tilde{\alpha}_-, \alpha_0)$ such that $\mathfrak{A}_0^{\hat{\alpha}} > \xi$ and

$$\Lambda_{\hat{\alpha}}(\tau + T, \tau) \geq 1 - \frac{K_- LT}{\xi}. \tag{6.19}$$

For arbitrary $\tau \in \mathbb{I}$, the solution $\mu_\tau(\cdot) := \lambda_{\hat{\alpha}}(\cdot, \tau, \xi)$ of $(6.10)_{\hat{\alpha}}$ is also a solution of the inhomogeneous linear differential equation

$$\dot{x} = a(t, \hat{\alpha})x + b(t, \hat{\alpha})(\mu_\tau(t))^2 + r(t, \mu_\tau(t), \hat{\alpha}). \tag{6.20}$$

Since $\mathfrak{A}_0^{\hat{\alpha}} > \xi$, we have

$$\mu_\tau(\tau + T) < \xi. \tag{6.21}$$

Moreover, from the definition of $K_-$ and (6.18), we directly get

$$\mu_\tau(\tau + t) \geq K_- \xi \quad \text{for all } t \in [0, T]. \tag{6.22}$$

We distinguish two cases.
*Case 1. There exists a $\bar{t} \in (0, T]$ such that*

$$\mu_\tau(\tau + \bar{t}) = \frac{\xi}{K_-}.$$

We choose $\bar{t}$ maximal with this property. Due to (6.21), this means that $\mu(\tau + t) \leq \xi / K_-$ for all $t \in [\bar{t}, T]$. Then the variation of constants formula (cf. Proposition A.6), applied to (6.20), implies the relation

$$\mu_\tau(\tau + T)$$

$$= \Lambda_{\hat{\alpha}}(\tau + T, \tau + \bar{t}) \frac{\xi}{K_-} +$$

$$+ \int_{\tau + \bar{t}}^{\tau + T} \Lambda_{\hat{\alpha}}(\tau + T, t) \big( b(t, \hat{\alpha})(\mu_\tau(t))^2 + r(t, \mu_\tau(t), \hat{\alpha}) \big) \, dt$$

$$\overset{(6.18)}{\geq} \xi + K_- L(T - \bar{t}) > \xi.$$

This contradicts (6.21).

*Case 2.* For all $t \in (0, T]$, we have

$$\mu_\tau(\tau + \bar{t}) < \frac{\xi}{K_-}.$$

In this case, the variation of constants formula, applied to (6.20), yields

$$\mu_\tau(\tau + T)$$

$$= \Lambda_{\hat{\alpha}}(\tau + T, \tau)\xi +$$

$$+ \int_\tau^{\tau + T} \Lambda_{\hat{\alpha}}(\tau + T, t) \big( b(t, \hat{\alpha})(\mu_\tau(t))^2 + r(t, \mu_\tau(t), \hat{\alpha}) \big) \, dt$$

$$\overset{(6.18), (6.19)}{\geq} \left( 1 - \frac{K_- L T}{\xi} \right) \xi + K_- L T = \xi.$$

This contradicts (6.21) also, and thus, $\lim_{\alpha \nearrow \alpha_0} \mathfrak{A}_0^\alpha = 0$ is proved. Analogously, one can show $\lim_{\alpha \searrow \alpha_0} \mathfrak{R}_0^\alpha = 0$ and treat the case (6.14). $\qquad \square$

*Remark 6.5.*

(i) The hypothesis on the linear part implies that the $(\tau, T)$-dichotomy spectrum of the linearization $\dot{x} = a(t, \alpha)x$ converges to $\{0\}$ in Hausdorff distance in the limit $\alpha \to \alpha_0$.

(ii) Condition (6.16) is only used to obtain the attractivity or repulsivity of the trivial solution by applying Theorem 5.10. Alternatively, one can directly postulate that the trivial solution changes their stability at the parameter value $\alpha_0$ from, say, attractivity to repulsivity.

## 6.2 Nonautonomous Pitchfork Bifurcation

In this section, nonautonomous generalizations of the classical pitchfork bifurcation are derived. First, the case of unbounded time domains is treated.

**Theorem 6.6 (Nonautonomous pitchfork bifurcation, part I).** *Let $x_- < 0 < x_+$ and $\alpha_- < \alpha_+$ be in $\overline{\mathbb{R}}$ and $\mathbb{I}$ be an unbounded interval of the* *form $\mathbb{R}_\kappa^-$, $\mathbb{R}_\kappa^+$ or $\mathbb{R}$, respectively, and consider the nonautonomous differential equation*

$$\dot{x} = a(t,\alpha)x + b(t,\alpha)x^3 + r(t,x,\alpha) \tag{6.23}_\alpha$$

*with continuous functions $a : \mathbb{I} \times (\alpha_-, \alpha_+) \to \mathbb{R}$, $b : \mathbb{I} \times (\alpha_-, \alpha_+) \to \mathbb{R}$ and $r : \mathbb{I} \times (x_-, x_+) \times (\alpha_-, \alpha_+) \to \mathbb{R}$ fulfilling $r(\cdot, 0, \cdot) \equiv 0$. Let $\Lambda_\alpha : \mathbb{I} \times \mathbb{I} \to \mathbb{R}$ be the transition operator of the linearized equation $\dot{x} = a(t,\alpha)x$, and assume, there exists an $\alpha_0 \in (\alpha_-, \alpha_+)$ such that the following hypotheses hold:*

- *Hypothesis on linear part. There exist two functions $\beta_1, \beta_2 : (\alpha_-, \alpha_+) \to \mathbb{R}$ which are either both monotone increasing or both monotone decreasing and $K \geq 1$ such that $\lim_{\alpha \to \alpha_0} \beta_1(\alpha) = \lim_{\alpha \to \alpha_0} \beta_2(\alpha) = 0$ and*

$$\Lambda_\alpha(t,s) \leq Ke^{\beta_1(\alpha)(t-s)} \quad \text{for all } \alpha \in (\alpha_-, \alpha_+) \text{ and } t,s \in \mathbb{I} \text{ with } t \geq s,$$
$$\Lambda_\alpha(t,s) \leq Ke^{\beta_2(\alpha)(t-s)} \quad \text{for all } \alpha \in (\alpha_-, \alpha_+) \text{ and } t,s \in \mathbb{I} \text{ with } t \leq s.$$

- *Hypothesis on nonlinearity. The cubic term either fulfills*

$$0 < \liminf_{\alpha \to \alpha_0} \inf_{t \in \mathbb{I}} b(t,\alpha) \leq \limsup_{\alpha \to \alpha_0} \sup_{t \in \mathbb{I}} b(t,\alpha) < \infty \tag{6.24}$$

*or*

$$-\infty < \liminf_{\alpha \to \alpha_0} \inf_{t \in \mathbb{I}} b(t,\alpha) \leq \limsup_{\alpha \to \alpha_0} \sup_{t \in \mathbb{I}} b(t,\alpha) < 0, \tag{6.25}$$

*and the remainder satisfies*

$$\lim_{x \to 0} \sup_{\alpha \in (\alpha_0 - x^2, \alpha_0 + x^2)} \sup_{t \in \mathbb{I}} \frac{|r(t,x,\alpha)|}{|x|^3} = 0 \tag{6.26}$$

*and*

$$\limsup_{\alpha \to \alpha_0} \limsup_{x \to 0} \sup_{t \in \mathbb{I}} \frac{2K|r(t,x,\alpha)|}{|x| \max\{-\beta_1(\alpha), \beta_2(\alpha)\}} < 1.$$

*Then there exist $\hat{\alpha}_- < 0 < \hat{\alpha}_+$ such that the following statements are fulfilled:*

(i) *In case (6.24) and the functions $\beta_1$ and $\beta_2$ are monotone increasing, the trivial solution is past (future, all-time, respectively) attractive for $\alpha \in (\hat{\alpha}_-, \alpha_0)$ and past (future, all-time, respectively) repulsive for $\alpha \in (\alpha_0, \hat{\alpha}_+)$. The differential equation (6.23)$_\alpha$ admits a past (future, all-time, respectively) bifurcation, since the corresponding radii of past (future, all-time, respectively) attraction satisfy*

$$\lim_{\alpha \nearrow \alpha_0} \mathfrak{A}_0^\alpha = 0.$$

*If, in addition, $\mathbb{I} = \mathbb{R}_\kappa^+$, then, for $\alpha \in (\hat{\alpha}_-, \alpha_0)$, there exists a nontrivial future repeller $R_\alpha \subset \mathbb{I} \times \mathbb{R}$, and we have a future repeller transition, since*

$$\lim_{\alpha \nearrow \alpha_0} d_H(R_\alpha(t), \{0\}) = 0 \quad \text{for all } t \in \mathbb{I}.$$

(ii) *In case (6.25) and the functions $\beta_1$ and $\beta_2$ are monotone increasing, the trivial solution is all-time (past, future, respectively) attractive for $\alpha \in (\hat{\alpha}_-, \alpha_0)$ and past (future, all-time, respectively) repulsive for $\alpha \in (\alpha_0, \hat{\alpha}_+)$. The differential equation $(6.23)_\alpha$ admits a past (future, all-time, respectively) bifurcation, since the corresponding radii of past (future, all-time, respectively) repulsion satisfy*

$$\lim_{\alpha \searrow \alpha_0} \mathfrak{R}_0^\alpha = 0.$$

*If, in addition, $\mathbb{I} = \mathbb{R}_\kappa^-$, then, for $\alpha \in (\alpha_0, \hat{\alpha}_+)$, there exists a nontrivial past attractor $A_\alpha \subset \mathbb{I} \times \mathbb{R}$, and we have a past attractor transition, since*

$$\lim_{\alpha \searrow \alpha_0} d_H\big(A_\alpha(t), \{0\}\big) = 0 \quad \text{for all } t \in \mathbb{I}.$$

(iii) *In case (6.24) and the functions $\beta_1$ and $\beta_2$ are monotone decreasing, the trivial solution is past (future, all-time, respectively) repulsive for $\alpha \in (\hat{\alpha}_-, \alpha_0)$ and past (future, all-time, respectively) attractive for $\alpha \in (\alpha_0, \hat{\alpha}_+)$. The differential equation $(6.23)_\alpha$ admits a past (future, all-time, respectively) bifurcation, since the corresponding radii of past (future, all-time, respectively) attraction satisfy*

$$\lim_{\alpha \searrow \alpha_0} \mathfrak{A}_0^\alpha = 0.$$

*If, in addition, $\mathbb{I} = \mathbb{R}^+$, then, for $\alpha \in (\alpha_0, \hat{\alpha}_+)$, there exists a nontrivial future repeller $R_\alpha \subset \mathbb{I} \times \mathbb{R}$, and we have a future repeller transition, since*

$$\lim_{\alpha \searrow \alpha_0} d_H\big(R_\alpha(t), \{0\}\big) = 0 \quad \text{for all } t \in \mathbb{I}.$$

(iv) *In case (6.25) and the functions $\beta_1$ and $\beta_2$ are monotone decreasing, the trivial solution is past (future, all-time, respectively) repulsive for $\alpha \in (\hat{\alpha}_-, \alpha_0)$ and past (future, all-time, respectively) attractive for $\alpha \in (\alpha_0, \hat{\alpha}_+)$. The differential equation $(6.23)_\alpha$ admits a past (future, all-time, respectively) bifurcation, since the corresponding radii of past (future, all-time, respectively) repulsion satisfy*

$$\lim_{\alpha \nearrow \alpha_0} \mathfrak{R}_0^\alpha = 0.$$

*If, in addition, $\mathbb{I} = \mathbb{R}^-$, then, for $\alpha \in (\hat{\alpha}_-, \alpha_0)$, there exists a nontrivial past attractor $A_\alpha \subset \mathbb{I} \times \mathbb{R}$, and we have a past attractor transition, since*

$$\lim_{\alpha \nearrow \alpha_0} d_H\big(A_\alpha(t), \{0\}\big) = 0 \quad \text{for all } t \in \mathbb{I}.$$

*Proof.* The first part of this theorem concerning the bifurcation of the attraction or repulsion areas, respectively, can be proved using the same methods as in the proof of Theorem 6.1. We write $\tilde{\alpha}_-$ and $\tilde{\alpha}_+$ for the constants $\hat{\alpha}_-$ and

$\hat{\alpha}_+$ used in this proof. For the proof of the attractor and repeller transitions, w.l.o.g., we only consider the case (ii), i.e., $\mathbb{I} = \mathbb{R}_\kappa^-$, condition (6.25) holds and the functions $\beta_1$ and $\beta_2$ are monotone increasing. We denote the general solution of (6.23)$_\alpha$ by $\lambda_\alpha$ and define

$$b_+ := \frac{1}{2} \sup_{t \in \mathbb{I},\, \alpha \in (\tilde{\alpha}_-, \tilde{\alpha}_+)} b(t, \alpha) < 0.$$

Due to (6.26), there exists a $\rho > 0$ such that

$$|r(t, x, \alpha)| \leq -b_+ |x|^3 \quad \text{for all } x \in [-\rho, \rho], \alpha \in (\alpha_0 - x^2, \alpha_0 + x^2) \text{ and } t \in \mathbb{I}.$$

The remaining proof is divided into two steps.

*Step 1.* For given $x_1, x_2, x_3 \leq \rho$ such that $0 < x_1 \leq x_2 \leq x_3/(2K)$, there exists a uniquely determined constant

$$\alpha^* = \alpha^*(x_1, x_2, x_3) \in \left(\alpha_0, \min\left\{\alpha_0 + x_1^2, \tilde{\alpha}_+\right\}\right]$$

with the following properties:

- We have $\lambda_\alpha(t, \tau, [-x_2, x_2]) \subset (-x_3, x_3)$ for all $\tau \leq t \leq \kappa$ and $\alpha \in (\alpha_0, \alpha^*)$,
- there exists a $T^* > 0$ such that for all $\alpha \in (\alpha_0, \alpha^*)$ and $\tau \leq \kappa - T^*$, there exist $t_+, t_- \in [0, T^*]$ with

$$\lambda_\alpha(\tau + t_+, \tau, x_2) = x_1 \quad \text{and} \quad \lambda_\alpha(\tau + t_-, \tau, -x_2) = -x_1,$$

- $\alpha^*$ is chosen maximal, i.e., for all bigger $\alpha^*$, one of the two above properties is violated.

We will only prove the existence of a constant $\alpha^*$ such that

(a)  for all $\tau \leq t \leq \kappa$ and $\alpha \in (\alpha_0, \alpha^*)$, we have $\lambda_\alpha(t, \tau, x_2) \leq x_3$,

(b)  there exists a $T^* > 0$ such that for all $\alpha \in (\alpha_0, \alpha^*)$ and $\tau \leq \kappa - T^*$, there exists a $t_+ \in [0, T^*]$ with $\lambda_\alpha(\tau + t_+, \tau, x_2) = x_1$,

since the extension to the above assertion follows similarly and by taking the supremum of all such $\alpha^*$. We first note that for arbitrary $\tau \in \mathbb{I}$, the solution $\mu_\tau(\cdot) := \lambda_\alpha(\cdot, \tau, x_2)$ of (6.23)$_\alpha$ is also a solution of the inhomogeneous linear differential equation

$$\dot{x} = a(t, \alpha)x + b(t, \alpha)(\mu_\tau(t))^3 + r(t, \mu_\tau(t), \alpha). \tag{6.27}$$

Concerning the expression

$$s(\alpha, T) := K e^{\beta_1(\alpha)T} x_2 + \frac{b_+ x_1^3}{K} T \quad \text{for all } \alpha \in (\alpha_0, \tilde{\alpha}_+) \text{ and } T \geq 0,$$

there exist $\alpha^* \in \left(\alpha_0, \min\left\{\alpha_0 + x_1^2, \tilde{\alpha}_+\right\}\right]$ and $T^* > 0$ such that for $\alpha \in (\alpha_0, \alpha^*]$, we have

$$s(\alpha, T^*) < 0 \quad \text{and} \quad s(\alpha, T) \leq 2Kx_2 \text{ for all } T \in [0, T^*].$$

This follows by choosing $T^*$ such that $(b_+ x_1^3/K)T^* \leq -2Kx_2$ and $\alpha^*$ such that $\exp(\beta_1(\alpha^*)T^*) \leq 2$. Choose $\alpha \in (\alpha_0, \alpha^*]$ and $\tau, \tau^* \leq \kappa$ with $\tau \leq \tau^*$. Assume that $x_1 \leq \mu_\tau(t) \leq x_3$ for all $t \in [\tau, \tau^*]$. Then the variation of constants formula (cf. Proposition A.6), applied to (6.27), yields the relation

$$\mu_\tau(\tau^*) = \Lambda_\alpha(\tau^*, \tau)x_2 +$$

$$+ \int_\tau^{\tau^*} \underbrace{\Lambda_\alpha(\tau^*, s)}_{\geq \frac{1}{K}\exp(\beta_2(\alpha)(\tau^*-s))} \underbrace{\left( b(s, \alpha)(\mu_\tau(s))^3 + r(s, \mu_\tau(s), \alpha) \right)}_{\leq b_+ x_1^3 < 0} ds$$

$$\leq Ke^{\beta_1(\alpha)(\tau^*-\tau)}x_2 + \int_\tau^{\tau^*} \frac{1}{K}e^{\beta_2(\alpha)(\tau^*-s)}b_+ x_1^3\, ds$$

$$= Ke^{\beta_1(\alpha)(\tau^*-\tau)}x_2 + \frac{b_+ x_1^3}{K\beta_2(\alpha)}\left(e^{\beta_2(\alpha)(\tau^*-\tau)} - 1\right)$$

$$\leq Ke^{\beta_1(\alpha)(\tau^*-\tau)}x_2 + \frac{b_+ x_1^3}{K}(\tau^* - \tau) = s(\alpha, \tau^* - \tau).$$

Since $s(\alpha, T) \leq 2Kx_2 \leq x_3$ for all $T \in [0, T^*]$, the assumption $\mu_\tau(t) \leq x_3$ for all $t \in [\tau, \tau^*]$ is justified. This proves (a). Because of $s(\alpha, T^*) < 0$, also (b) is fulfilled.

*Step 2. There exists an $\hat{\alpha}_+ \in (\alpha_0, \tilde{\alpha}_+)$ such that for all $\alpha \in (\alpha_0, \hat{\alpha}_+)$, there exists a nontrivial past attractor $A_\alpha \subset \mathbb{I} \times \mathbb{R}$ of (6.23)$_\alpha$ which fulfills*

$$\lim_{\alpha \searrow \alpha_0} d_H\big(A_\alpha(t), \{0\}\big) = 0 \quad \text{for all } t \in \mathbb{I}.$$

For $x_3 := \rho/K$ and $x_2 := x_3/(2K)$, we consider $\gamma : (0, x_2) \to (\alpha_0, \alpha_+)$, defined by

$$\gamma(x_1) := \alpha^*(x_1, x_2, x_3) \quad \text{for all } x_1 \in (0, x_2),$$

where $\alpha^*$ stems from Step 1. We set $\bar{\alpha} := \gamma(x_2/2)$ and define

$$\delta(\alpha) := \inf\big\{x_1 \in (0, x_2) : \gamma(x_1) \geq \alpha\big\} \quad \text{for all } \alpha \in (\alpha_0, \bar{\alpha}].$$

Due to $\alpha_0 < \alpha^*(x_1, x_2, x_3) \leq \alpha_0 + x_1^2$, we have $\lim_{x_1 \to 0} \gamma(x_1) = \alpha_0$, and since $\gamma$ is monotone increasing, this implies that $\delta$ is monotone increasing, $\delta(\alpha) > 0$ for all $\alpha \in (\alpha_0, \bar{\alpha}]$ and

$$\lim_{\alpha \searrow \alpha_0} \delta(\alpha) = 0. \tag{6.28}$$

We define

$$\bar{x}_3(\alpha) := 3K\delta(\alpha) \quad \text{and} \quad \bar{x}_2(\alpha) := \bar{x}_1(\alpha) := \frac{3}{2}\delta(\alpha) \quad \text{for all } \alpha \in (\alpha_0, \bar{\alpha}]$$

and consider the function $\bar{\gamma} : (\alpha_0, \bar{\alpha}] \to (\alpha_0, \alpha_+)$, defined by

$$\bar{\gamma}(\alpha) := \alpha^*\big(\bar{x}_1(\alpha), \bar{x}_2(\alpha), \bar{x}_3(\alpha)\big) \quad \text{for all } \alpha \in (\alpha_0, \bar{\alpha}],$$

where $\alpha^*$ is taken from Step 1 again. Moreover, we define

$$M := [-x_2, x_2] \quad \text{and} \quad B_\alpha := [-\bar{x}_3(\alpha), \bar{x}_3(\alpha)] \quad \text{for all } \alpha \in (\alpha_0, \bar{\alpha}]$$

and fix a $\beta \in (\alpha_0, \bar{\alpha}]$ and an $\alpha \in (\alpha_0, \min\{\bar{\gamma}(\beta), \beta\})$. Since $\alpha \leq \beta$ and $x_2 \geq 3\delta(\beta)/2$ and due to the definition of $\delta$, there exists a $T^* > 0$ such that for all $\tau \leq \kappa - T^*$, there exist $t_+, t_- \in [0, T^*]$ with

$$\lambda_\alpha(t^+, \tau, x_2) = \frac{3}{2}\delta(\beta) = \bar{x}_2(\beta) \quad \text{and} \quad \lambda_\alpha(t^-, \tau, -x_2) = -\frac{3}{2}\delta(\beta) = -\bar{x}_2(\beta).$$

Moreover, since $\alpha < \bar{\gamma}(\beta)$, we have

$$\lambda_\alpha\big(t, \tau, [-\bar{x}_2(\beta), \bar{x}_2(\beta)]\big) \subset (-\bar{x}_3(\beta), \bar{x}_3(\beta)) \quad \text{for all } \tau \leq t \leq \kappa.$$

This means that, considering equation $(6.23)_\alpha$, $B_\beta \times \mathbb{I}$ is past absorbing with respect to $\{M \times \mathbb{I}\}$. Then, due to Theorem 2.35 (i), there exists a past attractor $A_\alpha \subset B_\beta \times \mathbb{I}$. The past attractor is nontrivial due to Lemma 2.39. The limit relation

$$\lim_{\alpha \nearrow \alpha_0} d_H\big(A_\alpha(t), \{0\}\big) = 0 \quad \text{for all } t \in \mathbb{I}$$

follows from $A_\alpha \subset B_\beta \times \mathbb{I}$ for all $\alpha < \min\{\bar{\gamma}(\beta), \beta\}$ and (6.28). By setting $\hat{\alpha}_+ := \bar{\gamma}(\bar{\alpha})$, all assertions of this theorem are proved. $\qquad \square$

*Remark 6.7.*

(i) In the limit $\alpha \to \alpha_0$, the attractivity or repulsivity of the trivial solution is lost in both directions, i.e., no situation as described in Remark 6.2 (i) can occur. This means that nonautonomous pitchfork bifurcations are total bifurcations.

(ii) The hypothesis on the linear part implies that the past (future, all-time, respectively) dichotomy spectrum of the linearization $\dot{x} = a(t, \alpha)x$ converges to $\{0\}$ in Hausdorff distance in the limit $\alpha \to \alpha_0$.

(iii) As in Example 6.3, one can show that Theorem 6.6 is a proper generalization of the well-known autonomous pitchfork bifurcation (see, e.g., WIGGINS [182, p. 267 f.] and AULBACH [14, Satz 7.10.8]).

(iv) Please note that the above bifurcation result is essentially the combination of two scenarios which are independent of each other. This means that it is possible to consider $(6.23)_\alpha$ only for $\alpha > \alpha_0$ or $\alpha < \alpha_0$, respectively, in order to obtain the results which apply for these parameter values.

We now compare case (i) of the above theorem with the equivalent autonomous bifurcation.

*Example 6.8.* Let $x_- < 0 < x_+$ and $\alpha_- < 0 < \alpha_+$ be in $\overline{\mathbb{R}}$, and consider the autonomous differential equation

$$\dot{x} = f(x, \alpha), \tag{6.29}$$

where the $C^4$-function $f : (x_-, x_+) \times (\alpha_-, \alpha_+) \to \mathbb{R}$ satisfies the following assumptions:

(i)   $f(0, \alpha) = 0$   for all $\alpha \in (\alpha_-, \alpha_+)$,

(ii)  $D_1 f(0, 0) = 0$,

(iii) $D_1 D_2 f(0, 0) > 0$,

(iv)  $D_1^2 f(0, 0) = 0$,

(v)   $D_1^3 f(0, 0) > 0$.

Then (6.29) admits an autonomous pitchfork bifurcation (see, e.g., WIGGINS [182, p. 268 f.], and see Figure 1.1 for the bifurcation diagram). There exist a neighborhood $U \times V$ of $(0, 0)$ in $\mathbb{R}^2$ and a $C^2$-function $h : U \to V$ with $h(0) = 0$ and

$$f(x, h(x)) = 0 \quad \text{for all } x \in U.$$

Except the trivial equilibria and the equilibria described by $h$, there are no other equilibria in $U \times V$, and the function $h$ is maximal at $x = 0$. It can be verified that this situation fits into case (i) of Theorem 6.6: The functions $\beta_1$ and $\beta_2$ can be chosen to be increasing, since $D_1 D_2 f(0, 0) > 0$ by (iii), and (6.24) holds, since $D_1^3 f(0, 0) > 0$ by (v). Due to (iii), the trivial equilibrium of (6.29) is attractive for $\alpha < 0$ and repulsive for $\alpha > 0$, and this carries over to nonautonomous notions of attractivity and repulsivity. The function $h$ describes repulsive equilibria of (6.29), and these equilibria are the boundary of the domain of attraction of the trivial equilibria. Since $\lim_{x \to 0} h(x) = 0$, we have a nonautonomous bifurcation in form of a shrinking domain of attraction.

The case of compact time intervals is treated in the last theorem of this chapter.

**Theorem 6.9 (Nonautonomous pitchfork bifurcation, part II).** *Let $x_- < 0 < x_+$ and $\alpha_- < \alpha_+$ be in $\overline{\mathbb{R}}$ and $\mathbb{I} := [\tau, \tau + T]$, and consider the nonautonomous differential equation*

$$\dot{x} = a(t, \alpha)x + b(t, \alpha)x^3 + r(t, x, \alpha) \tag{6.30}_\alpha$$

*with continuous functions $a : \mathbb{I} \times (\alpha_-, \alpha_+) \to \mathbb{R}$, $b : \mathbb{I} \times (\alpha_-, \alpha_+) \to \mathbb{R}$ and $r : \mathbb{I} \times (x_-, x_+) \times (\alpha_-, \alpha_+) \to \mathbb{R}$ fulfilling $r(\cdot, 0, \cdot) \equiv 0$. Let $\Lambda_\alpha : \mathbb{I} \times \mathbb{I} \to \mathbb{R}$ denote the transition operator of the linearized equation $\dot{x} = a(t, \alpha)x$. We define*

$$K(\alpha) := \sup \{\Lambda_\alpha(t, s) : t, s \in \mathbb{I}\} \quad \text{for all } \alpha \in (\alpha_-, \alpha_+)$$

*and assume, there exists an $\alpha_0 \in (\alpha_-, \alpha_+)$ such that the following hypotheses hold:*

- Hypothesis on linear part. *We either have*

$$\Lambda_\alpha(\tau + T, \tau) < 1 \quad \text{for all } \alpha \in (\alpha_-, \alpha_0) \text{ and} \\ \Lambda_\alpha(\tau + T, \tau) > 1 \quad \text{for all } \alpha \in (\alpha_0, \alpha_+)$$  (6.31)

*or*

$$\Lambda_\alpha(\tau + T, \tau) > 1 \quad \text{for all } \alpha \in (\alpha_-, \alpha_0) \text{ and} \\ \Lambda_\alpha(\tau + T, \tau) < 1 \quad \text{for all } \alpha \in (\alpha_0, \alpha_+).$$  (6.32)

- Hypothesis on nonlinearity. *The cubic term either fulfills*

$$\liminf_{\alpha \to \alpha_0} \inf_{t \in \mathbb{I}} b(t, \alpha) > 0$$  (6.33)

*or*

$$\limsup_{\alpha \to \alpha_0} \sup_{t \in \mathbb{I}} b(t, \alpha) < 0,$$  (6.34)

*and the remainder satisfies*

$$\lim_{x \to 0} \sup_{\alpha \in (\alpha_0 - x^2, \alpha_0 + x^2)} \sup_{t \in \mathbb{I}} \frac{|r(t, x, \alpha)|}{|x|^3} = 0$$  (6.35)

*and*

$$\limsup_{\alpha \to \alpha_0} \limsup_{x \to 0} \sup_{t \in \mathbb{I}} - \frac{T K(\alpha) |r(t, x, \alpha)|}{|x| \ln\left(\min\left\{\Lambda_\alpha(\tau + T, \tau), \Lambda_\alpha(\tau, \tau + T)\right\}\right)} < 1.$$  (6.36)

*Then there exist $\hat{\alpha}_- < 0 < \hat{\alpha}_+$ such that the following statements are fulfilled:*

(i) *In case (6.31) and (6.33) is fulfilled, the trivial solution is $(\tau, T)$-attractive for $\alpha \in (\hat{\alpha}_-, \alpha_0)$ and $(\tau, T)$-repulsive for $\alpha \in (\alpha_0, \hat{\alpha}_+)$. The differential equation $(6.30)_\alpha$ admits a $(\tau, T)$-bifurcation, since the corresponding radii of $(\tau, T)$-attraction satisfy*

$$\lim_{\alpha \nearrow \alpha_0} \mathfrak{A}_0^\alpha = 0.$$

(ii) *In case (6.31) and (6.34) is fulfilled, the trivial solution is $(\tau, T)$-attractive for $\alpha \in (\hat{\alpha}_-, \alpha_0)$ and $(\tau, T)$-repulsive for $\alpha \in (\alpha_0, \hat{\alpha}_+)$. The differential equation $(6.30)_\alpha$ admits a $(\tau, T)$-bifurcation, since the corresponding radii of $(\tau, T)$-repulsion satisfy*

$$\lim_{\alpha \searrow \alpha_0} \mathfrak{R}_0^\alpha = 0.$$

(iii) *In case (6.32) and (6.33) is fulfilled, the trivial solution is $(\tau, T)$-repulsive for $\alpha \in (\hat{\alpha}_-, \alpha_0)$ and $(\tau, T)$-attractive for $\alpha \in (\alpha_0, \hat{\alpha}_+)$. The differential equation $(6.30)_\alpha$ admits a $(\tau, T)$-bifurcation, since the corresponding radii of $(\tau, T)$-attraction satisfy*

$$\lim_{\alpha \searrow \alpha_0} \mathfrak{A}_0^\alpha = 0.$$

(iv) *In case (6.32) and (6.34) is fulfilled, the trivial solution is $(\tau, T)$-repulsive for $\alpha \in (\hat{\alpha}_-, \alpha_0)$ and $(\tau, T)$-attractive for $\alpha \in (\alpha_0, \hat{\alpha}_+)$. The differential equation $(6.30)_\alpha$ admits a $(\tau, T)$-bifurcation, since the corresponding radii of $(\tau, T)$-repulsion satisfy*

$$\lim_{\alpha \nearrow \alpha_0} \mathfrak{R}_0^\alpha = 0 \,.$$

*Proof.* This theorem can be proved using the same methods as in the proof of Theorem 6.6.                                                              □

*Remark 6.10.*

 (i) The hypothesis on the linear part implies that the $(\tau, T)$-dichotomy spectrum of the linearization $\dot{x} = a(t, \alpha)x$ converges to $\{0\}$ in Hausdorff distance in the limit $\alpha \to \alpha_0$.

(ii) Condition (6.36) is only used to obtain the attractivity or repulsivity of the trivial solution by applying Theorem 5.10. Alternatively, one can directly postulate that the trivial solution changes the stability at the parameter value $\alpha_0$ from, say, attractivity to repulsivity.

# Bifurcations of Asymptotically Autonomous Systems

A nonautonomous differential equation

$$\dot{x} = f(t, x) \tag{7.1}$$

is called *past (future, respectively) asymptotically autonomous* with *limiting equation*

$$\dot{x} = g(x) \tag{7.2}$$

if $\lim_{t \to -\infty} f(t, x) = g(x)$ ($\lim_{t \to \infty} f(t, x) = g(x)$, respectively) holds uniformly for every element $x$ of the domain of the function $g$. This chapter deals with the question of transferring bifurcation phenomena from the autonomous differential equation (7.2) to the nonautonomous differential equation (7.1).

The study of asymptotically autonomous differential equations goes back to MARKUS [115]. Markus discusses properties of nonautonomous $\omega$-limit sets and generalizes the Theorem of Poincaré & Bendixson (see, e.g., PALIS & DE MELO [124] and HIRSCH & SMALE [81, Chapter 11]) to asymptotically autonomous planar systems. His work has stimulated the qualitative theory of nonautonomous differential equations (see, e.g., SELL [165, 166, 167]). Further fundamental work on asymptotically autonomous systems was achieved by STRAUSS & YORKE [178], ARTSTEIN [10, 11], THIEME [179] and MISCHAIKOW & SMITH & THIEME [121] in the context of differential equations (see also KATO & MARTYNYUK & SHESTAKOV [90]); for difference equations, we refer to SCHÖNEFUSS [163].

It is not clear a priori under which assumptions certain behavior carries over from the autonomous to the nonautonomous system. In fact, in THIEME [179], several examples of asymptotically autonomous systems are studied that behave quite differently from the limiting equations. In LANGA & ROBINSON & SUÁREZ [104], however, it is shown that the pullback and forward behavior of a special asymptotically autonomous Lotka-Volterra system is consistent to the underlying autonomous system.

In the first section of this chapter, some basic properties of asymptotically autonomous systems are prepared for later use. In Section 7.2, one-dimensional bifurcations such as the pitchfork, transcritical and saddle node bifurcation are discussed. Section 7.3 is devoted to study the Hopf bifurcation scenario.

Whenever considering a nonautonomous differential equation in this chapter, its general solution is denoted by $\lambda$. For the flow of an autonomous differential equation, we write $\phi$.

## 7.1 Basic Properties of Asymptotically Autonomous Systems

In this section, some useful lemmata are derived for asymptotically autonomous differential equations. The first two lemmata deal with the question of controlling the distances of the time evolutions of both systems on compact time intervals.

**Lemma 7.1.** *Consider an open set $D \subset \mathbb{R}^N$, a nonautonomous differential equation*

$$\dot{x} = f(t, x)$$

*with a $C^1$-function $f : (-\infty, 0) \times D \to \mathbb{R}^N$ and an autonomous differential equation*

$$\dot{x} = g(x)$$

*with a $C^1$-function $g : D \to \mathbb{R}^N$. We assume that*

$$\lim_{t \to -\infty} f(t, x) = g(x) \quad \text{uniformly for } x \in D. \tag{7.3}$$

*Furthermore, let $K \subset D$ be a compact and convex set. Then the following statements are fulfilled:*

(i) *For all $T > 0$ and $\varepsilon > 0$, there exists a $\tau_0 < -T$ such that for all $T' \leq T$ and $x \in K$ with*

$$\phi(t, x) \in K \quad \text{for all } t \in [0, T'],$$

*the relation*

$$\|\lambda(\tau + t, \tau, x) - \phi(t, x)\| \leq \varepsilon \quad \text{for all } \tau \leq \tau_0 \text{ and } t \in [0, T']$$

*is fulfilled.*

(ii) *For all $T > 0$ and $\varepsilon > 0$, there exists a $\tau_0 < 0$ such that for all $T' \leq T$ and $x \in K$ with*

$$\phi(-t, x) \in K \quad \text{for all } t \in [0, T'],$$

the relation

$$\|\lambda(\tau - t, \tau, x) - \phi(-t, x)\| \leq \varepsilon \quad \text{for all } \tau \leq \tau_0 \text{ and } t \in [0, T']$$

is fulfilled.

*Proof.* (i) Since $D$ is open, there exist a compact and convex set $\tilde{K}$ and an $\eta > 0$ such that $U_\eta(K) \subset \tilde{K}$. We choose $T > 0$ and $\varepsilon > 0$ arbitrarily and define $M := \max_{x \in \tilde{K}} \|Dg(x)\|$. Due to (7.3), there exists a $\tau_0 < -T$ with

$$\|f(t + T, x) - g(x)\| \leq \frac{\min\{\varepsilon, \eta\}}{Te^{MT}} \quad \text{for all } t \leq \tau_0 \text{ and } x \in D.$$

For the rest of this proof, we fix arbitrary numbers $\tau \leq \tau_0$, $T' \leq T$ and $x \in K$ fulfilling

$$\phi(t, x) \in K \quad \text{for all } t \in [0, T'].$$

Since for all $t \in [0, T']$, we have

$$\lambda(t + \tau, \tau, x) - \phi(t, x) = \int_0^t \left( f(s + \tau, \lambda(s + \tau, \tau, x)) - g(\phi(s, x)) \right) ds,$$

it follows from the mean value inequality (see, e.g., ABRAHAM & MARSDEN & RATIU [1, Theorem 2.4.8, p. 87]) that

$$\|\lambda(t + \tau, \tau, x) - \phi(t, x)\|$$
$$\leq \int_0^t \|f(s + \tau, \lambda(s + \tau, \tau, x)) - g(\phi(s, x))\| \, ds$$
$$\leq \int_0^t \left( \|f(s + \tau, \lambda(s + \tau, \tau, x)) - g(\lambda(s + \tau, \tau, x))\| + \right.$$
$$\left. \|g(\lambda(s + \tau, \tau, x)) - g(\phi(s, x))\| \right) ds$$
$$\leq \frac{t \min\{\varepsilon, \eta\}}{Te^{MT}} + M \int_0^t \|\lambda(s + \tau, \tau, x) - \phi(s, x)\| \, ds.$$

Assume, there exists a $t \in (0, T')$ with $\|\lambda(t + \tau, \tau, x) - \phi(t, x)\| \geq \min\{\varepsilon, \eta\}$. We define

$$T^* := \min\left\{t \in (0, T') : \|\lambda(t + \tau, \tau, x) - \phi(t, x)\| \geq \min\{\varepsilon, \eta\}\right\} < T'.$$

Hence, from Gronwall's inequality (Lemma A.8), we obtain

$$\|\lambda(T^* + \tau, \tau, x) - \phi(T^*, x)\| \leq \frac{T^* \min\{\varepsilon, \eta\}}{Te^{MT}} e^{MT^*} < \min\{\varepsilon, \eta\}.$$

This is a contradiction and finishes the proof of this lemma.
(ii) See proof of (i).    □

**Lemma 7.2.** *Consider an open set $D \subset \mathbb{R}^N$, a nonautonomous differential equation*

$$\dot{x} = f(t, x)$$

*with a $C^1$-function $f : (0, \infty) \times D \to \mathbb{R}^N$ and an autonomous differential equation*

$$\dot{x} = g(x)$$

*with a $C^1$-function $g : D \to \mathbb{R}^N$. We assume that*

$$\lim_{t \to \infty} f(t, x) = g(x) \quad \text{uniformly for } x \in D.$$

*Furthermore, let $K \subset D$ be a compact and convex set. Then the following statements are fulfilled:*

(i) *For all $T > 0$ and $\varepsilon > 0$, there exists a $\tau_0 > 0$ such that for all $T' \leq T$ and $x \in K$ with*

$$\phi(t, x) \in K \quad \text{for all } t \in [0, T'],$$

*the relation*

$$\|\lambda(\tau + t, \tau, x) - \phi(t, x)\| \leq \varepsilon \quad \text{for all } \tau \geq \tau_0 \text{ and } t \in [0, T']$$

*is fulfilled.*

(ii) *For all $T > 0$ and $\varepsilon > 0$, there exists a $\tau_0 > T$ such that for all $T' \leq T$ and $x \in K$ with*

$$\phi(-t, x) \in K \quad \text{for all } t \in [0, T'],$$

*the relation*

$$\|\lambda(\tau - t, \tau, x) - \phi(t, x)\| \leq \varepsilon \quad \text{for all } \tau \geq \tau_0 \text{ and } t \in [0, T']$$

*is fulfilled.*

*Proof.* See proof of Lemma 7.1. □

In case of the classical autonomous bifurcations for ordinary differential equations (such as pitchfork, transcritical, saddle node and Hopf bifurcation), *after* the bifurcation, the phase space can be separated into three invariant parts. Therefore, we restrict attention to the following situation: Let $D \subset \mathbb{R}^N$ be an open and convex set and

$$\dot{x} = g(x)$$

be an autonomous differential equation with a $C^1$-function $g : D \to \mathbb{R}^N$. We suppose that $D$ is the disjoint union of

- a bounded and open set $S^i$ (inner area),

**Fig. 7.1.** The situation in case of a (a) pitchfork bifurcation, (b) transcritical or saddle node bifurcation, (c) Hopf bifurcation.

- an open set $S^o$ (outer area),
- a compact set $S = \partial S^i = \partial S^o$ with int $S = \emptyset$.

The occurrence of one of the above mentioned autonomous bifurcations means that (exactly) one of the following two hypotheses holds:

- *Hypothesis ($H_1$).* The following conditions are fulfilled:
  (i)   The inner area $S^i$ is *forward invariant*, i.e.,

  $$\phi(t, x) \in S^i \quad \text{for all } t \geq 0 \text{ and } x \in S^i,$$

  and there exists an attractive equilibrium $x_0 \in S^i$ such that for all compact sets $K \subset S^i$, we have

  $$\lim_{t \to \infty} d\big(\phi(t, K)\big|\{x_0\}\big) = 0.$$

  (ii)  The outer area $S^o$ is *backward invariant*, i.e.,

  $$\phi(t, x) \in S^o \quad \text{for all } t \leq 0 \text{ and } x \in S^o,$$

  and $S$ is a *repeller*, i.e., there exists an $\eta > 0$ with

  $$\lim_{t \to \infty} d\big(\phi(-t, U_\eta(S))\big|S\big) = 0.$$

  (iii) $S$ is *invariant*, i.e.,

  $$\phi(t, x) \in S \quad \text{for all } t \in \mathbb{R} \text{ and } x \in S.$$

- *Hypothesis ($H_2$).* The following conditions are fulfilled:
  (i)   The inner area $S^i$ is backward invariant, and there exists a repulsive equilibrium $x_0 \in S^i$ such that for all compact sets $K \subset S^i$, we have

  $$\lim_{t \to \infty} d\big(\phi(-t, K)\big|\{x_0\}\big) = 0.$$

  (ii)  The outer area $S^o$ is forward invariant, and $S$ is an *attractor*, i.e., there exists an $\eta > 0$ with

  $$\lim_{t \to \infty} d\big(\phi(t, U_\eta(S))\big|S\big) = 0.$$

(iii) $S$ is invariant.

Some easy consequences of Hypothesis $(H_1)$ are derived in the following lemma.

**Lemma 7.3.** *Under Hypothesis $(H_1)$, the following statements hold:*

(i) *For all $\delta > 0$, there exists a $T > 0$ such that for $x \in D$ with $d(x, S) \geq \delta$, there exists a $\hat{T} \in [0, T]$ with*

$$d\big(\phi(\hat{T}, x), S\big) \geq \eta.$$

(ii) *For all $\gamma > 0$, there exists a $\delta > 0$ with*

$$\phi\big(-t, U_\delta(S^i)\big) \subset U_\gamma(S^i) \quad \text{for all } t \geq 0.$$

*Proof.* (i) We choose $\delta > 0$ arbitrarily. Due to the hypotheses, there exists a $T > 0$ with

$$\phi\big(-T, U_\eta(S)\big) \subset U_{\delta/2}(S).$$

This implies the assertion.

(ii) We choose $\gamma > 0$ arbitrarily. Since $S$ is repulsive and $S = \partial S^i$, there exists a $T > 0$ with

$$\phi\big(-t, U_\eta(S^i)\big) \subset U_\gamma(S^i) \quad \text{for all } t > T.$$

Arguing negatively, we assume that for all $n \in \mathbb{N}$, there exist $t_n \in [0, T]$ and $x_n \in U_{1/n}(S^i)$ with

$$d\big(\phi(-t_n, x_n), S^i\big) = d\big(\phi(-t_n, x_n), S\big) \geq \gamma.$$

Since $S$ is compact, we assume w.l.o.g. that the sequence $\{x_n\}_{n \in \mathbb{N}}$ is convergent with limit $x \in S$. Due to the continuity of the flow $\phi$ and the invariance of $S$, there exists a $\beta > 0$ such that for all $y \in U_\beta(x)$ and $t \in [0, T]$, we have

$$d\big(\phi(-t, y), S\big) < \frac{\gamma}{2}.$$

This is a contradiction and finishes the proof of this lemma.     $\square$

*Remark 7.4.* In case the outer area is forward invariant, statement (i) of the above lemma can be simplified as follows: For all $\delta > 0$, there exists a $T > 0$ such that for all $x \in D$ with $d(x, S) \geq \delta$,

$$d\big(\phi(T, x), S\big) \geq \eta$$

is fulfilled.

The following two lemmata deal with the question of determining past (future, respectively) attraction areas of past (future, respectively) attractive solutions.

**Lemma 7.5 (Attraction areas of past attractive solutions).** *We suppose that Hypothesis ($H_1$) is fulfilled and consider the nonautonomous differential equation*

$$\dot{x} = f(t, x) \tag{7.4}$$

*with a $C^1$-function $f : (-\infty, 0) \times D \to \mathbb{R}^N$ such that*

$$\lim_{t \to -\infty} f(t, x) = g(x) \quad \text{uniformly for all } x \in D.$$

*Furthermore, for some $\tau < 0$, let $\mu : (-\infty, \tau) \to \mathbb{R}^N$ be a past attractive solution of (7.4) with $\lim_{t \to -\infty} \mu(t) = x_0$. Then we have*

$$\mathcal{A}_\mu^\leftarrow = S^i - x_0.$$

*If, in addition, $S^i$ is bounded, then there exist $s < \tau$ and a past repeller $R \subset (-\infty, s) \times D$ with*

$$S^i \subset \liminf_{t \to -\infty} R(t) \subset \limsup_{t \to -\infty} R(t) \subset \text{cls } S^i.$$

*Proof.* The proof of this lemma is divided into four steps.
*Step 1.* $\mathcal{A}_\mu^\leftarrow \supset S^i - x_0$.
Since $\mu$ is past attractive, there exists a $\gamma > 0$ such that for all $s < \tau$, we have

$$\lim_{t \to -\infty} d\big(\lambda\big(s, t, U_\gamma(\mu(t))\big)\big|\mu(s)\big) = 0. \tag{7.5}$$

We choose $y \in S^i$ arbitrarily. Let $C$ be a neighborhood of $y$ such that there exists a $\delta > 0$ with $\text{cls } U_\delta(C) \subset S^i$. Since $\lim_{t \to -\infty} \mu(t) = x_0$, there exists a $t_1 < \tau$ such that

$$\mu(t) \in U_{\min\{\gamma/3, \delta\}}(x_0) \quad \text{for all } t \le t_1.$$

Due to the attractivity of $x_0$, there exists a $T > 0$ such that

$$d\big(\phi(T, U_\delta(C))\big|\{x_0\}\big) < \frac{\gamma}{3}.$$

Since it is possible to choose a compact and convex superset $K \subset D$ of $S^i$ ($D$ is convex), Lemma 7.1 (i) implies that there exists a $t_2 < t_1 - T$ with

$$\|\lambda(t + T, t, x) - \phi(T, x)\| \le \frac{\gamma}{3} \quad \text{for all } t \le t_2 \text{ and } x \in \underbrace{\mu(t) + C - x_0}_{\subset U_\delta(C)}.$$

Hence, for all $t \le t_2$ and $x \in \mu(t) + C - x_0$, we have

$$\|\lambda(t + T, t, x) - \mu(t + T)\|$$
$$\le \|\lambda(t + T, t, x) - \phi(T, x)\| + \|\phi(T, x) - x_0\| + \|x_0 - \mu(t + T)\|$$
$$< \frac{\gamma}{3} + \frac{\gamma}{3} + \frac{\gamma}{3} = \gamma.$$

Thus,

$$\lim_{t \to -\infty} d\big(\lambda\big(t_2, t, \mu(t) + C - x_0\big) \big| \{\mu(t_2)\}\big)$$

$$= \lim_{t \to -\infty} d\big(\lambda\big(t_2, t + T, \underbrace{\lambda(t + T, t, \mu(t) + C - x_0)}_{\subset U_\gamma(\mu(t+T))}\big) \big| \{\mu(t_2)\}\big) \overset{(7.5)}{=} 0\,.$$

This implies $y - x_0 \in \mathcal{A}_\mu^\leftarrow$, and since $y \in S^i$ has been chosen arbitrarily, we have $\mathcal{A}_\mu^\leftarrow \supset S^i - x_0$.

*Step 2.* $\mathcal{A}_\mu^\leftarrow \subset S^i - x_0$.

We choose $y \in S^o$ and $\beta > 0$ with $U_\beta(x_0) \subset S^i$ and define $\delta := d(y, S) > 0$. Due to Lemma 7.3 (i), there exists a $T > 0$ such that for all $x \in S^o$ with $d(x, S) \geq \min\{\delta/2, \eta/3\}$, there exists a $\hat{T} \in [0, T]$ with

$$d\big(\phi(\hat{T}, x), S\big) \geq \eta \quad \text{and} \quad \phi(\hat{T}, x) \in S^o\,. \tag{7.6}$$

Moreover, there exists a $t_1 < \tau$ with

$$\mu(t) \in U_{\min\{\delta/2, \eta/3, \beta/4\}}(x_0) \subset S^i \quad \text{for all } t \leq t_1\,.$$

Let $K$ be a compact and convex superset of $U_\eta(S)$. Then, because of Lemma 7.1 (i), there exists a $t_2 < t_1$ such that for all $\tilde{T} \in [0, T]$ and $x \in K$ with $\phi(t, x) \in K$ for all $t \in [0, \tilde{T}]$, we have

$$\big\|\lambda(\hat{t} + t, \hat{t}, x) - \phi(t, x)\big\| \leq \min\left\{\frac{\beta}{2}, \frac{\eta}{3}\right\} \quad \text{for all } \hat{t} \leq t_2 \text{ and } t \in [0, \tilde{T}]\,. \tag{7.7}$$

We argue negatively and suppose that

$$\lim_{t \to -\infty} \big\|\lambda\big(t_2, t, y - x_0 + \mu(t)\big) - \mu(t_2)\big\| = 0$$

holds. Therefore, since $\mu(t_2) \in U_{\beta/4}(x_0)$, there exists a $t_3 < t_2$ with

$$\lambda\big(t_2, t_3, y - x_0 + \mu(t_3)\big) \in U_{\beta/2}(x_0)\,. \tag{7.8}$$

We define

$$s := \max\left\{t \in [t_3, t_2] : d\big(\lambda\big(t, t_3, y - x_0 + \mu(t_3)\big), S\big) \geq \min\left\{\frac{\eta}{2}, \frac{\delta}{2}\right\}\right.$$

$$\left. \text{and} \qquad \lambda\big(t, t_3, y - x_0 + \mu(t_3)\big) \in S^o\right\}\,.$$

This implies the relations $d\big(\lambda\big(s, t_3, y - x_0 + \mu(t_3)\big), S\big) = \min\{\eta/2, \delta/2\}$ and $\lambda\big(s, t_3, y - x_0 + \mu(t_3)\big) \in S^o$.

We distinguish two cases.

*Case 1.* $t_2 - s \leq T$.

*Case 1.1. For all $t \in [0, t_2 - s]$, we have $\phi(t, \lambda(s, t_3, y - x_0 + \mu(t_3))) \in K$.*
Due to (7.7), we have for all $t \in [0, t_2 - s]$,

$$\left\| \phi(t, \lambda(s, t_3, y - x_0 + \mu(t_3))) - \lambda(t + s, t_3, y - x_0 + \mu(t_3)) \right\| \leq \frac{\beta}{2}.$$

Since $\phi(t, \lambda(s, t_3, y - x_0 + \mu(t_3))) \in S^o$ for all $t \in [0, t_2 - s]$, this leads to

$$\begin{aligned}
&\left\| \lambda(t_2, t_3, y - x_0 + \mu(t_3)) - x_0 \right\| \\
&\geq \left\| \phi(t_2 - s, \lambda(s, t_3, y - x_0 + \mu(t_3))) - x_0 \right\| - \\
&\quad \left\| \lambda(t_2, t_3, y - x_0 + \mu(t_3)) - \phi(t_2 - s, \lambda(s, t_3, y - x_0 + \mu(t_3))) \right\| \\
&\geq \beta - \frac{\beta}{2} = \frac{\beta}{2}.
\end{aligned}$$

This is a contradiction to (7.8).
*Case 1.2. There exists a $\tilde{t} \in [0, t_2 - s]$ with $\phi(\tilde{t}, \lambda(s, t_3, y - x_0 + \mu(t_3))) \notin K$.*
By defining

$$\hat{s} := \inf \left\{ t \in [0, t_2 - s] : \phi(t, \lambda(s, t_3, y - x_0 + \mu(t_3))) \notin K \right\} > 0,$$

we obtain $d(\phi(\hat{s}, \lambda(s, t_3, y - x_0 + \mu(t_3))), S) \geq \eta$. Due to (7.7), the relation

$$\left\| \lambda(\hat{s} + s, t_3, y - x_0 + \mu(t_3)) - \phi(\hat{s}, \lambda(s, t_3, y - x_0 + \mu(t_3))) \right\| \leq \frac{\eta}{3}$$

holds, and hence, we have both $d(\lambda(\hat{s} + s, t_3, y - x_0 + \mu(t_3)), S) \geq 2\eta/3$ and
$\lambda(\hat{s} + s, t_3, y - x_0 + \mu(t_3)) \in S^o$. This is a contradiction to the definition of $s$.
*Case 2. $t_2 - s > T$.*
*Case 2.1. For all $t \in [0, T]$, we have $\phi(t, \lambda(s, t_3, y - x_0 + \mu(t_3))) \in K$.*
Because of $d(\lambda(s, t_3, y - x_0 + \mu(t_3)), S) = \min\{\delta/2, \eta/2\} \geq \min\{\delta/2, \eta/3\}$ and
(7.6), there exists a $\hat{T} \in [0, T]$ with

$$d(\phi(\hat{T}, \lambda(s, t_3, y - x_0 + \mu(t_3))), S) \geq \eta,$$

and (7.7) yields

$$\left\| \phi(\hat{T}, \lambda(s, t_3, y - x_0 + \mu(t_3))) - \lambda(\hat{T} + s, t_3, y - x_0 + \mu(t_3)) \right\| \leq \frac{\eta}{3}.$$

Together, this implies

$$d(\lambda(\hat{T} + s, t_3, y - x_0 + \mu(t_3)), S) \geq \frac{2}{3}\eta$$

and

$$\lambda(\hat{T} + s, t_3, y - x_0 + \mu(t_3)) \in S^o.$$

This is a contradiction to the definition of $s$.
*Case 2.2. There exists a $\tilde{t} \in [0, T]$ with $\phi(\tilde{t}, \lambda(s, t_3, y - x_0 + \mu(t_3))) \notin K$.*

This case is treated analogously to Case 1.2 (by writing $T$ instead of $t_2 - s$). Consequently, we have $y - x_0 \notin \mathcal{A}_\mu^\leftarrow$. This leads to the assertion, since $S = \partial S^i$, $\text{int } S = \emptyset$ and $\mathcal{A}_\mu^\leftarrow$ is open.

*Step 3.* For all $\kappa \leq \eta$, there exist $T > 0$ and $t_1 < \tau$ such that for all $t_2 < t_1$ and $t > T$, we have

$$\lambda\left(t_2 - t, t_2, U_{5\kappa/6}\left(\mathcal{A}_\mu^\leftarrow + \mu(t_2)\right)\right) \subset U_{2\kappa/3}\left(\mathcal{A}_\mu^\leftarrow + \mu(t_2 - t)\right).$$

We choose $\kappa \leq \eta$ arbitrarily. By applying Lemma 7.3 (ii), there exists a $\delta > 0$ with

$$\phi\left(-t, U_\delta(S^i)\right) \subset U_{\kappa/4}(S^i) \quad \text{for all } t \geq 0. \tag{7.9}$$

Due to the repulsivity of $S$, there exists a $T > 0$ with

$$\phi\left(-t, U_\kappa(S^i)\right) \subset U_{\delta/2}(S^i) \quad \text{for all } t > T.$$

By choosing $K$ as a convex and compact superset of $\cup_{t \in [0,T]} \phi(-t, U_\kappa(S^i))$, we can apply Lemma 7.1 (ii), and we therefore get a $t_1 < \tau$ with

$$\|\lambda(t_2 - t, t_2, x) - \phi(-t, x)\| \leq \frac{\delta}{2} \quad \text{for all } x \in U_\kappa(S^i), t_2 < t_1 \text{ and } t \in [0, T]$$

$$\text{and} \quad \|\mu(t) - x_0\| \leq \frac{\kappa}{6} \quad \text{for all } t < t_1. \tag{7.10}$$

Thus,

$$\lambda\left(t_2 - T, t_2, U_\kappa(S^i)\right) \subset U_\delta(S^i) \subset U_{\kappa/4}(S^i) \quad \text{for all } t_2 < t_1 \tag{7.11}$$

is fulfilled. Because of (7.9), this leads to

$$\phi\left(-t, \lambda\left(t_2 - T, t_2, U_\kappa(S^i)\right)\right) \subset U_{\kappa/4}(S^i) \quad \text{for all } t_2 < t_1 \text{ and } t \geq 0.$$

Due to (7.10) and $\delta/2 < \kappa/4$, we have

$$\lambda\left(t_2 - t, t_2, U_\kappa(S^i)\right) \subset U_{\kappa/2}(S^i) \quad \text{for all } t_2 < t_1 \text{ and } t \in [T, 2T].$$

Suppose now, there exist $\hat{t} > 2T$ and $\hat{t}_2 < t_1$ with

$$d\left(\lambda\left(\hat{t}_2 - \hat{t}, \hat{t}_2, U_\kappa(S^i)\right) \big| S^i\right) \geq \frac{\kappa}{2}.$$

We define

$$s := \inf\left\{t > 2T : d\left(\lambda\left(\hat{t}_2 - t, \hat{t}_2, U_\kappa(S^i)\right) \big| S^i\right) \geq \frac{\kappa}{2}\right\} > 2T$$

and set $t_2 := \hat{t}_2 - s + T < t_1$. Consequently,

$$\lambda\left(\hat{t}_2 - s, \hat{t}_2, U_\kappa(S^i)\right) = \lambda\left(t_2 - T, t_2, \underbrace{\lambda\left(t_2, \hat{t}_2, U_\kappa(S^i)\right)}_{\subset U_{\kappa/2}(S^i)}\right) \overset{(7.11)}{\in} U_{\kappa/4}(S^i)$$

holds. This is a contradiction, i.e., for all $t_2 < t_1$ and $t > T$, we have

$$\lambda\big(t_2 - t, t_2, U_\kappa(S^i)\big) \subset U_{\kappa/2}(S^i).$$

Since $\mu(t) \in U_{\kappa/6}(x_0)$ for all $t < t_1$, the relation

$$\lambda\left(t_2 - t, t_2, U_{5\kappa/6}\big(\mathcal{A}_\mu^\leftarrow + \mu(t_2)\big)\right) \subset U_{2\kappa/3}\big(\mathcal{A}_\mu^\leftarrow + \mu(t_2 - t)\big)$$

is fulfilled for all $t_2 < t_1$ and $t > T$.

*Step 4. Existence of the past repeller.*
Repeated usage of Step 3 implies

$$\lim_{t \to -\infty} d\left(\lambda\left(t, \tau, U_{5\eta/6}\big(\mathcal{A}_\mu^\leftarrow + \mu(\tau)\big)\right) \,\Big|\, \mathcal{A}_\mu^\leftarrow + \mu(t)\right) = 0 \quad \text{for all } \tau < t_1.$$

Due to Theorem 2.41 (i), there exist $s < \tau$ and a past repeller $R \subset (-\infty, s) \times D$ with

$$\mathcal{A}_\mu^\leftarrow \subset \liminf_{t \to -\infty} \big(R(t) - \mu(t)\big) \subset \limsup_{t \to -\infty} \big(R(t) - \mu(t)\big) \subset \operatorname{cls} \mathcal{A}_\mu^\leftarrow.$$

Since $\lim_{t \to -\infty} \mu(t) = x_0$, we have

$$S^i \subset \liminf_{t \to -\infty} R(t) \subset \limsup_{t \to -\infty} R(t) \subset \operatorname{cls} S^i.$$

This finishes the proof of this lemma.    □

**Lemma 7.6 (Attraction areas of future attractive solutions).** *We suppose that Hypothesis ($H_1$) is fulfilled and consider the nonautonomous differential equation*

$$\dot{x} = f(t, x) \tag{7.12}$$

*with a $C^1$-function $f : (0, \infty) \times D \to \mathbb{R}^N$ such that*

$$\lim_{t \to \infty} f(t, x) = g(x) \quad \text{uniformly for all } x \in D.$$

*Furthermore, for some $\tau > 0$, let $\mu : (\tau, \infty) \to \mathbb{R}^N$ be a future attractive solution of (7.12) with $\lim_{t \to \infty} \mu(t) = x_0$. Then we have*

$$\mathcal{A}_\mu^\rightarrow = S^i - x_0.$$

*If, in addition, $S^i$ is bounded, then there exist $s > \tau$ and a future repeller $R \subset (s, \infty) \times D$ with*

$$S^i \subset \liminf_{t \to \infty} R(t) \subset \limsup_{t \to \infty} R(t) \subset \operatorname{cls} S^i.$$

*Proof.* The proof of this lemma is quite similar to that of Lemma 7.5, but this is not clear a priori and there are important differences. In the following, the entire proof is therefore written down.

*Step 1.* $\mathcal{A}_\mu^\rightarrow \supset S^i - x_0$.

Since $\mu$ is future attractive, there exists a $\gamma > 0$ such that for all $s > \tau$, we have

$$\lim_{t \to \infty} d\big(\lambda\big(t, s, U_\gamma(\mu(s))\big) \big| \mu(t)\big) = 0. \tag{7.13}$$

We choose $y \in S^i$ arbitrarily. Let $C$ be a neighborhood of $y$ such that there exists a $\delta > 0$ with $\text{cls}\, U_\delta(C) \subset S^i$. Since $\lim_{t \to \infty} \mu(t) = x_0$, there exists a $t_1 > \tau$ such that

$$\mu(t) \in U_{\min\{\gamma/3, \delta\}}(x_0) \quad \text{for all } t \geq t_1.$$

Due to the attractivity of $x_0$, there exists a $T > 0$ such that

$$d\big(\phi(T, U_\delta(C)) \big| \{x_0\}\big) < \frac{\gamma}{3}.$$

Since it is possible to choose a compact and convex superset $K \subset D$ of $S^i$ ($D$ is convex), Lemma 7.2 (i) implies that there exists a $t_2 > t_1$ with

$$\|\lambda(t + T, t, x) - \phi(T, x)\| \leq \frac{\gamma}{3} \quad \text{for all } t \geq t_2 \text{ and } x \in \underbrace{\mu(t) + C - x_0}_{\subset U_\delta(C)}.$$

Hence, for all $t \geq t_2$ and $x \in \mu(t) + C - x_0$, we have

$$
\begin{aligned}
&\|\lambda(t + T, t, x) - \mu(t + T)\| \\
&\leq \|\lambda(t + T, t, x) - \phi(T, x)\| + \|\phi(T, x) - x_0\| + \|x_0 - \mu(t + T)\| \\
&< \frac{\gamma}{3} + \frac{\gamma}{3} + \frac{\gamma}{3} = \gamma.
\end{aligned}
$$

Thus, for all $s \geq t_2$, we have the relation

$$
\begin{aligned}
&\lim_{t \to \infty} d\big(\lambda\big(t, s, \mu(s) + C - x_0\big) \big| \{\mu(s)\}\big) \\
&= \lim_{t \to \infty} d\big(\lambda\big(t, s + T, \underbrace{\lambda(s + T, t, \mu(s) + C - x_0)}_{\subset U_\gamma(\mu(s+T))}\big) \big| \{\mu(t)\}\big) \overset{(7.13)}{=} 0.
\end{aligned}
$$

This implies $y - x_0 \in \mathcal{A}_\mu^\rightarrow$, and since $y \in S^i$ has been chosen arbitrarily, we have $\mathcal{A}_\mu^\rightarrow \supset S^i - x_0$.

*Step 2.* $\mathcal{A}_\mu^\rightarrow \subset S^i - x_0$.

We choose $y \in S^o$ and $\beta > 0$ with $U_\beta(x_0) \subset S^i$ and define $\delta := d(y, S) > 0$. Due to Lemma 7.3 (i), there exists a $T > 0$ such that for all $x \in S^o$ with $d(x, S) \geq \min\{\delta/2, \eta/3\}$, there exists a $\hat{T} \in [0, T]$ with

$$d\big(\phi(\hat{T},x),S\big) \geq \eta \quad \text{and} \quad \phi(\hat{T},x) \in S^o. \tag{7.14}$$

Moreover, there exists a $t_1 > \tau$ with

$$\mu(t) \in U_{\min\{\delta/2,\eta/3,\beta/4\}}(x_0) \subset S^i \quad \text{for all } t \geq t_1.$$

Let $K$ be a compact and convex superset of $U_\eta(S)$. Then, because of Lemma 7.2 (i), there exists a $t_2 > t_1$ such that for all $\tilde{T} \in [0,T]$ and $x \in K$ with $\phi(t,x) \in K$ for all $t \in [0,\tilde{T}]$, we have

$$\big\|\lambda(\hat{t}+t,\hat{t},x)-\phi(t,x)\big\| \leq \min\left\{\frac{\beta}{2},\frac{\eta}{3}\right\} \quad \text{for all } \hat{t} \geq t_2 \text{ and } t \in [0,\tilde{T}]. \tag{7.15}$$

We argue negatively and suppose, there exists a $t_3 \geq t_2$ such that the relation

$$\lim_{t\to\infty} \big\|\lambda\big(t,t_3,y-x_0+\mu(t_3)\big) - \mu(t)\big\| = 0$$

holds. Therefore, since $\lim_{t\to\infty}\mu(t) = x_0$, there exists a $t_4 > t_3$ with

$$\lambda\big(t_4,t_3,y-x_0+\mu(t_3)\big) \in U_{\beta/2}(x_0). \tag{7.16}$$

We define

$$s := \max\left\{t \in [t_3,t_4] : d\big(\lambda(t,t_3,y-x_0+\mu(t_3)),S\big) \geq \min\left\{\frac{\eta}{2},\frac{\delta}{2}\right\}\right.$$

$$\text{and} \qquad \lambda(t,t_3,y-x_0+\mu(t_3)) \in S^o \Bigg\}.$$

This implies the relations $d\big(\lambda(s,t_3,y-x_0+\mu(t_3)),S\big) = \min\{\eta/2,\delta/2\}$ and $\lambda\big(s,t_3,y-x_0+\mu(t_3)\big) \in S^o$.
We distinguish two cases.
*Case 1.* $t_4 - s \leq T$.
*Case 1.1.* For all $t \in [0,t_4-s]$, we have $\phi\big(t,\lambda(s,t_3,y-x_0+\mu(t_3))\big) \in K$.
Due to (7.15), we have for all $t \in [0,t_4-s]$,

$$\big\|\phi\big(t,\lambda(s,t_3,y-x_0+\mu(t_3))\big) - \lambda\big(t+s,t_3,y-x_0+\mu(t_3)\big)\big\| \leq \frac{\beta}{2}.$$

Since $\phi\big(t,\lambda(s,t_3,y-x_0+\mu(t_3))\big) \in S^o$ for all $t \in [0,t_4-s]$, this leads to

$$\big\|\lambda(t_4,t_3,y-x_0+\mu(t_3)) - x_0\big\|$$
$$\geq \big\|\phi(t_4-s,\lambda(s,t_3,y-x_0+\mu(t_3))) - x_0\big\| -$$
$$\big\|\lambda(t_4,t_3,y-x_0+\mu(t_3)) - \phi(t_4-s,\lambda(s,t_3,y-x_0+\mu(t_3)))\big\|$$
$$\geq \beta - \frac{\beta}{2} = \frac{\beta}{2}.$$

This is a contradiction to (7.16).
*Case 1.2. There exists a $\tilde{t} \in [0, t_4 - s]$ with $\phi\big(\tilde{t}, \lambda\big(s, t_3, y - x_0 + \mu(t_3)\big)\big) \notin K$.*
By defining

$$\hat{s} := \inf\big\{t \in [0, t_4 - s] : \phi\big(t, \lambda\big(s, t_3, y - x_0 + \mu(t_3)\big)\big) \notin K\big\} > 0,$$

we have $d\big(\phi\big(\hat{s}, \lambda\big(s, t_3, y - x_0 + \mu(t_3)\big)\big), S\big) \geq \eta$. Due to (7.15), the relation

$$\big\|\lambda\big(\hat{s} + s, t_3, y - x_0 + \mu(t_3)\big) - \phi\big(\hat{s}, \lambda\big(s, t_3, y - x_0 + \mu(t_3)\big)\big)\big\| \leq \frac{\eta}{3}$$

holds. Hence, we have both $d\big(\lambda\big(\hat{s} + s, t_3, y - x_0 + \mu(t_3)\big), S\big) \geq 2\eta/3$ and $\lambda\big(\hat{s} + s, t_3, y - x_0 + \mu(t_3)\big) \in S^o$. This is a contradiction to the definition of $s$.
*Case 2. $t_4 - s > T$.*
*Case 2.1. For all $t \in [0, T]$, we have $\phi\big(t, \lambda\big(s, t_3, y - x_0 + \mu(t_3)\big)\big) \in K$.*
Because of $d\big(\lambda\big(s, t_3, y - x_0 + \mu(t_3)\big), S\big) = \min\{\delta/2, \eta/2\} \geq \min\{\delta/2, \eta/3\}$ and
(7.14), there exists a $\hat{T} \in [0, T]$ with

$$d\big(\phi\big(\hat{T}, \lambda\big(s, t_3, y - x_0 + \mu(t_3)\big)\big), S\big) \geq \eta,$$

and (7.15) yields

$$\big\|\phi\big(\hat{T}, \lambda\big(s, t_3, y - x_0 + \mu(t_3)\big)\big) - \lambda\big(\hat{T} + s, t_3, y - x_0 + \mu(t_3)\big)\big\| \leq \frac{\eta}{3}.$$

Together, this implies

$$d\big(\lambda\big(\hat{T} + s, t_3, y - x_0 + \mu(t_3)\big), S\big) \geq \frac{2}{3}\eta$$

and

$$\lambda\big(\hat{T} + s, t_3, y - x_0 + \mu(t_3)\big) \in S^o.$$

This is a contradiction to the definition of $s$.
*Case 2.2. There exists a $\tilde{t} \in [0, T]$ with $\phi\big(\tilde{t}, \lambda\big(s, t_3, y - x_0 + \mu(t_3)\big)\big) \notin K$.*
This case is treated analogously to Case 1.2 (by writing $T$ instead of $t_4 - s$).
Consequently, we have $y - x_0 \notin \mathcal{A}_\mu^\to$. This leads to the assertion, since $S = \partial S^i$,
int $S = \emptyset$ and $\mathcal{A}_\mu^\to$ is open.
*Step 3. For all $\kappa \leq \eta$, there exist $T > 0$ and $t_1 > 0$ such that for all $t_2 > t_1$
and $t > T$, we have*

$$\lambda\big(t_2, t_2 + t, U_{5\kappa/6}\big(\mathcal{A}_\mu^\to + \mu(t_2 + t)\big)\big) \subset U_{2\kappa/3}\big(\mathcal{A}_\mu^\to + \mu(t_2)\big).$$

We choose $\kappa \leq \eta$ arbitrarily. By applying Lemma 7.3 (ii), there exists a $\delta > 0$
with

$$\phi\big(-t, U_\delta\big(S^i\big)\big) \subset U_{\kappa/4}\big(S^i\big) \quad \text{for all } t \geq 0. \tag{7.17}$$

Due to the repulsivity of $S$, there exists a $T > 0$ with

$$\phi\big(-t, U_\kappa(S^i)\big) \subset U_{\delta/2}(S^i) \quad \text{for all } t > T.$$

By choosing $K$ as a convex and compact superset of $\bigcup_{t \in [0,T]} \phi\big(-t, U_\kappa(S^i)\big)$, we can apply Lemma 7.2 (ii), and we therefore get a $t_1 > \tau$ with

$$\|\lambda(t_2, t_2 + t, x) - \phi(-t, x)\| \leq \frac{\delta}{2} \quad \text{for all } x \in U_\kappa(S^i), t_2 > t_1 \text{ and } t \in [0, T]$$

$$\text{and } \|\mu(t) - x_0\| \leq \frac{\kappa}{6} \quad \text{for all } t > t_1. \tag{7.18}$$

Thus,

$$\lambda\big(t_2, t_2 + T, U_\kappa(S^i)\big) \subset U_\delta(S^i) \subset U_{\kappa/4}(S^i) \quad \text{for all } t_2 > t_1 \tag{7.19}$$

is fulfilled. Because of (7.17), this leads to

$$\phi\big(-t, \lambda(t_2 - T, t_2, U_\kappa(S^i))\big) \subset U_{\kappa/4}(S^i) \quad \text{for all } t_2 > t_1 \text{ and } t \geq 0.$$

Due to (7.18) and $\delta/2 < \kappa/4$, we have

$$\lambda\big(t_2, t_2 + t, U_\kappa(S^i)\big) \subset U_{\kappa/2}(S^i) \quad \text{for all } t_2 > t_1 \text{ and } t \in [T, 2T].$$

Suppose now that there exist $\hat{t} > 2T$ and $\hat{t}_2 < t_1$ with

$$d\big(\lambda(\hat{t}_2, \hat{t}_2 + \hat{t}, U_\kappa(S^i))\big| S^i\big) \geq \frac{\kappa}{2}.$$

We define

$$s := \inf \left\{ t > 2T : d\big(\lambda(\hat{t}_2, \hat{t}_2 + t, U_\kappa(S^i))\big| S^i\big) \geq \frac{\kappa}{2} \right\} > 2T$$

and set $t_2 := \hat{t}_2 + s - T > t_1$. Consequently,

$$\lambda\big(\hat{t}_2, \hat{t}_2 + s, U_\kappa(S^i)\big) = \lambda\big(\hat{t}_2, t_2, \underbrace{\lambda(t_2, \hat{t}_2 + s, U_\kappa(S^i))}_{\overset{(7.19)}{\subset} U_{\kappa/2}(S^i)}\big) \in U_{\kappa/4}(S^i)$$

holds. This is a contradiction, i.e., for all $t_2 < t_1$ and $t > T$, we have

$$\lambda\big(t_2, t_2 + t, U_\kappa(S^i)\big) \subset U_{\kappa/2}(S^i).$$

Since $\mu(t) \in U_{\kappa/6}(x_0)$ for all $t > t_1$, the relation

$$\lambda\big(t_2, t_2 + t, U_{5\kappa/6}(S^i + \mu(t_2 + t))\big) \subset U_{2\kappa/3}(S^i + \mu(t_2))$$

is fulfilled for all $t_2 > t_1$ and $t > T$.

*Step 4. Existence of the future repeller.*
Repeated usage of Step 3 implies that for all $\varepsilon > 0$, there exists an $s > 0$ such that for all $\tau \geq s$, there exists a $T > 0$ with

$$\lambda\left(\tau, \tau + t, U_{\eta/2}\left(\mathcal{A}_{\mu}^{\rightarrow} + \mu(\tau + t)\right)\right) \subset U_{\varepsilon}\left(\mathcal{A}_{\mu}^{\rightarrow} + \mu(\tau)\right) \quad \text{for all } t \geq T.$$

Because of Theorem 2.41 (iv), there exists an $s < \tau$ and a future repeller $R \subset (s, \infty) \times D$ with

$$\mathcal{A}_{\mu}^{\rightarrow} \subset \liminf_{t \to \infty} \left(R(t) - \mu(t)\right) \subset \limsup_{t \to \infty} \left(R(t) - \mu(t)\right) \subset \text{cls } \mathcal{A}_{\mu}^{\rightarrow}.$$

Since $\lim_{t \to \infty} \mu(t) = x_0$, we have

$$S^i \subset \liminf_{t \to \infty} R(t) \subset \limsup_{t \to \infty} R(t) \subset \text{cls } S^i.$$

This finishes the proof of this lemma. $\qquad\qquad\qquad\qquad\qquad\qquad \square$

Please note that similar lemmata can be derived for the determination of past (future, respectively) repulsion areas of past (future, respectively) repulsive solutions.

## 7.2 Bifurcations in Dimension One

In this section, one-dimensional differential equations are studied which exhibit pitchfork, transcritical or saddle node bifurcations. It is shown that under special assumptions, this bifurcation behavior is transferred to asymptotically autonomous systems.

Let $-\infty \leq x_- < x_+ \leq \infty$ and $\alpha_0 < \alpha_1$, and consider an autonomous differential equation

$$\dot{x} = g(x, \alpha) \tag{7.20}_\alpha$$

depending on a parameter $\alpha$ with a $C^1$-function $g : (x_-, x_+) \times (\alpha_0, \alpha_1] \to \mathbb{R}$. We assume that there exists an $x_0 \in (x_-, x_+)$ with

$$g(x_0, \alpha) = 0 \quad \text{and} \quad D_1 g(x_0, \alpha) \neq 0 \quad \text{for all } \alpha \in (\alpha_0, \alpha_1].$$

In the next four lemmata, conditions for the existence of nonautonomous counterparts for the equilibrium $x_0$ are studied. In a first instance, we restrict the parameter area to compact subintervals of $(\alpha_0, \alpha_1]$.

**Lemma 7.7 (Existence of past attractive solutions).** *Let $\alpha_- \leq \alpha_+$ be in $(\alpha_0, \alpha_1]$, and consider the nonautonomous differential equation*

$$\dot{x} = f(t, x, \alpha) \tag{7.21}_\alpha$$

*with a $C^1$-function $f : (-\infty, 0) \times (x_-, x_+) \times [\alpha_-, \alpha_+] \to \mathbb{R}$. We assume that*

$$\lim_{t \to -\infty} f(t, x, \alpha) = g(x, \alpha) \quad \text{and} \quad \lim_{t \to -\infty} D_2 f(t, x, \alpha) = D_1 g(x, \alpha)$$

*hold uniformly for all $x \in (x_-, x_+)$ and $\alpha \in [\alpha_-, \alpha_+]$. Furthermore, we suppose that*

$$D_1 g(x_0, \alpha) < 0 \quad \text{for all } \alpha \in [\alpha_-, \alpha_+].$$

*Then there exist a $\tau < 0$ and a continuous function $\mu : (-\infty, \tau] \times [\alpha_-, \alpha_+] \to \mathbb{R}$ such that $\mu(\cdot, \alpha)$ is the uniquely determined past attractive solution of $(7.21)_\alpha$ which fulfills*

$$\lim_{t \to -\infty} \mu(t, \alpha) = x_0.$$

*In addition, for fixed $\alpha \in [\alpha_-, \alpha_+]$, the following statements are fulfilled:*

(i) *If there exist $x_0^- < x_0$ and $x_0^+ > x_0$ with*

$$g(x_0^-, \alpha) = g(x_0^+, \alpha) = 0, \quad D_1 g(x_0^-, \alpha) > 0 \quad \text{and} \quad D_1 g(x_0^+, \alpha) > 0,$$

*and $g(x, \alpha) \neq 0$ for all $x \in (x_0^-, x_0) \cup (x_0, x_0^+)$, we have*

$$\mathcal{A}_{\mu(\cdot, \alpha)}^{\leftarrow} = (x_0^- - x_0, x_0^+ - x_0).$$

*Furthermore, there exists a past repeller $R(\alpha)$ of $(7.21)_\alpha$ with*

$$(x_0^-, x_0^+) \subset \liminf_{t \to -\infty} R(\alpha, t) \subset \limsup_{t \to -\infty} R(\alpha, t) \subset [x_0^-, x_0^+].$$

(ii) *If there exists an $x_0^- < x_0$ with*

$$g(x_0^-, \alpha) = 0 \quad \text{and} \quad D_1 g(x_0^-, \alpha) > 0,$$

*and $g(x, \alpha) \neq 0$ for all $x \in (x_0^-, x_0) \cup (x_0, x_+)$, we have*

$$\mathcal{A}_{\mu(\cdot, \alpha)}^{\leftarrow} = (x_0^- - x_0, x_+ - x_0).$$

(iii) *If there exists an $x_0^+ > x_0$ with*

$$g(x_0^+, \alpha) = 0 \quad \text{and} \quad D_1 g(x_0^+, \alpha) > 0,$$

*and $g(x, \alpha) \neq 0$ for all $x \in (x_-, x_0) \cup (x_0, x_0^+)$, we have*

$$\mathcal{A}_{\mu(\cdot, \alpha)}^{\leftarrow} = (x_- - x_0, x_0^+ - x_0).$$

*Remark 7.8.* The statement (i) of above lemma corresponds to the autonomous pitchfork bifurcation, where after the bifurcation, there are three equilibria, and (ii) and (iii) describe the situation after a transcritical or saddle node bifurcation.

*Proof (Lemma 7.7).* The proof is divided into three steps.
*Step 1.* There exist a $\tau < 0$ and a continuous function $\mu : (-\infty, \tau] \times [\alpha_-, \alpha_+] \to \mathbb{R}$ such that $\mu(\cdot, \alpha)$ is the uniquely determined past attractive solution of $(7.21)_\alpha$ which fulfills $\lim_{t \to -\infty} \mu(t, \alpha) = x_0$.

Due to the hypotheses (please note that $[\alpha_-, \alpha_+]$ is compact and $g$ is uniformly continuous on compact sets), there exist $\beta > 0, \gamma < 0$ and $\tau < 0$ with

$$f(t, x_0 - \beta, \alpha) > 0, \; f(t, x_0 + \beta, \alpha) < 0 \quad \text{and} \quad D_2 f(t, x, \alpha) \leq \gamma$$

for all $x \in U_{2\beta}(x_0)$, $t \leq \tau$ and $\alpha \in [\alpha_-, \alpha_+]$. We fix an $\alpha \in [\alpha_-, \alpha_+]$ for the rest of this step. The sets

$$M_1 := \big\{ x \in \text{cls}\, U_\beta(x_0) : \text{There exists a } t < \tau \text{ with } \lambda(t, \tau, x, \alpha) < x_0 - \beta \big\}$$

and

$$M_2 := \big\{ x \in \text{cls}\, U_\beta(x_0) : \text{There exists a } t < \tau \text{ with } \lambda(t, \tau, x, \alpha) > x_0 + \beta \big\}$$

are obviously nonempty and due to the continuity of the general solution (cf. Proposition A.3) relatively open in $\text{cls}\, U_\beta(x_0)$. Hence, we have $M_1 \cup M_2 \subsetneq \text{cls}\, U_\beta(x_0)$. Therefore, there exists a $y \in U_\beta(x_0)$ such that $\mu(t, \alpha) := \lambda(t, \tau, y, \alpha) \in U_\beta(x_0)$ for all $t \leq \tau$. To show that this solution is past attractive, we study the differential equation of the perturbed motion

$$\dot{x} = h(t, x, \alpha) := f\big(t, x + \mu(t, \alpha), \alpha\big) - f\big(t, \mu(t, \alpha), \alpha\big),$$

whose general solution will be denoted by $\tilde{\lambda}$. Due to the mean value theorem (see, e.g., LANG [102, Theorem 4.2, p. 341]), we have

$$h(t, x, \alpha) = x \int_0^1 D_2 h(t, \theta x, \alpha)\, d\theta = x \int_0^1 D_2 f\big(t, \theta x + \mu(t, \alpha), \alpha\big)\, d\theta.$$

This implies

$$h(t, x, \alpha) \geq \gamma x \quad \text{for all } t \leq \tau \text{ and } x \in (-\beta, 0)$$

$$\text{and} \quad h(t, x, \alpha) \leq \gamma x \quad \text{for all } t \leq \tau \text{ and } x \in (0, \beta).$$

We therefore obtain

$$\lim_{t \to -\infty} d\big(\tilde{\lambda}\big(\tau, t, U_{\beta/2}(0), \alpha\big) \big| \{0\}\big) = 0,$$

and consequently,

$$\lim_{t \to -\infty} d\big(\lambda\big(\tau, t, U_{\beta/2}(\mu(t, \alpha)), \alpha\big) \big| \{\mu(\tau, \alpha)\}\big) = 0$$

holds (cf. Proposition A.7). Thus, the solution $\mu(\cdot, \alpha)$ is past attractive. Moreover, the limit relation $\lim_{t \to -\infty} \mu(t, \alpha) = x_0$ is obviously fulfilled. The uniqueness of $\mu(\cdot, \alpha)$ follows directly from Proposition 2.37 (i).

*Step 2. $\mu$ is continuous.*

First, we consider the function $d : [\alpha_-, \alpha_+] \to (x_-, x_+)$, defined by

$$d(\alpha) := \mu(\tau, \alpha) \quad \text{for all } \alpha \in [\alpha_-, \alpha_+].$$

We suppose that there exist a $\tilde{\alpha} \in [\alpha_-, \alpha_+]$ and a sequence $\{\tilde{\alpha}_n\}_{n \in \mathbb{N}}$ with $\lim_{n \to \infty} \tilde{\alpha}_n = \tilde{\alpha}$ such that $\{d(\tilde{\alpha}_n)\}_{n \in \mathbb{N}}$ does not converge to $d(\tilde{\alpha})$. Since this sequence is bounded, we assume w.l.o.g. that it is convergent with the limit $\tilde{x} \in \operatorname{cls} U_\beta(x_0)$, $\tilde{x} \neq d(\tilde{\alpha})$. Because of Step 1, there exists a $\tilde{t} < \tau$ such that $\lambda(\tilde{t}, \tau, \tilde{x}, \tilde{\alpha}) \notin \operatorname{cls} U_\beta(x_0)$. The continuity of the general solution implies the existence of a neighborhood $V$ of $(\tilde{x}, \tilde{\alpha})$ such that

$$\lambda(\tilde{t}, \tau, x, \alpha) \notin \operatorname{cls} U_\beta(x_0) \quad \text{for all } (x, \alpha) \in V.$$

In particular, there exists an $n \in \mathbb{N}$ with $(d(\tilde{\alpha}_n), \tilde{\alpha}_n) \in V$. This implies

$$\mu(\tilde{t}, \tilde{\alpha}_n) = \lambda(\tilde{t}, \tau, \mu(\tau, \tilde{\alpha}_n), \tilde{\alpha}_n) = \lambda(\tilde{t}, \tau, d(\tilde{\alpha}_n), \tilde{\alpha}_n) \notin \operatorname{cls} U_\beta(x_0).$$

This is a contradiction, and therefore, the function $d$ is continuous. To prove the continuity of $\mu$, we choose a sequence $\{(\hat{t}_n, \hat{\alpha}_n)\}_{n \in \mathbb{N}}$ in $(-\infty, \tau] \times [\alpha_-, \alpha_+]$ with $\lim_{n \to \infty} (\hat{t}_n, \hat{\alpha}_n) = (\hat{t}, \hat{\alpha})$. The continuity of $\mu$ follows from

$$\begin{aligned}
\lim_{n \to \infty} \mu(\hat{t}_n, \hat{\alpha}_n) &= \lim_{n \to \infty} \lambda(\hat{t}_n, \tau, \mu(\tau, \hat{\alpha}_n), \hat{\alpha}_n) \\
&= \lim_{n \to \infty} \lambda(\hat{t}_n, \tau, d(\hat{\alpha}_n), \hat{\alpha}_n) \\
&= \lambda(\hat{t}, \tau, d(\hat{\alpha}), \hat{\alpha}) = \mu(\hat{t}, \hat{\alpha}).
\end{aligned}$$

*Step 3. The statements (i), (ii) and (iii) are fulfilled.*
The asserted relations for $\mathcal{A}_{\mu(\cdot, \alpha)}^{\leftarrow}$ and the existence of a past repeller follow directly from Lemma 7.5 if we define the repulsive set $S$ as $\{x_0^-, x_0^+\}$ in case (i), $\{x_0^-\}$ in case (ii) or $\{x_0^+\}$ in case (iii), respectively (cf. also Figure 7.1).

□

**Lemma 7.9 (Existence of future attractive solutions).** *Let $\alpha_- \leq \alpha_+$ be in $(\alpha_0, \alpha_1]$, and consider the nonautonomous differential equation*

$$\dot{x} = f(t, x, \alpha) \tag{7.22}_\alpha$$

*with a $C^1$-function $f : (0, \infty) \times (x_-, x_+) \times [\alpha_-, \alpha_+] \to \mathbb{R}$. We assume that*

$$\lim_{t \to \infty} f(t, x, \alpha) = g(x, \alpha) \quad \text{and} \quad \lim_{t \to \infty} D_2 f(t, x, \alpha) = D_1 g(x, \alpha)$$

*hold uniformly for all $x \in (x_-, x_+)$ and $\alpha \in [\alpha_-, \alpha_+]$. Furthermore, we suppose that*

$$D_1 g(x_0, \alpha) < 0 \quad \text{for all } \alpha \in [\alpha_-, \alpha_+].$$

*Then there exist $\tau > 0$ and $\beta > 0$ such that every solution $\lambda(\cdot, \tau, x, \alpha)$ for $x \in U_\beta(x_0)$ and $\alpha \in [\alpha_-, \alpha_+]$ is future attractive with*

$$\lim_{t \to \infty} \lambda(t, \tau, x, \alpha) = x_0.$$

*Let $\nu : [\tau, \infty) \to \mathbb{R}$ be such a solution of $(7.22)_\alpha$ for fixed $\alpha \in [\alpha_-, \alpha_+]$. Then the following statements are fulfilled:*

(i) *If there exist $x_0^- < x_0$ and $x_0^+ > x_0$ with*

$$g(x_0^-, \alpha) = g(x_0^+, \alpha) = 0, \quad D_1 g(x_0^-, \alpha) > 0 \quad and \quad D_1 g(x_0^+, \alpha) > 0,$$

*and $g(x, \alpha) \neq 0$ for all $x \in (x_0^-, x_0) \cup (x_0, x_0^+)$, we have*

$$\mathcal{A}_\nu^\rightarrow = (x_0^- - x_0, x_0^+ - x_0).$$

*Furthermore, there exists a future repeller $R(\alpha)$ of $(7.22)_\alpha$ with*

$$(x_0^-, x_0^+) \subset \liminf_{t \to \infty} R(\alpha, t) \subset \limsup_{t \to \infty} R(\alpha, t) \subset [x_0^-, x_0^+].$$

(ii) *If there exists an $x_0^- < x_0$ with*

$$g(x_0^-, \alpha) = 0 \quad and \quad D_1 g(x_0^-, \alpha) > 0,$$

*and $g(x, \alpha) \neq 0$ for all $x \in (x_0^-, x_0) \cup (x_0, x_+)$, we have*

$$\mathcal{A}_\nu^\rightarrow = (x_0^- - x_0, x_+ - x_0).$$

(iii) *If there exists an $x_0^+ > x_0$ with*

$$g(x_0^+, \alpha) = 0 \quad and \quad D_1 g(x_0^+, \alpha) > 0,$$

*and $g(x, \alpha) \neq 0$ for all $x \in (x_-, x_0) \cup (x_0, x_0^+)$, we have*

$$\mathcal{A}_\nu^\rightarrow = (x_- - x_0, x_0^+ - x_0).$$

*Proof.* Due to the hypotheses (please note that $[\alpha_-, \alpha_+]$ is compact and $g$ is uniformly continuous on compact sets), there exist $\beta > 0, \gamma < 0$ and $\tau > 0$ with

$$f(t, x_0 - \beta, \alpha) > 0, \ f(t, x_0 + \beta, \alpha) < 0 \quad and \quad D_2 f(t, x, \alpha) \leq \gamma$$

for all $x \in U_{2\beta}(x_0)$, $t \geq \tau$ and $\alpha \in [\alpha_-, \alpha_+]$. We fix $\hat{x} \in U_\beta(x_0)$ and $\alpha \in [\alpha_-, \alpha_+]$ and consider for the rest of the proof in particular the solution $\nu(\cdot) := \lambda(\cdot, \tau, \hat{x}, \alpha)$ of $(7.22)_\alpha$ on the interval $[\tau, \infty)$. Obviously, the relation $\lim_{t \to \infty} \nu(t) = x_0$ holds. To show that this solution is future attractive, we study the differential equation of the perturbed motion

$$\dot{x} = h(t, x, \alpha) := f(t, x + \nu(t), \alpha) - f(t, \nu(t), \alpha),$$

whose general solution will be denoted by $\tilde{\lambda}$. Due to the mean value theorem, we have

$$h(t, x, \alpha) = x \int_0^1 D_2 h(t, \theta x, \alpha) \, d\theta = x \int_0^1 D_2 f(t, \theta x + \nu(t), \alpha) \, d\theta.$$

This implies

$$h(t, x, \alpha) \geq \gamma x \quad \text{for all } t \geq \tau \text{ and } x \in (-\beta, 0)$$
$$\text{and} \quad h(t, x, \alpha) \leq \gamma x \quad \text{for all } t \geq \tau \text{ and } x \in (0, \beta).$$

We therefore obtain

$$\lim_{t \to \infty} d\big(\tilde{\lambda}(t, s, U_{\beta/2}(0), \alpha)\big|\{0\}\big) = 0 \quad \text{for all } s \geq \tau,$$

and consequently,

$$\lim_{t \to \infty} d\big(\lambda\big(t, s, U_{\beta/2}(\nu(s)), \alpha\big)\big|\{\nu(t)\}\big) = 0 \quad \text{for all } s \geq \tau$$

holds (cf. Proposition A.7). Thus, the solution $\nu$ is future attractive. The asserted relations for $\mathcal{A}_\nu^{\to}$ and the existence of a future repeller follow directly from Lemma 7.6 if we define the repulsive set $S$ as $\{x_0^-, x_0^+\}$ in case (i), $\{x_0^-\}$ in case (ii) or $\{x_0^+\}$ in case (iii), respectively (cf. also Figure 7.1). $\qquad\square$

Under the assumption $D_1 g(x_0, \alpha) > 0$ for all $\alpha \in [\alpha_-, \alpha_+]$, analogous statements are obtained for past (future, respectively) repulsive solutions.

**Lemma 7.10 (Existence of past repulsive solutions).** *Let $\alpha_- \leq \alpha_+$ be in $(\alpha_0, \alpha_1]$, and consider the nonautonomous differential equation*

$$\dot{x} = f(t, x, \alpha) \qquad (7.23)_\alpha$$

*with a $C^1$-function $f : (-\infty, 0) \times (x_-, x_+) \times [\alpha_-, \alpha_+] \to \mathbb{R}$. We assume that*

$$\lim_{t \to -\infty} f(t, x, \alpha) = g(x, \alpha) \quad \text{and} \quad \lim_{t \to -\infty} D_2 f(t, x, \alpha) = D_1 g(x, \alpha)$$

*hold uniformly for all $x \in (x_-, x_+)$ and $\alpha \in [\alpha_-, \alpha_+]$. Furthermore, we suppose that*

$$D_1 g(x_0, \alpha) > 0 \quad \text{for all } \alpha \in [\alpha_-, \alpha_+].$$

*Then there exist $\tau < 0$ and $\beta > 0$ such that every solution $\lambda(\cdot, \tau, x, \alpha)$ for $x \in U_\beta(x_0)$ and $\alpha \in [\alpha_-, \alpha_+]$ is past repulsive with*

$$\lim_{t \to -\infty} \lambda(t, \tau, x, \alpha) = x_0.$$

*Let $\nu : (-\infty, \tau] \to \mathbb{R}$ be such a solution of $(7.23)_\alpha$ for fixed $\alpha \in [\alpha_-, \alpha_+]$. Then the following statements are fulfilled:*

*(i) If there exist $x_0^- < x_0$ and $x_0^+ > x_0$ with*

$$g\big(x_0^-, \alpha\big) = g\big(x_0^+, \alpha\big) = 0, \quad D_1 g\big(x_0^-, \alpha\big) < 0 \quad \text{and} \quad D_1 g\big(x_0^+, \alpha\big) < 0,$$

*and $g(x, \alpha) \neq 0$ for all $x \in (x_0^-, x_0) \cup (x_0, x_0^+)$, we have*

$$\mathcal{R}_\nu^\leftarrow = \left(x_0^- - x_0, x_0^+ - x_0\right).$$

*Furthermore, there exists a past attractor $A(\alpha)$ of $(7.23)_\alpha$ with*

$$\left(x_0^-, x_0^+\right) \subset \liminf_{t \to -\infty} A(\alpha, t) \subset \limsup_{t \to -\infty} A(\alpha, t) \subset \left[x_0^-, x_0^+\right].$$

(ii) *If there exists an $x_0^- < x_0$ with*

$$g\left(x_0^-, \alpha\right) = 0 \quad and \quad D_1 g\left(x_0^-, \alpha\right) < 0,$$

*and $g(x, \alpha) \neq 0$ for all $x \in \left(x_0^-, x_0\right) \cup (x_0, x_+)$, we have*

$$\mathcal{R}_\nu^\leftarrow = \left(x_0^- - x_0, x_+ - x_0\right).$$

(iii) *If there exists an $x_0^+ > x_0$ with*

$$g\left(x_0^+, \alpha\right) = 0 \quad and \quad D_1 g\left(x_0^+, \alpha\right) < 0,$$

*and $g(x, \alpha) \neq 0$ for all $x \in \left(x_-, x_0\right) \cup \left(x_0, x_0^+\right)$, we have*

$$\mathcal{R}_\nu^\leftarrow = \left(x_- - x_0, x_0^+ - x_0\right).$$

*Proof.* The assertions follow from Proposition 2.32 and Lemma 7.9.     □

**Lemma 7.11 (Existence of future repulsive solutions).** *Let $\alpha_- \leq \alpha_+$ be in $(\alpha_0, \alpha_1]$, and consider the nonautonomous differential equation*

$$\dot{x} = f(t, x, \alpha) \tag{7.24$_\alpha$}$$

*with a $C^1$-function $f : (0, \infty) \times (x_-, x_+) \times [\alpha_-, \alpha_+] \to \mathbb{R}$. We assume that*

$$\lim_{t \to \infty} f(t, x, \alpha) = g(x, \alpha) \quad and \quad \lim_{t \to \infty} D_2 f(t, x, \alpha) = D_1 g(x, \alpha)$$

*hold uniformly for all $x \in (x_-, x_+)$ and $\alpha \in [\alpha_-, \alpha_+]$. Furthermore, we suppose that*

$$D_1 g(x_0, \alpha) > 0 \quad for \ all \ \alpha \in [\alpha_-, \alpha_+].$$

*Then there exist a $\tau > 0$ and a continuous function $\mu : [\tau, \infty) \times [\alpha_-, \alpha_+] \to \mathbb{R}$ such that $\mu(\cdot, \alpha)$ is the uniquely determined future repulsive solution of $(7.24)_\alpha$ which fulfills*

$$\lim_{t \to \infty} \mu(t, \alpha) = x_0.$$

*In addition, for fixed $\alpha \in [\alpha_-, \alpha_+]$, the following statements are fulfilled:*

(i) *If there exist $x_0^- < x_0$ and $x_0^+ > x_0$ with*

$$g\left(x_0^-, \alpha\right) = g\left(x_0^+, \alpha\right) = 0, \quad D_1 g\left(x_0^-, \alpha\right) < 0 \quad and \quad D_1 g\left(x_0^+, \alpha\right) < 0,$$

*and $g(x,\alpha) \neq 0$ for all $x \in \left(x_0^-, x_0\right) \cup \left(x_0, x_0^+\right)$, we have*

$$\mathcal{R}_{\overrightarrow{\mu(\cdot,\alpha)}} = \left(x_0^- - x_0, x_0^+ - x_0\right).$$

*Furthermore, there exists a future attractor $A(\alpha)$ of $(7.24)_\alpha$ with*

$$\left(x_0^-, x_0^+\right) \subset \liminf_{t\to\infty} A(\alpha,t) \subset \limsup_{t\to\infty} A(\alpha,t) \subset \left[x_0^-, x_0^+\right].$$

(ii) *If there exists an $x_0^- < x_0$ with*

$$g\left(x_0^-, \alpha\right) = 0 \quad and \quad D_1 g\left(x_0^-, \alpha\right) < 0,$$

*and $g(x,\alpha) \neq 0$ for all $x \in \left(x_0^-, x_0\right) \cup \left(x_0, x_+\right)$, we have*

$$\mathcal{R}_{\overrightarrow{\mu(\cdot,\alpha)}} = \left(x_0^- - x_0, x_+ - x_0\right).$$

(iii) *If there exists an $x_0^+ > x_0$ with*

$$g\left(x_0^+, \alpha\right) = 0 \quad and \quad D_1 g\left(x_0^+, \alpha\right) < 0,$$

*and $g(x,\alpha) \neq 0$ for all $x \in \left(x_-, x_0\right) \cup \left(x_0, x_0^+\right)$, we have*

$$\mathcal{R}_{\overrightarrow{\mu(\cdot,\alpha)}} = \left(x_- - x_0, x_0^+ - x_0\right).$$

*Proof.* The assertions follow from Proposition 2.32 and Lemma 7.7.    □

In the following, we observe that pitchfork bifurcations of $(7.20)_\alpha$ give rise to total nonautonomous bifurcations. Transcritical and saddle node bifurcations, however, lead to partial nonautonomous bifurcations.

First, the attention is restricted to the situation that the autonomous differential equation $(7.20)_\alpha$ admits a supercritical pitchfork bifurcation at $(x_0, \alpha_0)$. More precisely, there exist a monotone increasing continuous function $h_1 : (\alpha_0, \alpha_1] \to (x_-, x_+)$ and a monotone decreasing continuous function $h_2 : (\alpha_0, \alpha_1] \to (x_-, x_+)$ such that for all $\alpha \in (\alpha_0, \alpha_1]$, we have

$$h_1(\alpha) < x_0 < h_2(\alpha),$$
$$g(h_1(\alpha), \alpha) = g(x_0, \alpha) = g(h_2(\alpha), \alpha) = 0,$$
$$D_1 g(h_1(\alpha), \alpha) \neq 0, \; D_1 g(x_0, \alpha) \neq 0, \; D_1 g(h_2(\alpha), \alpha) \neq 0.$$

Moreover, for all $\alpha \in (\alpha_0, \alpha_1]$ and $x \in (h_1(\alpha), x_0) \cup (x_0, h_2(\alpha))$, $g(x,\alpha) \neq 0$ is satisfied, and we have $\lim_{\alpha\to\alpha_0} h_1(\alpha) = \lim_{\alpha\to\alpha_0} h_2(\alpha) = x_0$.

**Theorem 7.12 (Total past bifurcation).** *We suppose that $(7.20)_\alpha$ admits a pitchfork bifurcation as described above and consider the nonautonomous differential equation*

$$\dot{x} = f(t, x, \alpha) \tag{$7.25)_\alpha$}$$

with a $C^1$-function $f : (-\infty, 0) \times (x_-, x_+) \times (\alpha_0, \alpha_1] \to \mathbb{R}$. We assume that

$$\lim_{t \to -\infty} f(t, x, \alpha) = g(x, \alpha) \quad and \quad \lim_{t \to -\infty} D_2 f(t, x, \alpha) = D_1 g(x, \alpha)$$

hold uniformly for all $x \in (x_-, x_+)$ and $\alpha \in (\alpha_0, \alpha_1]$. Then the following statements are fulfilled:

(i) If $D_1 g(x_0, \alpha_1) < 0$ is satisfied, then there exists a continuous function $\mu : D \subset \mathbb{R} \times (\alpha_0, \alpha_1] \to \mathbb{R}$ such that $\mu(\cdot, \alpha)$ is a past attractive solution of $(7.25)_\alpha$. We have a total past bifurcation, since

$$\lim_{\alpha \searrow \alpha_0} d\left(\mathcal{A}^-_{\mu(\cdot,\alpha)} \big| \{0\}\right) = 0 \,.$$

Furthermore, for all $\alpha \in (\alpha_0, \alpha_1]$, there exists a past repeller $R(\alpha)$. Due to

$$\lim_{\alpha \searrow \alpha_0} d\left(\limsup_{t \to -\infty} R(\alpha, t) \big| \{x_0\}\right) = 0 \,,$$

we also have a past repeller transition.

(ii) If $D_1 g(x_0, \alpha_1) > 0$ is satisfied, then there exists a continuous function $\mu : D \subset \mathbb{R} \times (\alpha_0, \alpha_1] \to \mathbb{R}$ such that $\mu(\cdot, \alpha)$ is a past repulsive solution of $(7.25)_\alpha$. We have a total past bifurcation, since

$$\lim_{\alpha \searrow \alpha_0} d\left(\mathcal{R}^-_{\mu(\cdot,\alpha)} \big| \{0\}\right) = 0 \,.$$

Furthermore, for all $\alpha \in (\alpha_0, \alpha_1]$, there exists a past attractor $A(\alpha)$. Due to

$$\lim_{\alpha \searrow \alpha_0} d\left(\limsup_{t \to -\infty} A(\alpha, t) \big| \{x_0\}\right) = 0 \,,$$

we also have a past attractor transition.

*Proof.* We define the compact intervals $I_0 := \left\{\alpha \in (\alpha_0, \alpha_1] : h_1(\alpha) \le x_0 - 1\right\}$ and

$$I_n := \left\{\alpha \in (\alpha_0, \alpha_1] : h_1(\alpha) \in \left[x_0 - \frac{1}{n}, x_0 - \frac{1}{n+1}\right]\right\} \quad \text{for all } n \in \mathbb{N}.$$

(i) For all $n \in \mathbb{N}_0$, we restrict $(7.25)_\alpha$ to the parameter area $I_n$ and apply Lemma 7.7. Then there exists a continuous function $\mu_n : (-\infty, \tau_n] \times I_n \to \mathbb{R}$ which describes uniquely determined past attractive solutions. We define

$$\mu(t, \alpha) := \mu_n(t, \alpha) \quad \text{for all } t < 0 \text{ and } \alpha \in (\alpha_0, \alpha_1] \text{ with } \alpha \in I_n \text{ and } t \le \tau_n \,.$$

Due to the uniqueness of the $\mu_n$, the so-defined function $\mu : D \to \mathbb{R}$ for some $D \subset \mathbb{R} \times (\alpha_0, \alpha_1]$ is well defined, and the continuity of $\mu$ follows directly. The existence of the past repellers and the limit relations are consequences of

Lemma 7.7 (i).

(ii) For all $n \in \mathbb{N}_0$, we restrict $(7.25)_\alpha$ to the parameter area $I_n$ and apply Lemma 7.10. It is obvious that one can construct a continuous function $\mu : D \subset \mathbb{R} \times (\alpha_0, \alpha_1) \to \mathbb{R}$ which describes past repulsive solutions. The existence of the past attractors and the limit relations are consequences of Lemma 7.10 (i). $\qquad\square$

**Theorem 7.13 (Total future bifurcation).** *We suppose that* $(7.20)_\alpha$ *admits a pitchfork bifurcation as described above and consider the nonautonomous differential equation*

$$\dot{x} = f(t, x, \alpha) \qquad (7.26)_\alpha$$

*with a $C^1$-function $f : (0, \infty) \times (x_-, x_+) \times (\alpha_0, \alpha_1] \to \mathbb{R}$. We assume that*

$$\lim_{t\to\infty} f(t, x, \alpha) = g(x, \alpha) \quad and \quad \lim_{t\to\infty} D_2 f(t, x, \alpha) = D_1 g(x, \alpha)$$

*hold uniformly for all $x \in (x_-, x_+)$ and $\alpha \in (\alpha_0, \alpha_1]$. Then the following statements are fulfilled:*

(i) *If $D_1 g(x_0, \alpha_1) < 0$ is satisfied, then there exists a continuous function $\mu : D \subset \mathbb{R} \times (\alpha_0, \alpha_1] \to \mathbb{R}$ such that $\mu(\cdot, \alpha)$ is a future attractive solution of $(7.26)_\alpha$. We have a total future bifurcation, since*

$$\lim_{\alpha\searrow\alpha_0} d\left(\mathcal{A}_{\mu(\cdot,\alpha)}^{\to}\big|\{0\}\right) = 0.$$

*Furthermore, for all $\alpha \in (\alpha_0, \alpha_1]$, there exists a future repeller $R(\alpha)$. Due to*

$$\lim_{\alpha\searrow\alpha_0} d\left(\limsup_{t\to\infty} R(\alpha; t)\big|\{x_0\}\right) = 0,$$

*we also have a future repeller transition.*

(ii) *If $D_1 g(x_0, \alpha_1) > 0$ is satisfied, then there exists a continuous function $\mu : D \subset \mathbb{R} \times (\alpha_0, \alpha_1] \to \mathbb{R}$ such that $\mu(\cdot, \alpha)$ is a future repulsive solution of $(7.26)_\alpha$. We have a total future bifurcation, since*

$$\lim_{\alpha\searrow\alpha_0} d\left(\mathcal{R}_{\mu(\cdot,\alpha)}^{\to}\big|\{0\}\right) = 0.$$

*Furthermore, for all $\alpha \in (\alpha_0, \alpha_1]$, there exists a future attractor $A(\alpha)$. Due to*

$$\lim_{\alpha\searrow\alpha_0} d\left(\limsup_{t\to\infty} A(\alpha, t)\big|\{x_0\}\right) = 0,$$

*we also have a future attractor transition.*

*Proof.* See proof of Theorem 7.12. $\qquad\square$

In the following example, which is a special case of PÖTZSCHE & RASMUSSEN [142, Example 4.1], the center manifold reduction (see, e.g., CARR [38] and HALE [75, Chapter 4] in the autonomous context) is used to verify past and future bifurcations and transitions in an asymptotically autonomous version of the Lorenz system.

*Example 7.14.* We consider a nonautonomous version of the famous Lorenz equation (see, e.g., LORENZ [108] and KUZNETSOW [101, pp. 166, 249]), given by the three-dimensional system

$$\dot{x}_1 = \sigma_\alpha(t)(x_2 - x_1)$$
$$\dot{x}_2 = \rho_\alpha(t)x_1 - x_2 - x_1 x_3 \ .$$
$$\dot{x}_3 = -\beta_\alpha(t)x_3 + x_1 x_2$$

In our situation, $\sigma_\alpha, \rho_\alpha, \beta_\alpha$ are perturbed nonautonomously, i.e., we assume that the functions $\sigma_\alpha, \rho_\alpha, \beta_\alpha : \mathbb{R} \to (0, \infty)$ are given by

$$\sigma_\alpha(t) := \sigma_0 + \alpha\sigma(t), \quad \rho_\alpha(t) := 1 + \rho_0 + \alpha\rho(t), \quad \beta_\alpha(t) := \beta_0 + \alpha\beta(t)$$

with real constants $\sigma_0, \rho_0, \beta_0 > 0$, bounded $C^3$-functions $\sigma, \rho, \beta$ and $\alpha \in \mathbb{R}$, which will be the bifurcation parameter. It is our goal to study the stability of the equilibrium $x = 0$ for different values of $\alpha$. From the linearization of the trivial equilibrium, which is given by

$$\begin{pmatrix} -\sigma_0 & \sigma_0 & 0 \\ \rho_0 & -1 & 0 \\ 0 & 0 & -\beta_0 \end{pmatrix},$$

we see that in case $\alpha = 0$ (i.e., in case of the autonomous Lorenz system), the origin is attractive for $\rho_0 < 0$ and repulsive for $\rho_0 > 0$. More interesting is the nonhyperbolic case $\rho_0 = 0$, where an autonomous pitchfork bifurcation occurs as $\rho_0$ passes through 0 (see KUZNETSOW [101, p. 249]). To mimic this situation, we assume $\rho_0 = 0$ from now on. Before proceeding, we formally append the trivial equation $\dot{\alpha} = 0$ and—to simplify our calculations—apply the transformation

$$\begin{pmatrix} y_1 \\ y_2 \\ y_3 \\ y_4 \end{pmatrix} := \begin{pmatrix} -\sigma_0 & 0 & 1 & 0 \\ 1 & 0 & 1 & 0 \\ 0 & 1 & 0 & 0 \\ 0 & 0 & 0 & 1 \end{pmatrix} \begin{pmatrix} x_1 \\ x_2 \\ x_3 \\ \alpha \end{pmatrix}.$$

This implies the system

$$\dot{y} = Ay + F(t, y) \tag{7.27}$$

with $A := \mathrm{diag}(-\sigma_0 - 1, -\beta_0, 0, 0)$ and the nonlinearity

$$
F(t,y) := \begin{pmatrix} \frac{\sigma_0}{\sigma_0+1}y_1y_2 - \frac{\sigma(t)+\sigma_0(\sigma(t)+\rho(t))}{\sigma_0+1}y_1y_4 - \frac{1}{\sigma_0+1}y_2y_3 + \frac{\rho(t)}{\sigma_0+1}y_3y_4 \\ -\sigma_0 y_1^2 + (1-\sigma_0)y_1y_3 - \beta(t)y_2y_4 + y_3^2 + 2y_4^2 \\ \frac{\sigma_0^2}{\sigma_0+1}y_1y_2 + \frac{\sigma(t)+\sigma_0(\sigma(t)-\sigma_0\rho(t))}{\sigma_0+1}y_1y_4 - \frac{\sigma_0}{\sigma_0+1}y_2y_3 + \frac{\sigma_0\rho(t)}{\sigma_0+1}y_3y_4 \\ 0 \end{pmatrix}.
$$

Thus, we can apply Theorem 5.3 to (7.27) to show that there exists a local two-dimensional all-time center-unstable manifold $\mathcal{S}^-$, given as graph of a function $s^- : V \times \mathbb{R} \to \mathbb{R}^2$, where $V \subset \mathbb{R}^2$ is a neighborhood of 0. The ansatz

$$
s^-(y_3, y_4, t) = \sum_{i=0}^{2} y_3^{2-i} y_4^i \begin{pmatrix} s_{2-i,i}^1(t) \\ s_{2-i,i}^2(t) \end{pmatrix} + O\left(\sqrt{y_3^2 + y_4^2}^3\right)
$$

yields that the equation reduced to the all-time center-unstable manifold $\mathcal{S}^-$ is given by

$$
\dot{y}_3 = \frac{\sigma_0}{\sigma_0+1}\alpha\rho(t)y_3 - s_{2,0}^2(t)y_3^3 + O\left(\alpha y_3^2, \alpha^2 y_3, y_3^3\right).
$$

Using PÖTZSCHE & RASMUSSEN [142, Theorem 3.1], we obtain $s_{2,0}^2(t) \equiv 1/\beta_0$, and consequently, the one-dimensional bifurcation equation is given by

$$
\dot{y}_3 = \frac{\sigma_0}{\sigma_0+1}\alpha\rho(t)y_3 - \tfrac{1}{\beta_0}y_3^3 + O\left(\alpha y_3^2, \alpha^2 y_3, y_3^3\right). \tag{7.28}
$$

We henceforth assume that our system is past (future, respectively) asymptotically autonomous, i.e., the limits $t \to \pm\infty$ of the functions $\sigma, \rho$ and $\beta$ exist. We define

$$
\sigma^\pm := \lim_{t\to\pm\infty} \sigma(t), \quad \rho^\pm := \lim_{t\to\pm\infty} \rho(t) \quad \text{and} \quad \beta^\pm := \lim_{t\to\pm\infty} \beta(t).
$$

The autonomous limiting equations of the bifurcation equation are then given by

$$
\dot{y}_3 = \frac{\sigma_0}{\sigma_0+1}\alpha\rho^\pm y_3 - \tfrac{1}{\beta_0}y_3^3 + O\left(\alpha y_3^2, \alpha^2 y_3, y_3^3\right). \tag{7.29}
$$

It is easy to check that this equation admits a pitchfork bifurcation, i.e., the equilibrium 0 is attractive for $\alpha < 0$ and repulsive for $\alpha > 0$. For small $\alpha > 0$, there are two additional attractive equilibria branching from the origin. One can show that the convergence of the right hand side in (7.28) is uniform in a neighborhood of 0, and also the derivative with respect to $y_3$ of (7.28) converges locally uniformly to the corresponding derivative in (7.29). Thus, Theorem 7.12 or Theorem 7.13, respectively, is applicable, and therefore, system (7.28) admits a total past or future bifurcation and a past or future attractor transition, respectively. Please note that not only the reduced equation (7.28) gives rise to a nonautonomous transition but also the nonautonomous Lorenz equation itself. This is due to the fact that there exists an asymptotic phase for the center manifold (see AULBACH & WANNER [22, Theorem 3.3]), i.e., every solution approaches a solution lying in the center manifold in forward time exponentially. Therefore, for small $\alpha > 0$, there also exists a past or future attractor of the three-dimensional system, respectively, which shrinks down in the limit $\alpha \searcol 0$. However, the three-dimensional nonautonomous Lorenz equation does not admit a past or future bifurcation, since due to the asymptotic phase, the trivial solution is not past repulsive for $\alpha > 0$.

To obtain partial nonautonomous bifurcations, we assume that the differential equation $(7.20)_\alpha$ admits a supercritical transcritical or saddle node bifurcation at $(x_0, \alpha_0)$. This means, there exists a strictly increasing continuous function $h : (\alpha_0, \alpha_1) \to (x_-, x_+)$ such that for all $\alpha \in (\alpha_0, \alpha_1)$, we have

$$h(\alpha) < x_0 \,,$$
$$g(h(\alpha), \alpha) = g(x_0, \alpha) = 0 \,,$$
$$D_1 g(h(\alpha), \alpha) \neq 0, \ D_1 g(x_0, \alpha) \neq 0 \,.$$

Moreover, for all $\alpha \in (\alpha_0, \alpha_1)$ and $x \in (h(\alpha), x_0)$, the relation $g(x, \alpha) \neq 0$ is satisfied, and we have $\lim_{\alpha \to \alpha_0} h(\alpha) = x_0$. Please note that in case of a saddle node bifurcation, one has to transform the greater equilibrium into $x_0$. In case of a transcritical bifurcation, we assume that the bigger equilibrium equals $x_0$. This can be also reached by a transformation.

**Theorem 7.15 (Partial past bifurcation).** *We suppose that $(7.20)_\alpha$ admits a transcritical or saddle node bifurcation as described above and consider the nonautonomous differential equation*

$$\dot{x} = f(t, x, \alpha) \qquad\qquad (7.30)_\alpha$$

*with a $C^1$-function $f : (-\infty, 0) \times (x_-, x_+) \times (\alpha_0, \alpha_1] \to \mathbb{R}$. We assume that*

$$\lim_{t \to -\infty} f(t, x, \alpha) = g(x, \alpha) \quad and \quad \lim_{t \to -\infty} D_2 f(t, x, \alpha) = D_1 g(x, \alpha)$$

*hold uniformly for all $x \in (x_-, x_+)$ and $\alpha \in (\alpha_0, \alpha_1]$. Then the following statements are fulfilled:*

(i) *If $D_1 g(x_0, \alpha_1) < 0$ is satisfied, then there exists a continuous function $\mu : D \subset \mathbb{R} \times (\alpha_0, \alpha_1] \to \mathbb{R}$ such that $\mu(\cdot, \alpha)$ is a past attractive solution of $(7.30)_\alpha$. We have a partial bifurcation, since*

$$\lim_{\alpha \searrow \alpha_0} \mathfrak{A}^{\leftarrow}_{\mu(\cdot, \alpha)} = 0 \,.$$

(ii) *If $D_1 g(x_0, \alpha_1) > 0$ is satisfied, then there exists a continuous function $\mu : D \subset \mathbb{R} \times (\alpha_0, \alpha_1] \to \mathbb{R}$ such that $\mu(\cdot, \alpha)$ is a past repulsive solution of $(7.30)_\alpha$. We have a partial bifurcation, since*

$$\lim_{\alpha \searrow \alpha_0} \mathfrak{R}^{\leftarrow}_{\mu(\cdot, \alpha)} = 0 \,.$$

*Proof.* See proof of Theorem 7.12. $\hspace{5cm}$ □

**Theorem 7.16 (Partial future bifurcation).** *We suppose that* $(7.20)_\alpha$ *admits a transcritical or saddle node bifurcation as described above and consider the nonautonomous differential equation*

$$\dot{x} = f(t, x, \alpha) \qquad\qquad (7.31)_\alpha$$

*with a* $C^1$*-function* $f : (0, \infty) \times (x_-, x_+) \times (\alpha_0, \alpha_1] \to \mathbb{R}$. *We assume that*

$$\lim_{t \to \infty} f(t, x, \alpha) = g(x, \alpha) \quad and \quad \lim_{t \to \infty} D_2 f(t, x, \alpha) = D_1 g(x, \alpha)$$

*hold uniformly for all* $x \in (x_-, x_+)$ *and* $\alpha \in (\alpha_0, \alpha_1]$. *Then the following statements are fulfilled:*

(i) *If* $D_1 g(x_0, \alpha_1) < 0$ *is satisfied, then there exists a continuous function* $\mu : D \subset \mathbb{R} \times (\alpha_0, \alpha_1] \to \mathbb{R}$ *such that* $\mu(\cdot, \alpha)$ *is a future attractive solution of* $(7.31)_\alpha$. *We have a partial bifurcation, since*

$$\lim_{\alpha \searrow \alpha_0} \overrightarrow{\mathfrak{A}}_{\mu(\cdot,\alpha)} = 0 \,.$$

(ii) *If* $D_1 g(x_0, \alpha_1) > 0$ *is satisfied, then there exists a continuous function* $\mu : D \subset \mathbb{R} \times (\alpha_0, \alpha_1] \to \mathbb{R}$ *such that* $\mu(\cdot, \alpha)$ *is a future repulsive solution of* $(7.31)_\alpha$. *We have a partial bifurcation, since*

$$\lim_{\alpha \searrow \alpha_0} \overrightarrow{\mathfrak{R}}_{\mu(\cdot,\alpha)} = 0 \,.$$

*Proof.* See proof of Theorem 7.12. $\hspace{5cm}$ □

## 7.3 Bifurcations in Dimension Two

In this section, two-dimensional differential equations which exhibit Hopf bifurcations are studied (see, e.g., MARSDEN & MCCRACKEN [117]). As in the previous section, this bifurcation behavior is transferred to asymptotically autonomous systems.

More precisely, we consider the autonomous differential equation

$$\begin{aligned} \dot{x} &= g_1(x, y, \alpha) \\ \dot{y} &= g_2(x, y, \alpha) \end{aligned} \qquad\qquad (7.32)_\alpha$$

with a $C^1$-function $g : (x_-, x_+) \times (y_-, y_+) \times (\alpha_0, \alpha_1] \to \mathbb{R}^2$ which admits a supercritical Hopf bifurcation at $(x_0, y_0, \alpha_0)$, i.e., for all $\alpha \in (\alpha_0, \alpha_1]$, we have

$$g(x_0, y_0, \alpha) = 0 \quad \text{and} \quad D_{(1,2)}g(x_0, y_0, \alpha) = \begin{pmatrix} a(\alpha) & -b(\alpha) \\ b(\alpha) & a(\alpha) \end{pmatrix}$$

with continuous functions $a : (\alpha_0, \alpha_1] \to \mathbb{R}$ and $b : (\alpha_0, \alpha_1] \to \mathbb{R}$ which fulfill $a(\alpha) \neq 0$ and $b(\alpha) \neq 0$. Furthermore, let $S(\alpha)$ be an attractive (in case $a(\alpha) < 0$) or a repulsive (in case $a(\alpha) > 0$) periodic orbit of $(7.32)_\alpha$, respectively, which depends continuously on $\alpha$ with respect to the Hausdorff distance and converges to $(x_0, y_0)$ in the limit $\alpha \to \alpha_0$. We denote the inner area of $S(\alpha)$ by $S^i(\alpha)$.

As in the previous section, in the next four lemmata, conditions for the existence of nonautonomous counterparts for the equilibrium $(x_0, y_0)$ are studied. In the first instance, the parameter area is restricted to compact subintervals of $(\alpha_0, \alpha_1]$.

**Lemma 7.17 (Existence of past attractive solutions).** *Consider the nonautonomous differential equation*

$$\begin{aligned} \dot{x} &= f_1(t, x, y, \alpha) \\ \dot{y} &= f_2(t, x, y, \alpha) \end{aligned} \qquad (7.33)_\alpha$$

*with a $C^1$-function $f : (-\infty, 0) \times (x_-, x_+) \times (y_-, y_+) \times [\alpha_-, \alpha_+] \to \mathbb{R}^2$. We assume that*

$$\lim_{t \to -\infty} f(t, x, y, \alpha) = g(x, y, \alpha), \quad \lim_{t \to -\infty} D_{(2,3)}f(t, x, y, \alpha) = D_{(1,2)}g(x, y, \alpha)$$

*hold uniformly for all $x \in (x_-, x_+)$, $y \in (y_-, y_+)$ and $\alpha \in [\alpha_-, \alpha_+]$. Furthermore, we suppose that*

$$a(\alpha) < 0 \quad \text{for all } \alpha \in [\alpha_-, \alpha_+].$$

*Then there exist $\tau < 0$ and a continuous function $\mu : (-\infty, \tau] \times [\alpha_-, \alpha_+] \to \mathbb{R}^2$ such that $\mu(\cdot, \alpha)$ is the uniquely determined past attractive solution of $(7.33)_\alpha$ which fulfills*

$$\lim_{t \to -\infty} \mu(t, \alpha) = (x_0, y_0).$$

*Moreover, we have*

$$\mathcal{A}^{\leftarrow}_{\mu(\cdot, \alpha)} = S^i(\alpha) - (x_0, y_0) \quad \text{for all } \alpha \in [\alpha_-, \alpha_+].$$

*Furthermore, there exists a past repeller $R(\alpha)$ of $(7.33)_\alpha$ with*

$$S^i(\alpha) \subset \liminf_{t \to -\infty} R(\alpha, t) \subset \limsup_{t \to -\infty} R(\alpha, t) \subset \text{cls}\, S^i(\alpha) \quad \text{for all } \alpha \in [\alpha_-, \alpha_+].$$

*Proof.* For simplicity, we assume w.l.o.g. that $(x_0, y_0) = (0, 0)$ in this proof. The proof is divided into three steps.
*Step 1. There exist a $\tau < 0$ and a continuous function $\mu : (-\infty, \tau] \times [\alpha_-, \alpha_+] \to$*

$\mathbb{R}^2$ such that $\mu(\cdot, \alpha)$ is the uniquely determined past attractive solution of $(7.33)_\alpha$ which fulfills $\lim_{t \to -\infty} \mu(t, \alpha) = (0, 0)$.

Due to the compactness of $[\alpha_-, \alpha_+]$ and the uniform continuity of $g$ on compact sets, there exist $\beta > 0$ and $\gamma < 0$ with

$$\frac{\partial g_1}{\partial x}(x, y, \alpha) \le 2\gamma, \quad \frac{\partial g_2}{\partial y}(x, y, \alpha) \le 2\gamma, \quad \left| \frac{\partial g_1}{\partial y}(x, y, \alpha) + \frac{\partial g_2}{\partial x}(x, y, \alpha) \right| \le -\gamma$$

for all $(x, y) \in \text{cls}\, U_{2\beta}((0,0))$ and $\alpha \in [\alpha_-, \alpha_+]$. This implies the existence of a $\tau < 0$ with

$$\frac{\partial f_1}{\partial x}(t, x, y, \alpha) \le \gamma, \quad \frac{\partial f_2}{\partial y}(t, x, y, \alpha) \le \gamma$$

and

$$\left| \frac{\partial f_1}{\partial y}(t, x, y, \alpha) + \frac{\partial f_2}{\partial x}(t, x, y, \alpha) \right| \le -\frac{\gamma}{2}$$

and

$$|f_1(t, 0, 0, \alpha)| + |f_2(t, 0, 0, \alpha)| \le -\frac{\gamma\beta}{4}$$

for all $t \le \tau$, $x, y \in \text{cls}\, U_{2\beta}((0,0))$ and $\alpha \in [\alpha_-, \alpha_+]$. For the rest of this step, we fix an $\alpha \in [\alpha_-, \alpha_+]$. For all $t \le \tau$ and $x, y \in \text{cls}\, U_{2\beta}((0,0))$, the mean value theorem implies

$$f(t, x, y, \alpha) = f(t, 0, 0, \alpha) + \int_0^1 D_{(2,3)} f(t, \theta x, \theta y, \alpha) \cdot (x, y)\, d\theta\,.$$

It follows that for all $t \le \tau$, $\alpha \in [\alpha_-, \alpha_+]$ and $x, y$ with $x^2 + y^2 = \beta^2$,

$$\langle f(t, x, y, \alpha), (x, y) \rangle = f_1(t, x, y, \alpha)x + f_2(t, x, y, \alpha)y$$
$$= f_1(t, 0, 0, \alpha)x + f_2(t, 0, 0, \alpha)y +$$
$$\int_0^1 \left( \frac{\partial f_1}{\partial x}(t, \theta x, \theta y, \alpha)x^2 + \frac{\partial f_2}{\partial y}(t, \theta x, \theta y, \alpha)y^2 + \right.$$
$$\left. \frac{\partial f_1}{\partial y}(t, \theta x, \theta y, \alpha)xy + \frac{\partial f_2}{\partial x}(t, \theta x, \theta y, \alpha)xy \right) d\theta$$
$$\le -\frac{\gamma\beta^2}{4} + \int_0^1 \left( \gamma x^2 + \gamma y^2 - \frac{\gamma}{2}|xy| \right) d\theta \le \frac{\gamma\beta^2}{4} < 0$$

holds. Therefore, the subset $\text{cls}\, U_\beta((0,0))$ of the phase space is forward invariant in the following sense: For all $t^- \le t^+ \le \tau$ and $\alpha \in [\alpha_-, \alpha_+]$, we have

$$\lambda\big(t^+, t^-, \text{cls}\, U_\beta((0,0)), \alpha\big) \subset \text{cls}\, U_\beta((0,0))\,.$$

Thus, the set

$$M := \{(x,y) \in \mathrm{cls}\, U_\beta((0,0)) : \text{There exists a } t \le \tau \text{ such that}$$
$$\|\lambda(t,\tau,x,y,\alpha)\| = \beta \text{ and}$$
$$\lambda(t,\tau,x,y,\alpha) \ne (\beta,0) \text{ and}$$
$$\lambda(t,\tau,x,y,\alpha) \ne (-\beta,0)\}$$

is nonempty and due to the continuity of the general solution relatively open in $\mathrm{cls}\, U_\beta((0,0))$. This means that $\tilde{M} := \mathrm{cls}\, U_\beta((0,0)) \setminus M$ is closed. The sets

$$M_1 := \{(x,y) \in \tilde{M} : \text{There exists a } t \le \tau \text{ such that}$$
$$\|\lambda(t,\tau,x,y,\alpha)\| = \beta \text{ and}$$
$$\lambda(t,\tau,x,y,\alpha) = (\beta,0)\}$$

and

$$M_2 := \{(x,y) \in \tilde{M} : \text{There exists a } t \le \tau \text{ such that}$$
$$\|\lambda(t,\tau,x,y,\alpha)\| = \beta \text{ and}$$
$$\lambda(t,\tau,x,y,\alpha) = (-\beta,0)\}$$

are obviously nonempty and due to the continuity of the general solution relatively open in $\tilde{M}$. This implies that $M_1 \cup M_2 \subsetneq \tilde{M}$. Therefore, there exists a $(\hat{x},\hat{y}) \in U_\beta(0,0)$ with

$$\mu(t,\alpha) := \lambda(t,\tau,\hat{x},\hat{y},\alpha) \in U_\beta((0,0)) \quad \text{for all } t \le \tau.$$

To show that $\mu$ is past attractive, we study the differential equation of the perturbed motion

$$\begin{aligned}
\dot{x} &= h_1(t,x,y,\alpha) := f_1\big(t,x+\mu_1(t,\alpha),y+\mu_2(t,\alpha),\alpha\big) - f_1\big(t,\mu(t,\alpha),\alpha\big) \\
\dot{y} &= h_2(t,x,y,\alpha) := f_2\big(t,x+\mu_1(t,\alpha),y+\mu_2(t,\alpha),\alpha\big) - f_2\big(t,\mu(t,\alpha),\alpha\big)
\end{aligned}.$$

Due to the mean value theorem, for all $t \le \tau$ and $(x,y) \in U_\beta((0,0))$,

$$\begin{aligned}
h(t,x,y,\alpha) &= \int_0^1 D_{(2,3)} h(t,\theta x, \theta y, \alpha) \cdot (x,y)\, d\theta \\
&= \int_0^1 D_{(2,3)} f\big(t,\theta x + \mu_1(t,\alpha), \theta y + \mu_2(t,\alpha), \alpha\big) \cdot (x,y)\, d\theta
\end{aligned}$$

is fulfilled. Thus, for all $(r,\phi) \in (0,\beta) \times [0,2\pi)$ and $t \le \tau$, we have

$$h_1(t,r\cos\phi,r\sin\phi,\alpha)\cos\phi + h_2(t,r\cos\phi,r\sin\phi,\alpha)\sin\phi$$
$$= \int_0^1 \Big( \frac{\partial f_1}{\partial x}(t,\theta r\cos\phi + \mu_1(t,\alpha), \theta r\sin\phi + \mu_2(t,\alpha), \alpha) r\cos^2\phi +$$
$$\frac{\partial f_2}{\partial y}(t,\theta r\cos\phi + \mu_1(t,\alpha), \theta r\sin\phi + \mu_2(t,\alpha), \alpha) r\sin^2\phi +$$

$$\frac{\partial f_1}{\partial y}(t, \theta r \cos\phi + \mu_1(t,\alpha), \theta r \sin\phi + \mu_2(t,\alpha), \alpha) r \cos\phi \sin\phi +$$

$$\left. \frac{\partial f_2}{\partial x}(t, \theta r \cos\phi + \mu_1(t,\alpha), \theta r \sin\phi + \mu_2(t,\alpha), \alpha) r \cos\phi \sin\phi \right) d\theta$$

$$\leq \int_0^1 \left( \gamma r \cos^2\phi + \gamma r \sin^2\phi - \frac{\gamma}{2} r \cos\phi \sin\phi \right) d\theta \leq r\frac{\gamma}{2}.$$

Applying polar coordinates (see AULBACH [14, Satz 5.2.1, p. 192]), we see that $\mu(\cdot,\alpha)$ is past attractive. Moreover, the relation $\lim_{t\to-\infty} \mu(t,\alpha) = (0,0)$ is obviously satisfied. The uniqueness of $\mu(\cdot,\alpha)$ follows directly from Proposition 2.37 (i).

*Step 2. $\mu$ is continuous.*
See Step 2 of the proof of Lemma 7.7.

*Step 3. The assertions concerning $\mathcal{A}^{\leftarrow}_{\mu(\cdot,\alpha)}$ and the past repellers are fulfilled.*
This follows directly from Lemma 7.5. $\qquad\qquad\qquad\qquad\qquad\qquad\qquad\square$

**Lemma 7.18 (Existence of future attractive solutions).** *Consider the nonautonomous differential equation*

$$\begin{aligned} \dot{x} &= f_1(t, x, y, \alpha) \\ \dot{y} &= f_2(t, x, y, \alpha) \end{aligned} \qquad (7.34)_\alpha$$

*with a $C^1$-function $f : (0,\infty) \times (x_-, x_+) \times (y_-, y_+) \times [\alpha_-, \alpha_+] \to \mathbb{R}^2$. We assume that*

$$\lim_{t\to\infty} f(t,x,y,\alpha) = g(x,y,\alpha), \quad \lim_{t\to\infty} D_{(2,3)}f(t,x,y,\alpha) = D_{(1,2)}g(x,y,\alpha)$$

*hold uniformly for all $x \in (x_-, x_+)$, $y \in (y_-, y_+)$ and $\alpha \in [\alpha_-, \alpha_+]$. Furthermore, we suppose that*

$$a(\alpha) < 0 \quad \text{for all } \alpha \in [\alpha_-, \alpha_+].$$

*Then there exist $\tau > 0$ and $\beta > 0$ such that for $(x,y) \in U_\beta((x_0, y_0))$ and $\alpha \in [\alpha_-, \alpha_+]$, the solution $\lambda(\cdot, \tau, x, y, \alpha)$ of $(7.34)_\alpha$ is future attractive with*

$$\lim_{t\to\infty} \lambda(t, \tau, x, y, \alpha) = (x_0, y_0)$$

*and*

$$\mathcal{A}^{\rightarrow}_{\lambda(\cdot,\tau,x,y,\alpha)} = S^i(\alpha) - (x_0, y_0).$$

*Furthermore, there exists a future repeller $R(\alpha)$ of $(7.34)_\alpha$ with*

$$S^i(\alpha) \subset \liminf_{t\to\infty} R(\alpha, t) \subset \limsup_{t\to\infty} R(\alpha, t) \subset \text{cls } S^i(\alpha) \quad \text{for all } \alpha \in [\alpha_-, \alpha_+].$$

*Proof.* For simplicity, we assume w.l.o.g. that $(x_0, y_0) = (0,0)$ in this proof. Due to the compactness of $[\alpha_-, \alpha_+]$ and the uniform continuity of $g$ on compact sets, there exist $\beta > 0$ and $\gamma < 0$ with

$$\frac{\partial g_1}{\partial x}(x,y,\alpha) \le 2\gamma, \quad \frac{\partial g_2}{\partial y}(x,y,\alpha) \le 2\gamma, \quad \left|\frac{\partial g_1}{\partial y}(x,y,\alpha) + \frac{\partial g_2}{\partial x}(x,y,\alpha)\right| \le -\gamma$$

for all $(x,y) \in \mathrm{cls}\,U_{2\beta}((0,0))$ and $\alpha \in [\alpha_-, \alpha_+]$. This implies the existence of a $\tau < 0$ with

$$\frac{\partial f_1}{\partial x}(t,x,y,\alpha) \le \gamma, \quad \frac{\partial f_2}{\partial y}(t,x,y,\alpha) \le \gamma$$

and

$$\left|\frac{\partial f_1}{\partial y}(t,x,y,\alpha) + \frac{\partial f_2}{\partial x}(t,x,y,\alpha)\right| \le -\frac{\gamma}{2}$$

and

$$|f_1(t,0,0,\alpha)| + |f_2(t,0,0,\alpha)| \le -\frac{\gamma\beta}{4}$$

for all $t \ge \tau$, $x,y \in \mathrm{cls}\,U_{2\beta}((0,0))$ and $\alpha \in [\alpha_-, \alpha_+]$. For the rest of this proof, we fix an $\alpha \in [\alpha_-, \alpha_+]$. For all $t \ge \tau$ and $x,y \in \mathrm{cls}\,U_{2\beta}((0,0))$, the mean value theorem implies

$$f(t,x,y,\alpha) = f(t,0,0,\alpha) + \int_0^1 f_x(t,\theta x,\theta y,\alpha) \cdot (x,y)\,\mathrm{d}\theta\,.$$

Thus, for all $t \ge \tau$ and $x,y$ with $x^2 + y^2 = \beta^2$, we have

$$\langle f(t,x,y,\alpha),(x,y)\rangle = f_1(t,x,y,\alpha)x + f_2(t,x,y,\alpha)y$$
$$= f_1(t,0,0,\alpha)x + f_2(t,0,0,\alpha)y +$$
$$\int_0^1 \left(\frac{\partial f_1}{\partial x}(t,\theta x,\theta y,\alpha)x^2 + \frac{\partial f_2}{\partial y}(t,\theta x,\theta y,\alpha)y^2 + \right.$$
$$\left. \frac{\partial f_1}{\partial y}(t,\theta x,\theta y,\alpha)xy + \frac{\partial f_2}{\partial x}(t,\theta x,\theta y,\alpha)xy\right)\,\mathrm{d}\theta$$
$$\le -\frac{\gamma\beta^2}{4} + \int_0^1 \left(\gamma x^2 + \gamma y^2 - \frac{\gamma}{2}|xy|\right)\,\mathrm{d}\theta \le \frac{\gamma\beta^2}{4} < 0\,.$$

Therefore, the subset $\mathrm{cls}\,U_\beta((0,0))$ of the phase space is forward invariant in the following sense: For all $t^+ \ge t^- \ge \tau$, we have

$$\lambda\bigl(t^+,t^-,\mathrm{cls}\,U_\beta((0,0)),\alpha\bigr) \subset \mathrm{cls}\,U_\beta((0,0))\,.$$

We choose $(\hat{x},\hat{y}) \in U_\beta((0,0))$ arbitrarily and consider for the rest of this proof in particular the solution $\nu(\cdot) := \lambda(\cdot,\tau,\hat{x},\hat{y},\alpha)$ on the interval $[\tau,\infty)$. It is obvious that $\lim_{t\to\infty} \nu(t) = (0,0)$ holds. To show that $\nu$ is future attractive, we study the differential equation of the perturbed motion

$$\dot{x} = h_1(t,x,y,\alpha) := f_1\bigl(t,x+\nu_1(t),y+\nu_2(t),\alpha\bigr) - f_1\bigl(t,\nu(t),\alpha\bigr)$$
$$\dot{y} = h_2(t,x,y,\alpha) := f_2\bigl(t,x+\nu_1(t),y+\nu_2(t),\alpha\bigr) - f_2\bigl(t,\nu(t),\alpha\bigr)\,.$$

Due to the mean value theorem, we have for all $t \ge \tau$ and $(x,y) \in U_\beta((0,0))$,

$$h(t, x, y, \alpha) = \int_0^1 D_{(2,3)} h(t, \theta x, \theta y, \alpha) \cdot (x, y) \, d\theta =$$

$$\int_0^1 D_{(2,3)} f\left(t, \theta x + \mu_1(t, \alpha), \theta y + \mu_2(t, \alpha), \alpha\right) \cdot (x, y) \, d\theta \, .$$

Thus, for all $(r, \phi) \in (0, \beta) \times [0, 2\pi)$ and $t \geq \tau$, we have

$$h_1(t, r \cos \phi, r \sin \phi, \alpha) \cos \phi + h_2(t, r \cos \phi, r \sin \phi, \alpha) \sin \phi$$

$$= \int_0^1 \left( \frac{\partial f_1}{\partial x} (t, \theta r \cos \phi + \mu_1(t, \alpha), \theta r \sin \phi + \mu_2(t, \alpha), \alpha) r \cos^2 \phi + \right.$$

$$\frac{\partial f_2}{\partial y} (t, \theta r \cos \phi + \mu_1(t, \alpha), \theta r \sin \phi + \mu_2(t, \alpha), \alpha) r \sin^2 \phi +$$

$$\frac{\partial f_1}{\partial y} (t, \theta r \cos \phi + \mu_1(t, \alpha), \theta r \sin \phi + \mu_2(t, \alpha), \alpha) r \cos \phi \sin \phi +$$

$$\left. \frac{\partial f_2}{\partial x} (t, \theta r \cos \phi + \mu_1(t, \alpha), \theta r \sin \phi + \mu_2(t, \alpha), \alpha) r \cos \phi \sin \phi \right) \, d\theta$$

$$\leq \int_0^1 \left( \gamma r \cos^2 \phi + \gamma r \sin^2 \phi - \frac{\gamma}{2} r \cos \phi \sin \phi \right) \, d\theta \leq r \frac{\gamma}{2} \, .$$

Applying polar coordinates, we see that $\nu$ is future attractive. The asserted relations concerning $\mathcal{A}_{\vec{\mu}(\cdot, \alpha)}^{\rightarrow}$ and the existence of the future repellers follow directly from Lemma 7.6. □

**Lemma 7.19 (Existence of past repulsive solutions).** *Consider the nonautonomous differential equation*

$$\begin{aligned} \dot{x} &= f_1(t, x, y, \alpha) \\ \dot{y} &= f_2(t, x, y, \alpha) \end{aligned} \qquad (7.35)_\alpha$$

*with a $C^1$-function $f : (-\infty, 0) \times (x_-, x_+) \times (y_-, y_+) \times [\alpha_-, \alpha_+] \to \mathbb{R}^2$. We assume that*

$$\lim_{t \to -\infty} f(t, x, y, \alpha) = g(x, y, \alpha), \qquad \lim_{t \to -\infty} D_{(2,3)} f(t, x, y, \alpha) = D_{(1,2)} g(x, y, \alpha)$$

*hold uniformly for all $x \in (x_-, x_+)$, $y \in (y_-, y_+)$ and $\alpha \in [\alpha_-, \alpha_+]$. Furthermore, we suppose that*

$$a(\alpha) > 0 \quad \text{for all } \alpha \in [\alpha_-, \alpha_+] \, .$$

*Then there exist $\tau < 0$ and $\beta > 0$ such that for $(x, y) \in U_\beta((x_0, y_0))$ and $\alpha \in [\alpha_-, \alpha_+]$, the solution $\lambda(\cdot, \tau, x, y, \alpha)$ of $(7.35)_\alpha$ is past repulsive with*

$$\lim_{t \to -\infty} \lambda(t, \tau, x, y, \alpha) = (x_0, y_0)$$

*and*

$$\mathcal{R}^{\leftarrow}_{\lambda(\cdot,\tau,x,y,\alpha)} = S^i(\alpha) - (x_0, y_0)\,.$$

Furthermore, there exists a past attractor $A(\alpha)$ of $(7.35)_\alpha$ with

$$S^i(\alpha) \subset \liminf_{t \to -\infty} A(\alpha, t) \subset \limsup_{t \to -\infty} A(\alpha, t) \subset \mathrm{cls}\, S^i(\alpha) \quad \text{for all } \alpha \in [\alpha_-, \alpha_+].$$

*Proof.* The assertions follow directly from Proposition 2.32 and Lemma 7.18.
□

**Lemma 7.20 (Existence of future repulsive solutions).** *Consider the nonautonomous differential equation*

$$\begin{aligned}
\dot{x} &= f_1(t, x, y, \alpha) \\
\dot{y} &= f_2(t, x, y, \alpha)
\end{aligned} \tag{7.36}_\alpha$$

*with a $C^1$-function $f : (0, \infty) \times (x_-, x_+) \times (y_-, y_+) \times [\alpha_-, \alpha_+] \to \mathbb{R}^2$. We assume that*

$$\lim_{t \to \infty} f(t, x, y, \alpha) = g(x, y, \alpha)\,, \quad \lim_{t \to \infty} D_{(2,3)} f(t, x, y, \alpha) = D_{(1,2)} g(x, y, \alpha)$$

*hold uniformly for all $x \in (x_-, x_+)$, $y \in (y_-, y_+)$ and $\alpha \in [\alpha_-, \alpha_+]$. Furthermore, we suppose that*

$$a(\alpha) > 0 \quad \text{for all } \alpha \in [\alpha_-, \alpha_+]\,.$$

*Then there exist a $\tau > 0$ and a continuous function $\mu : [\tau, \infty) \times [\alpha_-, \alpha_+] \to \mathbb{R}^2$ such that $\mu(\cdot, \alpha)$ is the uniquely determined future repulsive solution of $(7.36)_\alpha$ which fulfills $\lim_{t \to \infty} \mu(t, \alpha) = (x_0, y_0)$. Moreover, we have*

$$\mathcal{R}^{\rightarrow}_{\mu(\cdot, \alpha)} = S^i(\alpha) - (x_0, y_0) \quad \text{for all } \alpha \in [\alpha_-, \alpha_+]\,.$$

*Furthermore, there exists a future attractor $A(\alpha)$ of $(7.36)_\alpha$ with*

$$S^i(\alpha) \subset \liminf_{t \to \infty} A(\alpha, t) \subset \limsup_{t \to \infty} A(\alpha, t) \subset \mathrm{cls}\, S^i(\alpha) \quad \text{for all } \alpha \in [\alpha_-, \alpha_+].$$

*Proof.* The assertions follow directly from Proposition 2.32 and Lemma 7.17.
□

As in the previous subsection, these four lemmata lead to the existence of total nonautonomous bifurcations and transitions.

**Theorem 7.21 (Past Hopf bifurcation).** *We suppose that $(7.32)_\alpha$ admits a Hopf bifurcation as described above and consider the nonautonomous differential equation*

$$\begin{aligned}
\dot{x} &= f_1(t, x, y, \alpha) \\
\dot{y} &= f_2(t, x, y, \alpha)
\end{aligned} \tag{7.37}_\alpha$$

*with a $C^1$-function $f : (-\infty, 0) \times (x_-, x_+) \times (y_-, y_+) \times (\alpha_0, \alpha_1] \to \mathbb{R}^2$. We assume that*

$$\lim_{t \to -\infty} f(t, x, y, \alpha) = g(x, y, \alpha), \quad \lim_{t \to -\infty} D_{(2,3)} f(t, x, y, \alpha) = D_{(1,2)} g(x, y, \alpha)$$

*hold uniformly for all $x \in (x_-, x_+)$, $y \in (y_-, y_+)$ and $\alpha \in (\alpha_0, \alpha_1]$. Then the following statements are fulfilled:*

(i) *If $a(\alpha_1) < 0$, there exists a continuous function $\mu : D \subset \mathbb{R} \times (\alpha_0, \alpha_1] \to \mathbb{R}^2$ such that $\mu(\cdot, \alpha)$ is a past attractive solution of $(7.37)_\alpha$. We have a total past bifurcation, since*

$$\lim_{\alpha \searrow \alpha_0} d\left(\mathcal{A}^{\leftarrow}_{\mu(\cdot,\alpha)} | \{0\}\right) = 0.$$

*Furthermore, for all $\alpha \in (\alpha_0, \alpha_1]$, there exists a past repeller $R(\alpha)$. We also have a past repeller transition, since*

$$\lim_{\alpha \searrow \alpha_0} d\left(\limsup_{t \to -\infty} R(\alpha, t) | \{(x_0, y_0)\}\right) = 0.$$

(ii) *If $a(\alpha_1) > 0$, there exists a continuous function $\mu : D \subset \mathbb{R} \times (\alpha_0, \alpha_1] \to \mathbb{R}^2$ such that $\mu(\cdot, \alpha)$ is a past repulsive solution of $(7.37)_\alpha$. We have a total past bifurcation, since*

$$\lim_{\alpha \searrow \alpha_0} d\left(\mathcal{R}^{\leftarrow}_{\mu(\cdot,\alpha)} | \{0\}\right) = 0.$$

*Furthermore, for all $\alpha \in (\alpha_0, \alpha_1]$, there exists a past attractor $A(\alpha)$. We also have a past attractor transition, since*

$$\lim_{\alpha \searrow \alpha_0} d\left(\limsup_{t \to -\infty} A(\alpha, t) | \{(x_0, y_0)\}\right) = 0.$$

*Proof.* See proof of Theorem 7.12. □

**Theorem 7.22 (Future Hopf bifurcation).** *We suppose that $(7.32)_\alpha$ admits a Hopf bifurcation as described above and consider the nonautonomous differential equation*

$$\begin{aligned} \dot{x} &= f_1(t, x, y, \alpha) \\ \dot{y} &= f_2(t, x, y, \alpha) \end{aligned} \qquad (7.38)_\alpha$$

*with a $C^1$-function $f : (0, \infty) \times (x_-, x_+) \times (y_-, y_+) \times (\alpha_0, \alpha_1] \to \mathbb{R}^2$. We assume that*

$$\lim_{t \to \infty} f(t, x, y, \alpha) = g(x, y, \alpha), \quad \lim_{t \to \infty} D_{(2,3)} f(t, x, y, \alpha) = D_{(1,2)} g(x, y, \alpha)$$

*hold uniformly for all $x \in (x_-, x_+)$, $y \in (y_-, y_+)$ and $\alpha \in (\alpha_0, \alpha_1]$. Then the following statements are fulfilled:*

(i) *If $a(\alpha_1) < 0$, there exists a continuous function $\mu : D \subset \mathbb{R} \times (\alpha_0, \alpha_1] \to \mathbb{R}^2$ such that $\mu(\cdot, \alpha)$ is a future attractive solution of $(7.38)_\alpha$. We have a total future bifurcation, since*

$$\lim_{\alpha \searrow \alpha_0} d\left(\overrightarrow{\mathcal{A}_{\mu(\cdot,\alpha)}} \middle| \{0\}\right) = 0.$$

*Furthermore, for all $\alpha \in (\alpha_0, \alpha_1]$, there exists a future repeller $R(\alpha)$. We also have a future repeller transition, since*

$$\lim_{\alpha \searrow \alpha_0} d\left(\limsup_{t \to \infty} R(\alpha, t) \middle| \{(x_0, y_0)\}\right) = 0.$$

(ii) *If $a(\alpha_1) > 0$, there exists a continuous function $\mu : D \subset \mathbb{R} \times (\alpha_0, \alpha_1] \to \mathbb{R}^2$ such that $\mu(\cdot, \alpha)$ is a future repulsive solution of $(7.38)_\alpha$. We have a total future bifurcation, since*

$$\lim_{\alpha \searrow \alpha_0} d\left(\overrightarrow{\mathcal{R}_{\mu(\cdot,\alpha)}} \middle| \{0\}\right) = 0.$$

*Furthermore, for all $\alpha \in (\alpha_0, \alpha_1]$, there exists a future attractor $A(\alpha)$. We also have a future attractor transition, since*

$$\lim_{\alpha \searrow \alpha_0} d\left(\limsup_{t \to \infty} A(\alpha, t) \middle| \{(x_0, y_0)\}\right) = 0.$$

*Proof.* See proof of Theorem 7.12.    □

This chapter is concluded with the following famous example.

*Example 7.23.* We consider a nonautonomous version of the unforced Duffing-van der Pol equation

$$\dot{x}_1 = x_2$$
$$\dot{x}_2 = -x_1 + \alpha\beta(t)x_2 - x_1^2(x_1 + x_2)$$

depending on a real parameter $\alpha$. We assume that $\beta : \mathbb{R} \to \mathbb{R}^+$ is a $C^1$-function. This differential equation describes a nonlinear oscillator. It is well-known (see, e.g., HOLMES & RAND [82] or MARSDEN & MCCRACKEN [117]) that in case the function $\beta$ is constant and positive, i.e., the system is autonomous, the equilibrium $(0,0)$ is attractive for $\alpha < 0$. At $\alpha = 0$, the system undergoes a Hopf bifurcation: The equilibrium $(0,0)$ becomes repulsive and an attractive periodic orbit appears. As a consequence, we also have a bifurcation of autonomous attractors (see AULBACH & RASMUSSEN & SIEGMUND [16]): For values $\alpha \leq 0$, the singleton $\{(0,0)\}$ is an attractor. If $\alpha$ is small and positive, then the interior of the bifurcating periodic orbit is an attractor of the system.

We assume that the nonautonomous system is past (future, respectively) asymptotically autonomous, i.e., the function $\beta$ fulfills

$$\bar{\beta} := \lim_{t \to \pm\infty} \beta(t) > 0 \,.$$

Then the differential equation fulfills the hypotheses of Theorem 7.21 or Theorem 7.22 in some neighborhood of $(0,0)$, respectively, and we have a nonautonomous bifurcation and transition as described in these theorems.

# A

# Appendix

This supplementary appendix contains well-known definitions and results used in this book which—to provide reading fluency—are not stated before.

The first section of this appendix is devoted to fundamental facts about ordinary differential equations. In Section A.2, some useful lemmata are stated, and in the last section, basic properties of projective spaces are treated.

## A.1 Ordinary Differential Equations

We begin with the definition of an ordinary differential equation in the Euclidean space $\mathbb{R}^N$.

**Definition A.1 (Ordinary differential equation).** *For given* $N, M \in \mathbb{N}$, *let* $D \subset \mathbb{R} \times \mathbb{R}^N \times \mathbb{R}^M$ *be an open set and* $f : D \to \mathbb{R}^N$ *be a function. Then the equation*

$$\dot{x} = f(t, x, \alpha) \qquad (A.1)_\alpha$$

*is called* (nonautonomous) ordinary differential equation *which depends on a parameter* $\alpha$. *For fixed* $\hat{\alpha} \in \mathbb{R}^M$, *we say, a differentiable function* $\mu : \mathbb{I} \to \mathbb{R}^N$, $\mathbb{I}$ *an open interval, is a solution of* $(A.1)_{\hat{\alpha}}$ *if* $(t, \mu(t), \hat{\alpha}) \in D$ *for all* $t \in \mathbb{I}$ *and*

$$\dot{\mu}(t) := \frac{d\mu}{dt}(t) = f(t, \mu(t), \hat{\alpha}) \quad \text{for all } t \in \mathbb{I}$$

*is fulfilled. The combination of the differential equation* $(A.1)_{\hat{\alpha}}$ *and an* initial value condition $x(\tau) = \xi$ *is called* initial value problem. *We say, a solution* $\mu$ *of* $(A.1)_{\hat{\alpha}}$ *solves this initial value problem if* $\mu(\tau) = \xi$.

For the uniqueness of solutions of ordinary differential equations, the concept of Lipschitz continuity is appropriate.

**Definition A.2 (Lipschitz continuous function).** *For given $N, M \in \mathbb{N}$, let $D \subset \mathbb{R}^{1+N+M}$ and $g : D \to \mathbb{R}^N$ be a function. We say that $g$ is (globally) Lipschitz continuous if there exists a constant $L \geq 0$ with*

$$\|g(t,x,\alpha) - g(t,y,\alpha)\| \leq L\|x - y\| \quad \text{for all } (t,x,\alpha),(t,y,\alpha) \in D.$$

*$g$ is called locally Lipschitz continuous if for all $(t,x,\alpha) \in D$, there exist neighborhoods $V$ of $t$ and $W$ of $\alpha$ such that the restriction of $g$ to $V \times W \times \{\alpha\}$ is globally Lipschitz continuous.*

The proof of the following proposition can be found, e.g., in AULBACH [14, Definition 2.6.2, Satz 7.2.2].

**Proposition A.3 (General solution).** *Let $N, M \in \mathbb{N}$, $D \subset \mathbb{R} \times \mathbb{R}^N \times \mathbb{R}^M$ be open and $f : D \to \mathbb{R}^N$ be a locally Lipschitz continuous function, and consider the nonautonomous differential equation $(A.1)_\alpha$. Then there exist an open set $\Omega \subset \mathbb{R} \times \mathbb{R} \times \mathbb{R}^N \times \mathbb{R}^M$ and a continuous function $\lambda : \Omega \to \mathbb{R}^N$ such that for fixed $(\tau,\xi,\hat\alpha) \in D$, the function $\lambda(\cdot,\tau,\xi,\hat\alpha)$ is a non-continuable solution of the initial value problem $(A.1)_{\hat\alpha}$, $x(\tau) = \xi$. The function $\lambda$ is called the general solution of $(A.1)_\alpha$.*

*Remark A.4.* In case the differential equation $(A.1)_\alpha$ does not depend on $\alpha$, the fourth argument of the general solution is omitted.

**Definition A.5 (Transition operator).** *Let $\mathbb{I} \subset \mathbb{R}$ be an interval, and consider the nonautonomous linear differential equation*

$$\dot{x} = A(t)x \tag{A.2}$$

*with a continuous function $A : \mathbb{I} \to \mathbb{R}^{N \times N}$. The (uniquely determined) function $\Lambda : \mathbb{I} \times \mathbb{I} \to \mathbb{R}^{N \times N}$ with*

$$\Lambda(t,\tau)\xi = \lambda(t,\tau,\xi) \quad \text{for all } t,\tau \in \mathbb{I} \text{ and } \xi \in \mathbb{R}^N,$$

*where $\lambda$ denotes the general solution of (A.2), is called* transition operator *of (A.2). In case (A.2) is autonomous, i.e., $A = A(t)$ for all $t \in \mathbb{I} = \mathbb{R}$ with a matrix $A \in \mathbb{R}^{N \times N}$, the* matrix exponential function *$e^{A\cdot} : \mathbb{R} \to \mathbb{R}^{N \times N}$ is defined by*

$$e^{At} := \Lambda(t,0) \quad \text{for all } t \in \mathbb{R}.$$

Inhomogeneous linear differential equations are solved by the variation of constants formula.

**Proposition A.6 (Variation of constants formula).** *Let $\mathbb{I} \subset \mathbb{R}$ be an interval, and consider the nonautonomous inhomogeneous linear differential equation*

$$\dot{x} = A(t)x + b(t) \tag{A.3}$$

*with continuous functions* $A : \mathbb{I} \to \mathbb{R}^{N \times N}$ *and* $b : \mathbb{I} \to \mathbb{R}^N$. *Let* $\lambda$ *denote the general solution of* (A.3) *and* $\Lambda$ *denote the transition operator of* $\dot{x} = A(t)x$. *Then we have the representation*

$$\lambda(t, \tau, \xi) = \Lambda(t, \tau)\xi + \int_\tau^t \Lambda(t, s)b(s)\,\mathrm{d}s \quad \text{for all } t, \tau \in \mathbb{I} \text{ and } \xi \in \mathbb{R}^N.$$

*This equation is called the* variation of constants formula.

*Proof.* See, e.g., COPPEL [54, p. 45]. □

For the analysis in the vicinity of a given reference solution, the differential equation of perturbed motion is of great importance.

**Proposition A.7 (Differential equation of perturbed motion).** *For given* $D \subset \mathbb{R} \times \mathbb{R}^N$, *let* $f : D \to \mathbb{R}^N$ *be a locally Lipschitz continuous function, and consider the nonautonomous differential equation*

$$\dot{x} = f(t, x) \tag{A.4}$$

*with a solution* $\lambda : \mathbb{I} \to \mathbb{R}^N$, $\mathbb{I}$ *an interval. Then the so-called* differential equation of perturbed motion

$$\dot{x} = f\big(t, x + \lambda(t)\big) - f\big(t, \lambda(t)\big) \tag{A.5}$$

*has the following properties:*

(i) *If* $\nu : \mathbb{J} \to \mathbb{R}^N$ *is a solution of* (A.4) *and* $\mathbb{J} \subset \mathbb{I}$, *then* $\mu := \nu - \lambda$ *is a solution of* (A.5) *on* $\mathbb{J}$.

(ii) *If* $\mu : \mathbb{J} \to \mathbb{R}^N$ *is a solution of* (A.5) *and* $\mathbb{J} \subset \mathbb{I}$, *then* $\nu := \mu + \lambda$ *is a solution of* (A.4) *on* $\mathbb{J}$.

## A.2 Useful Lemmata

The following lemma, which goes back to GRONWALL [71], plays a central role in obtaining estimates for solutions of differential equations.

**Lemma A.8 (Gronwall's inequality).** *Let* $a \geq 0$ *and* $u, b : [\tau_-, \tau_+] \to \mathbb{R}_0^+$ *be continuous functions, and suppose that*

$$u(t) \leq a + \int_{\tau_-}^t b(s)u(s)\,\mathrm{d}s \quad \text{for all } t \in [\tau_-, \tau_+]$$

*is fulfilled. Then*

$$u(t) \leq a \exp\left(\int_{\tau_-}^t b(s)\,\mathrm{d}s\right) \quad \text{for all } t \in [\tau_-, \tau_+].$$

*Proof.* See, e.g., Abraham & Marsden & Ratiu [1, Theorem 4.1.7, p. 242].
□

The following lemma provides a triangle inequality for the Hausdorff-semi distance, which has been introduced in Section 2.1.

**Lemma A.9 (Triangle inequality for the Hausdorff semi-distance).**
*Let $X$ be a metric space and $d$ denote the Hausdorff semi-distance. Then, for all nonempty sets $A, B, C \subset X$, the relation*

$$d(A|C) \leq d(A|B) + d(B|C)$$

*is fulfilled.*

*Proof.* Obviously, for all nonempty sets $M_1, M_2 \subset X$, the Hausdorff semi-distance fulfills

$$d(M_1|M_2) = \inf \left\{ \delta > 0 : M_1 \subset U_\delta(M_2) \right\}.$$

Hence, for all $\varepsilon > 0$, we have

$$A \subset U_{d(A|B)+\varepsilon/2}(B) \quad \text{and} \quad B \subset U_{d(B|C)+\varepsilon/2}(C).$$

This implies $A \subset U_{d(A|B)+d(B|C)+\varepsilon}(C)$ and finishes the proof of this lemma.
□

**Lemma A.10.** *Let $A, B, C$ be linear subspaces of the $\mathbb{R}^N$ such that $A \supset C$. Then the relation*

$$A \cap (B + C) = (A \cap B) + C$$

*is fulfilled.*

*Proof.* See Siegmund [171, Hilfssatz 2.36, p. 58].
□

## A.3 Projective Spaces

In this section, the real projective space $\mathbb{P}^{N-1}$ of the vector space $\mathbb{R}^N$ is introduced, and some basic properties are derived. Here, the $\mathbb{R}^N$ is equipped with the Euclidean norm $\| \cdot \|$ and the Euclidean scalar product $\langle \cdot, \cdot \rangle$ (cf. Section 2.1). We say, two nonzero elements $x, y \in \mathbb{R}^N$ are equivalent if there exists a real number $c \in \mathbb{R}$ such that $x = cy$. The equivalence class of $x \in \mathbb{R}^N$ is denoted by $\mathbb{P}x$, and we call the set of all equivalence classes the *projective space* $\mathbb{P}^{N-1}$. Equipped with the metric $d_{\mathbb{P}} : \mathbb{P}^{N-1} \times \mathbb{P}^{N-1} \to [0, \sqrt{2}]$, given by

$$d_{\mathbb{P}}(\mathbb{P}v, \mathbb{P}w) = \min \left\{ \left\| \frac{v}{\|v\|} - \frac{w}{\|w\|} \right\|, \left\| \frac{v}{\|v\|} + \frac{w}{\|w\|} \right\| \right\} \quad \text{for all } v, w \in \mathbb{R}^N,$$

the projective space is a compact metric space. For any $v \in \mathbb{P}^{N-1}$, we define

$$\mathbb{P}^{-1}v := \left\{ x \in \mathbb{R}^N : \mathbb{P}x = v \right\} \cup \{0\}.$$

**Lemma A.11.** *For all $\varepsilon > 0$, there exists a $\delta \in (0,1)$ such that for all nonzero $v, w \in \mathbb{R}^N$ with*

$$\frac{\langle v, w \rangle^2}{\|v\|^2 \|w\|^2} \geq 1 - \delta \,,$$

*we have*

$$d_{\mathbb{P}}(\mathbb{P}v, \mathbb{P}w) \leq \varepsilon \,.$$

*Proof.* This is a direct consequence of COLONIUS & KLIEMANN [50, Lemma B.1.17., p. 538]. □

**Lemma A.12.** *Let $V, W \subset \mathbb{R}^N$ be linear subspaces of the $\mathbb{R}^N$ with $V \subsetneq W$. Then*

$$d_{\mathbb{P}}(\mathbb{P}W \,|\, \mathbb{P}V) = \sqrt{2} \,.$$

*Proof.* The linear subspace $V^{\perp} \cap W$, where

$$V^{\perp} := \left\{ x \in \mathbb{R}^N : \langle x, v \rangle = 0 \text{ for all } v \in V \right\},$$

is obviously nontrivial. Let $w$ be a nonzero element of $V^{\perp} \cap W$. Then, for all $v \in V$, we have

$$d_{\mathbb{P}}(\mathbb{P}w, \mathbb{P}v) = \min\left\{ \left\| \frac{v}{\|v\|} \pm \frac{w}{\|w\|} \right\| \right\}$$

$$= \min\left\{ \sqrt{ \underbrace{\left\langle \frac{v}{\|v\|}, \frac{v}{\|v\|} \right\rangle}_{=1} + \underbrace{\left\langle \frac{w}{\|w\|}, \frac{w}{\|w\|} \right\rangle}_{=1} \pm 2 \underbrace{\left\langle \frac{v}{\|v\|}, \frac{w}{\|w\|} \right\rangle}_{=0} } \right\}$$

$$= \sqrt{2} \,.$$

Since $d_{\mathbb{P}}(x, y) \leq \sqrt{2}$ for all $x, y \in \mathbb{P}^{N-1}$, this implies the assertion. □

# References

1. ABRAHAM, R. H., MARSDEN, J. E., AND RATIU, T. *Manifolds, Tensor Analysis, and Applications.* Springer, New York, 1988.
2. AGARWAL, R. P. *Difference Equations and Inequalities.* Marcel Dekker Inc., New York, 1992.
3. AKIN, E. *The General Topology of Dynamical Systems.* No. 1 in Graduate Studies in Mathematics. American Mathematical Society, Providence, Rhode Island, 1993.
4. ANDRONOV, A. A., AND PONTRYAGIN, L. Systemes grossiers. *Dokl. Akad. Nauk SSSR 14* (1937), 247–251.
5. ARNOLD, L. *Random Dynamical Systems.* Springer, Berlin Heidelberg New York, 1998.
6. ARNOLD, L. Recent Progress in Stochastic Bifurcation Theory. Report Nr. 439, Institut für Dynamische Systeme, Universität Bremen, 1999.
7. ARNOLD, L., AND SCHMALFUSS, B. Lyapunov's Second Method for Random Dynamical Systems. *Journal of Differential Equations 177*, 1 (2001), 235–265.
8. ARNOLD, L., SRI NAMACHCHIVAYA, N., AND SCHENK-HOPPÉ, K. R. Toward an Understanding of Stochastic Hopf Bifurcation: A Case Study. *International Journal of Bifurcation and Chaos 6*, 11 (1996), 1947–1975.
9. ARNOLD, L., AND XU, K. Invariant Measures for Random Dynamical Systems, and a Necessary Condition for Stochastic Bifurcation from a Fixed Point. *Random & Computational Dynamics 2*, 2 (1994), 165–182.
10. ARTSTEIN, Z. Limiting Equations and Stability of Nonautonomous Ordinary Differential Equations. In *J. P. LaSalle, The Stability of Dynamical Systems*, vol. 25 of *CBMS Regional Conference Series in Applied Mathematics*. SIAM, Philadelphia, 1976, pp. 57–76.
11. ARTSTEIN, Z. The Limiting Equations of Nonautonomous Ordinary Differential Equations. *Journal of Differential Equations 25* (1977), 184–202.
12. AUBIN, J. P., AND FRANKOWSKA, H. *Set-Valued Analysis*, vol. 2 of *Systems and Control: Foundations and Applications*. Birkhäuser, Boston, 1990.
13. AULBACH, B. A Reduction Principle for Nonautonomous Differential Equations. *Archiv der Mathematik 39* (1982), 217–232.
14. AULBACH, B. *Gewöhnliche Differenzialgleichungen.* Spektrum Akademischer Verlag, Heidelberg, 2004. (in German).

15. AULBACH, B., AND KALKBRENNER, J. Exponential Forward Splitting for Noninvertible Difference Equations. *Computers & Mathematics with Applications 42* (2001), 743–754.

16. AULBACH, B., RASMUSSEN, M., AND SIEGMUND, S. Approximation of Attractors of Nonautonomous Dynamical Systems. *Discrete and Continuous Dynamical Systems B 5*, 2 (2005), 215–238.

17. AULBACH, B., RASMUSSEN, M., AND SIEGMUND, S. Invariant Manifolds as Pullback Attractors of Nonautonomous Difference Equations. In *Proceedings of the Eighth International Conference on Difference Equations and Applications* (Boca Raton, 2005), S. Elaydi, G. Ladas, B. Aulbach, and O. Dosly, Eds., CRC Press.

18. AULBACH, B., RASMUSSEN, M., AND SIEGMUND, S. Invariant Manifolds as Pullback Attractors of Nonautonomous Differential Equations. *Discrete and Continuous Dynamical Systems 15*, 2 (2006), 579–596.

19. AULBACH, B., AND SIEGMUND, S. A Spectral Theory for Nonautonomous Difference Equations. In *Proceedings of the Fifth Conference on Difference Equations and Applications, Temuco/Chile 2000* (2000), Gordon & Breach Publishers.

20. AULBACH, B., AND SIEGMUND, S. The Dichotomy Spectrum for Noninvertible Systems of Linear Difference Equations. *Journal of Difference Equations and Applications 7*, 6 (2001), 895–913.

21. AULBACH, B., AND WANNER, T. Integral Manifolds for Carathéodory Type Differential Equations in Banach Spaces. In *Six Lectures on Dynamical Systems*, B. Aulbach and F. Colonius, Eds. World Scientific, Singapore, 1996.

22. AULBACH, B., AND WANNER, T. Invariant Foliations for Carathéodory Type Differential Equations in Banach Spaces. In *Advances of Stability Theory at the End of XX Century*, V. Lakshmikantham and A. A. Martynyuk, Eds. Gordon & Breach Publishers, 1999.

23. AUSLANDER, J., BHATIA, N. P., AND SEIBERT, P. Attractors in Dynamical Systems. *Boletín de la Sociedad Matemática Mexicana. Segunda Serie. 9* (1964), 55–66.

24. BARREIRA, L., AND PESIN, Y. B. *Lyapunov Exponents and Smooth Ergodic Theory*, vol. 23 of *University Lecture Series*. American Mathematical Society, Providence, Rhode Island, 2002.

25. BAXENDALE, P. Wiener Processes on Manifolds of Maps. *Proceedings of the Royal Society of Edinburgh. Section A 87* (1980), 127–152.

26. BEBUTOV, M. V. Sur les systèmes dynamiques dans l'espace des fonctions continues. *C. R. Acad. Sci. URSS 27* (1940), 904–906.

27. BENOÎT, E., Ed. *Dynamic Bifurcations*, vol. 1493 of *Springer Lecture Notes in Mathematics*. Springer, Berlin, 1991.

28. BERGER, A., AND SIEGMUND, S. On the Gap between Random Dynamical Systems and Continuous Skew Products. *Journal of Dynamics and Differential Equations 15*, 2–3 (2003), 237–279.

29. BERGLUND, N. Adiabatic Dynamical Systems and Hysteresis. Thesis EPFL no. 1800, 1998.

30. BHATIA, N. P., AND SZEGÖ, G. P. *Stability Theory of Dynamical Systems*. Springer, Berlin Heidelberg New York, 1970.

31. BIRKHOFF, G. D. *Dynamical Systems*, vol. 9 of *Colloquium Publications*. American Mathematical Society, New York, 1927.

32. BISMUT, J.-M. A Generalized Formula of Itô and some other Properties of Stochastic Flows. *Zeitschrift für Wahrscheinlichkeitstheorie und verwandte Gebiete 55* (1981), 331–350.

33. BRAAKSMA, B. L. J., AND BROER, H. W. On a Quasi-Periodic Hopf Bifurcation. *Annales de l'Institut Henri Poincaré – Analyse Non Linéaire 4*, 2 (1987), 115–168.

34. BRAGA BARROS, C. J., AND SAN MARTIN, L. A. B. Chain Transitive Sets for Flows on Flag Bundles. to appear in: Forum Mathematicum.

35. BROER, H. W. Quasi Periodicity in Local Bifurcation Theory. In *Bifurcation Theory, Mechanics and Physics* (1983), C. P. Bruter, A. Aragnol, and A. Lichnerowicz, Eds., D. Reidel Publishing Company, pp. 177–208.

36. BROER, H. W., HUITEMA, G. B., TAKENS, F., AND BRAAKSMA, B. L. J. *Unfoldings and Bifurcations of Quasi-Periodic Tori.* No. 421 in Memoirs of the AMS. American Mathematical Society, Providence, Rhode Island, 1990.

37. CARABALLO, T., AND LANGA, J. A. On the Upper Semicontinuity of Cocycle Attractors for Non-Autonomous and Random Dynamical Systems. *Dynamics of Continuous, Discrete and Impulsive Systems A 10*, 4 (2003), 491–513.

38. CARR, J. *Applications of Centre Manifold Theory*, vol. 35 of *Applied Mathematical Sciences*. Springer, Berlin Heidelberg New York, 1981.

39. CESARI, L. *Asymptotic Behavior and Stability Problems in Ordinary Differential Equations.* Springer, Berlin, 1963.

40. CHEBAN, D. N., KLOEDEN, P. E., AND SCHMALFUSS, B. Pullback Attractors in Dissipative Nonautonomous Differential Equations Under Discretization. *Journal of Dynamics and Differential Equations 13*, 1 (2001), 185–213.

41. CHEBAN, D. N., KLOEDEN, P. E., AND SCHMALFUSS, B. The Relationship between Pullback, Forward and Global Attractors of Nonautonomous Dynamical Systems. *Nonlinear Dynamics and Systems Theory 2*, 2 (2002), 125–144.

42. CHENCINER, A., AND IOOSS, G. Bifurcations de tores invariants. *Archive for Rational Mechanics and Analysis 69* (1979), 109–198.

43. CHENCINER, A., AND IOOSS, G. Persistance et bifurcation de tores invariants. *Archive for Rational Mechanics and Analysis 71* (1979), 301–306.

44. CHEPYZHOV, V. V., AND VISHIK, M. I. *Attractors for Equations of Mathematical Physics*, vol. 49 of *Colloquium Publications*. American Mathematical Society, Providence, Rhode Island, 2002.

45. CHICONE, C. *Ordinary Differential Equations with Applications*, vol. 34 of *Texts in Applied Mathematics*. Springer, New York, 1999.

46. CHOW, S.-N., AND HALE, J. K. *Methods of Bifurcation Theory*, vol. 251 of *Grundlehren der mathematischen Wissenschaften*. Springer, Berlin Heidelberg New York, 1996.

47. CHOW, S.-N., LI, C., AND WANG, D. *Normal Forms and Bifurcation of Planar Vector Fields.* Cambridge University Press, 1994.

48. CODDINGTON, E. A., AND LEVINSON, N. *Theory of Ordinary Differential Equations.* McGraw-Hill Book Company, New York Toronto London, 1955.

49. COLONIUS, F., AND KLIEMANN, W. The Morse Spectrum of Linear Flows on Vector Bundles. *Transactions of the American Mathematical Society 348*, 11 (1996), 4355–4388.

50. COLONIUS, F., AND KLIEMANN, W. *The Dynamics of Control.* Birkhäuser, 2000.

51. COLONIUS, F., AND KLIEMANN, W. Morse Decompositions and Spectra on Flag Bundles. *Journal of Dynamics and Differential Equations 14*, 4 (2002), 719–741.

52. COLONIUS, F., KLOEDEN, P. E., AND SIEGMUND, S., Eds. *Foundations of Nonautonomous Dynamical Systems* (2004), Special Issue of *Stochastics and Dynamics 4*, 3.

53. CONLEY, C. C. *Isolated Invariant Sets and the Morse Index*. No. 38 in Regional Conference Series in Mathematics. American Mathematical Society, Providence, Rhode Island, 1978.

54. COPPEL, W. A. *Stability and Asymptotic Behavior of Differential Equations*. Heath, Boston, 1965.

55. COPPEL, W. A. *Dichotomies in Stability Theory*, vol. 629 of *Springer Lecture Notes in Mathematics*. Springer, Berlin Heidelberg New York, 1978.

56. CRAUEL, H., DEBUSSCHE, A., AND FLANDOLI, F. Random Attractors. *Journal of Dynamics and Differential Equations 9*, 2 (1997), 307–341.

57. CRAUEL, H., DUC, L. H., AND SIEGMUND, S. Towards a Morse Theory for Random Dynamical Systems. *Stochastics and Dynamics 4*, 3 (2004), 277–296.

58. CRAUEL, H., AND FLANDOLI, F. Attractors for Random Dynamical Systems. *Probability Theory and Related Fields 100*, 3 (1994), 365–393.

59. CRAUEL, H., IMKELLER, P., AND STEINKAMP, M. Bifurcation of One-Dimensional Stochastic Differential Equation. In *Stochastic Dynamics*, H. Crauel and M. Gundlach, Eds. Springer, Berlin Heidelberg New York, 1999, pp. 27–47.

60. CRAWFORD, J. Introduction to Bifurcation Theory. *Reviews of Modern Physics 63* (1991), 991–1037.

61. DALECKIĬ, J. L., AND KREĬN, M. G. *Stability of Solutions of Differential Equations in Banach Spaces*, vol. 43 of *Translations of Mathematical Monographs*. American Mathematical Society, Providence, Rhode Island, 1974.

62. DIECI, L., AND VAN VLECK, E. S. Lyapunov and Other Spectra: A Survey. In *Collected Lectures on the Preservation of Stability under Discretization*. SIAM, 2002, pp. 197–218.

63. DIECI, L., AND VAN VLECK, E. S. Lyapunov Spectral Intervals: Theory and Computation. *SIAM Journal on Numerical Analysis 40*, 2 (2002), 516–542.

64. ELSTRODT, J. *Maß- und Integrationstheorie*. Springer, Berlin Heidelberg New York, 1996. (in German).

65. ELWORTHY, K. D. Stochastic Dynamical Systems and Their Flows. In *Stochastic Analysis* (New York, 1978), A. Friedman and M. Pinsky, Eds., Academic Press, pp. 79–95.

66. FABBRI, R., AND JOHNSON, R. A. On a Saddle-Node Bifurcation in a Problem of Quasi-Periodic Harmonic Forcing. Preprint, 2005.

67. FABBRI, R., JOHNSON, R. A., AND MANTELLINI, F. A Nonautonomous Saddle-Node Bifurcation Pattern. *Stochastics and Dynamics 4*, 3 (2004), 335–350.

68. FLANDOLI, F., AND SCHMALFUSS, B. Random Attractors for the 3-D Stochastic Navier-Stokes Equation with Mulitiplicative White Noise. *Stochastics and Stochastics Reports 59*, 1–2 (1996), 21–45.

69. FRANKS, J. A Variation on the Poincaré-Birkhoff Theorem. In *Hamiltonian Dynamical Systems* (1988), vol. 81 of *Contemporary Mathematics*, pp. 111–117.

70. GLENDINNING, P. Non-Smooth Pitchfork Bifurcations. *Discrete and Continuous Dynamical Systems B 4*, 2 (2004), 457–464.

71. GRONWALL, T. H. Note on the Derivatives with respect to a Parameter of the Solutions of a System of Differential Equations. *Annals of Mathematics 20*, 2 (1919), 292–296.

72. GUCKENHEIMER, J., AND HOLMES, P. *Nonlinear Oscillation, Dynamical Systems, and Bifurcations of Vector Fields*. Springer, New York, 1983.

73. HADAMARD, J. Sur l'itération et les solutions asymptotiques des équations différentielles. *Bulletin de la Société Mathématique de France 29* (1901), 224–228.

74. HAHN, W. *Stability of Motion*. Springer, Berlin, 1967.

75. HALE, J. K. *Topics in Dynamic Bifurcation Theory*. No. 47 in Regional Conference Series in Mathematics. American Mathematical Society, Providence, Rhode Island, 1981.

76. HALE, J. K. Introduction to Dynamic Bifurcation. In *Bifurcation Theory and Applications*, vol. 1057 of *Springer Lecture Notes in Mathematics*. Springer, Berlin Heidelberg New York, 1984, pp. 106–151.

77. HALE, J. K. *Asymptotic Behavior of Dissipative Systems*, vol. 25 of *Mathematical Surveys and Monographs*. American Mathematical Society, Providence, Rhode Island, 1988.

78. HALE, J. K., AND KOÇAK, H. *Dynamics and Bifurcations*. Springer, New York, 1991.

79. HENRY, D. *Geometric Theory of Semilinear Parabolic Equations*, vol. 840 of *Springer Lecture Notes in Mathematics*. Springer, Berlin Heidelberg New York, 1981.

80. HIRSCH, M. W., PUGH, C. C., AND SHUB, M. *Invariant Manifolds*, vol. 583 of *Springer Lecture Notes in Mathematics*. Springer, Berlin Heidelberg New York, 1977.

81. HIRSCH, M. W., AND SMALE, S. *Differential Equations, Dynamical Systems, and Linear Algebra*. Pure and Applied Mathematics. Academic Press, 1974.

82. HOLMES, P., AND RAND, D. Phase Portraits and Bifurcations of the Nonlinear Oscillator $\ddot{x} + (\alpha + \gamma x^2)\dot{x} + \beta x + \delta x^3 = 0$. *International Journal of Non-Linear Mechanics 15* (1980), 449–458.

83. IKEDA, N., AND WATANABE, S. *Stochastic Differential Equations and Diffusion Processes*. North Holland-Kodansha, Tokyo, 1981.

84. JOHNSON, R. A. An Application of Topological Dynamics to Bifurcation Theory. In *Topological Dynamics and Applications. A volume in honor of Robert Ellis*, M. G. Nerurkar, D. P. Dokken, and D. B. Ellis, Eds., vol. 215 of *Contemporary Mathematics*. American Mathematical Society, Providence, Rhode Island, 1998, pp. 323–334.

85. JOHNSON, R. A., KLOEDEN, P. E., AND PAVANI, R. Two-Step Transition in Nonautonomous Bifurcations: An Explanation. *Stochastics and Dynamics 2*, 1 (2002), 67–92.

86. JOHNSON, R. A., AND MANTELLINI, F. A Nonautonomous Transcritical Bifurcation Problem with an Application to Quasi-Periodic Bubbles. *Discrete and Continuous Dynamical Systems 9*, 1 (2003), 209–224.

87. JOHNSON, R. A., AND YI, Y. Hopf Bifurcation from Non-Periodic Solutions of Differential Equations, II. *Journal of Differential Equations 107*, 2 (1994), 310–340.

88. KALKBRENNER, J. Nichthyperbolische exponentielle Dichotomie. Diploma Thesis, University of Augsburg, 1992. (in German).

89. KALKBRENNER, J. Exponentielle Dichotomie und chaotische Dynamik nicht-invertierbarer Differenzengleichungen. Ph. D. Thesis, University of Augsburg, 1994. (in German).

90. KATO, J., MARTYNYUK, A. A., AND SHESTAKOV, A. A. *Stability of Motion of Nonautonomous Systems (Method of Limiting Equations)*, vol. 3 of *Stability and Control: Theory, Methods and Applications*. Gordon & Breach Publishers, Philadelphia, 1996.

91. KELLEY, A. Stability of the Center-Stable Manifold. *Journal of Mathematical Analysis and Applications 18* (1967), 336–344.

92. KELLEY, A. The Stable, Center-Stable, Center, Center-Unstable, Unstable Manifolds. *Journal of Differential Equations 3* (1967), 546–570.

93. KIRCHGRABER, U., AND PALMER, K. J. *Geometry in the Neigborhood of Invariant Manifolds of Maps and Flows and Linearization*, vol. 233 of *Pitman Research Notes in Mathematical Series*. Longman, Burnt Mill, 1990.

94. KLOEDEN, P. E. Lyapunov Functions for Cocycle Attractors in Nonautonomous Difference Equations. *Izvetsiya Akad Nauk Rep Moldovia Mathematika 26* (1998), 32–42.

95. KLOEDEN, P. E. A Lyapunov Function for Pullback Attractors of Nonautonomous Differential Equations. In *Conference 05* (2000), Electronic Journal of Differential Equations, pp. 91–102.

96. KLOEDEN, P. E. Pullback Attractors in Nonautonomous Difference Equations. *Journal of Difference Equations and Applications 6*, 1 (2000), 91–102.

97. KLOEDEN, P. E. Pitchfork and Transcritical Bifurcations in Systems with Homogeneous Nonlinearities and an Almost Periodic Time Coefficient. *Communications on Pure and Applied Analysis 3*, 2 (2004), 161–173.

98. KLOEDEN, P. E., KELLER, H., AND SCHMALFUSS, B. Towards a Theory of Random Numerical Dynamics. In *Stochastic Dynamics*, H. Crauel and M. Gundlach, Eds. Springer, Berlin Heidelberg New York, 1999.

99. KLOEDEN, P. E., AND SIEGMUND, S. Bifurcations and Continuous Transitions of Attractors in Autonomous and Nonautonomous Systems. *International Journal of Bifurcation and Chaos 15*, 3 (2005), 743–762.

100. KUNITA, H. On the Decomposition of Solutions of Stochastic Differential Equations. In *Stochastic Integrals* (1981), D. Williams, Ed., vol. 851 of *Springer Lecture Notes in Mathematics*, pp. 213–255.

101. KUZNETSOW, Y. A. *Elements of Applied Bifurcation Theory*, vol. 112 of *Applied Mathematical Sciences*. Springer, New York, 1995.

102. LANG, S. *Real and Functional Analysis*. Springer, New York, 1993.

103. LANGA, J. A., ROBINSON, J. C., AND SUÁREZ, A. Stability, Instability and Bifurcation Phenomena in Non-Autonomous Differential Equations. *Nonlinearity 15*, 3 (2002), 887–903.

104. LANGA, J. A., ROBINSON, J. C., AND SUÁREZ, A. Forwards and Pullback Behaviour of a Non-Autonomous Lotka-Volterra System. *Nonlinearity 16*, 4 (2003), 1277–1293.

105. LANGA, J. A., ROBINSON, J. C., AND SUÁREZ, A. Bifurcations in Non-Autonomous Scalar Equations. *Journal of Differential Equations 221*, 1 (2006), 1–35.

106. LEBOVITZ, N. R., AND SCHAAR, R. J. Exchange of Stabilities in Autonomous Systems. *Studies in Applied Mathematics 54* (1975), 229–260.

107. LEBOVITZ, N. R., AND SCHAAR, R. J. Exchange of Stabilities in Autonomous Systems II, Vertical Bifurcation. *Studies in Applied Mathematics 56* (1977), 1–50.

108. LORENZ, E. N. Deterministic Nonperiodic Flow. *Journal of the Atmospheric Sciences 20* (1963), 130–141.

109. LUO, WANG, ZHU, AND HAN. *Bifurcation Theory and Methods of Dynamical Systems.* World Scientific, Singapore, 1997.

110. LYAPUNOV, A. M. *The General Problem of the Stability of Motion.* Mathematical Society of Kharkov, Kharkov, 1892. (in Russian).

111. LYAPUNOV, A. M. *Sur les figures d'equilibre peu differentes des ellipsodies d'une masse liquide homogène donnee d'un mouvement de rotation.* Academy of Science St. Petersburg, St. Petersburg, 1906. (in French).

112. LYAPUNOV, A. M. Problème générale de la stabilité de mouvement. *Annales de la Faculte des Sciences de Toulouse 9* (1907), 203–474. (in French).

113. LYAPUNOV, A. M. *Stability of Motion,* vol. 30 of *Mathematics in Science and Engineering.* Academic Press, New York, London, 1966. Translated from Russian by F. Abramovici and M. Shimshoni.

114. MA, T., AND WANG, S. Attractor Bifurcation Theory and its Applications to Rayleigh-Bénard Convection. *Communications on Pure and Applied Analysis 2,* 4 (2003), 591–599.

115. MARKUS, L. Asymptotically Autonomous Differential Systems. In *Contributions to the Theory of Nonlinear Oscillations III, Annals of Mathematical Studies,* S. Lefschetz, Ed., vol. 36. Princeton University Press, 1956, pp. 17–29.

116. MARSDEN, J. E., AND HUGHES, T. J. R. *Mathematical Foundations of Elasticity.* Prentice-Hall, Englewood Cliffs, New Jersey, 1983.

117. MARSDEN, J. E., AND MCCRACKEN, M. *The Hopf Bifurcation and its Applications,* vol. 19 of *Applied Mathematical Sciences.* Springer, Berlin Heidelberg New York, 1976.

118. MASSERA, J. L., AND SCHÄFFER, J. J. *Linear Differential Equations and Function Spaces.* Academic Press, New York, London, 1966.

119. MENDELSON, P. On Unstable Attractors. *Boletín de la Sociedad Matemática Mexicana. Segunda Serie. 5* (1960), 270–276.

120. MILLER, R. K. Almost Periodic Differential Equations as Dynamical Systems with Applications to the Existence of Almost Periodic Solutions. *Journal of Differential Equations 1,* 3 (1965), 337–345.

121. MISCHAIKOW, K., SMITH, H., AND THIEME, H. R. Asymptotically Autonomous Semiflows: Chain Recurrence and Lyapunov Functions. *Transactions of the American Mathematical Society 347,* 5 (1995), 1669–1685.

122. NORTON, D. E. The Fundamental Theorem of Dynamical Systems. *Commentationes Mathematicae Universitatis Carolinae 36* (1995), 585–597.

123. OCHS, G. Weak Random Attractors. Report Nr. 449, Institut für Dynamische Systeme, Universität Bremen, 1999.

124. PALIS, J., AND DE MELO, W. *Geometric Theory of Dynamical Systems. An Introduction.* Springer, Berlin Heidelberg New York, 1982.

125. PALMER, K. Exponential Dichotomy, Integral Separation and Diagonalizability of Linear Systems of Ordinary Differential Equations. *Journal of Differential Equations 43* (1982), 184–203.

126. PALMER, K. Exponential Separation, Exponential Dichotomy and Spectral Theory for Linear Systems of Ordinary Differential Equations. *Journal of Differential Equations 46* (1982), 324–345.

127. PALMER, K. Exponential Dichotomies and Transversal Homoclinic Points. *Journal of Differential Equations 55* (1984), 225–256.

128. PALMER, K., AND SIEGMUND, S. Generalized Attractor-Repeller Pairs, Diagonalizability and Integral Separation. *Advanced Nonlinear Studies 4* (2004), 189–207.

129. PAPASCHINOPOULOS, G. Exponential Separation, Exponential Dichotomy, and Almost Periodicity of Linear Difference Equations. *Journal of Mathematical Analysis and Applications 120* (1986), 276–287.

130. PERRON, O. Über Stabilität und asymptotisches Verhalten der Integrale von Differentialgleichungssystemen. *Mathematische Zeitschrift 29* (1928), 129–160.

131. PERRON, O. Die Stabilitätsfrage bei Differentialgleichungen. *Mathematische Zeitschrift 32* (1930), 703–728.

132. PLISS, V. A. Principle Reduction in the Theory of the Stability of Motion. *Izv. Akad. Nauk SSSR, Mat. Ser. 28* (1964), 1297–1323. (in Russian).

133. PLISS, V. A., AND SELL, G. R. Robustness of Exponential Dichotomies in Infinite-Dimensional Dynamical Systems. *Journal of Dynamics and Differential Equations 11*, 3 (1999), 471–513.

134. POINCARÉ, H. Sur les propriétés des fonctions définies par les équations aux différences partielles. Thèse, Gauthier-Villars, Paris, 1879. (in French).

135. POINCARÉ, H. Sur l'equilibre d'une masse fluids animes d'un mouvement de rotation. *Acta Mathematica 7* (1885), 259–380. (in French).

136. POINCARÉ, H. Mémoire sur les courbes définie par une équation différentielle IV. *Journal de Mathématiques pures et appliquées 2* (1886), 151–217. (in French).

137. POINCARÉ, H. *Les méthodes nouvelles de la mécanique céleste.* Gauthier-Villars, Paris, 1892–1899. (3 volumes, in French).

138. PÖTZSCHE, C. Extended Hierarchies of Invariant Fiber Bundles for Dynamic Equations on Measure Chains. Preprint.

139. PÖTZSCHE, C. Exponential Dichotomies of Linear Dynamic Equations on Measure Chains under Slowly Varying Coefficients. *Journal of Mathematical Analysis and Applications 289*, 1 (2004), 317–335.

140. PÖTZSCHE, C., AND RASMUSSEN, M. Local Approximation of Invariant Fiber Bundles: An Algorithmic Approach. In *Difference Equations and Discrete Dynamical Systems* (Singapore, 2005), L. Allen, B. Aulbach, S. Elaydi, and R. Sacker, Eds., World Scientific.

141. PÖTZSCHE, C., AND RASMUSSEN, M. Taylor Approximation of Invariant Fiber Bundles of Nonautonomous Difference Equations. *Nonlinear Analysis. Theory, Methods & Applications 60*, 7 (2005), 1303–1330.

142. PÖTZSCHE, C., AND RASMUSSEN, M. Taylor Approximation of Integral Manifolds. *Journal of Dynamics and Differential Equations 18*, 2 (2006), 427–460.

143. RASMUSSEN, M. Morse Decompositions of Nonautonomous Dynamical Systems. to appear in: Transactions of the American Mathematical Society.

144. RASMUSSEN, M. Approximation von Attraktoren und Mannigfaltigkeiten nichtautonomer Systeme. Diploma Thesis, University of Augsburg, 2002. (in German).

145. RASMUSSEN, M. Towards a Bifurcation Theory for Nonautonomous Difference Equations. *Journal of Difference Equations and Applications 12*, 3–4 (2006), 297–312.

146. ROBINSON, C. *Dynamical Systems. Stability, Symbolic Dynamics and Chaos,* 2 ed. CRC Press, Boca Raton, 1999.

147. RUELLE, D. Small Random Perturbations of Dynamical Systems and the Definition of Attractors. *Communications in Mathematical Physics 82* (1981), 137–151.

148. RUELLE, D., AND TAKENS, F. On the Nature of Turbulence. *Communications in Mathematical Physics 20* (1971), 167–192.

149. RYBAKOWSKI, K. P. *The Homotopy Index and Partial Differential Equations.* Springer, Berlin Heidelberg New York, 1987.

150. SACKER, R. J. Existence of Dichotomies and Invariant Splittings for Linear Differential Systems IV. *Journal of Differential Equations 27* (1978), 106–137.

151. SACKER, R. J., AND SELL, G. R. Existence of Dichotomies and Invariant Splittings for Linear Differential Systems I. *Journal of Differential Equations 15* (1974), 429–458.

152. SACKER, R. J., AND SELL, G. R. Existence of Dichotomies and Invariant Splittings for Linear Differential Systems II. *Journal of Differential Equations 22* (1976), 478–496.

153. SACKER, R. J., AND SELL, G. R. Existence of Dichotomies and Invariant Splittings for Linear Differential Systems III. *Journal of Differential Equations 22* (1976), 497–522.

154. SACKER, R. J., AND SELL, G. R. A Spectral Theory for Linear Differential Systems. *Journal of Differential Equations 27* (1978), 320–358.

155. SALAMON, D., AND ZEHNDER, E. Flows on Vector Bundles and Hyperbolic Sets. *Transactions of the American Mathematical Society 306*, 2 (1988), 623–649.

156. SCHENK-HOPPÉ, K. R. Bifurcation Scenarios of the Noisy Duffing-van der Pol Oscillator. *Nonlinear Dynamics 11* (1996), 255–274.

157. SCHENK-HOPPÉ, K. R. Stochastic Hopf Bifurcation: An Example. *International Journal of Non-Linear Mechanics 31*, 5 (1996), 685–692.

158. SCHENK-HOPPÉ, K. R. The Stochastic Duffing-Van der Pol Equation. Ph. D. Thesis, Fachbereich Mathematik, Universität Bremen, 1996.

159. SCHENK-HOPPÉ, K. R. Random Attractors – General Properties, Existence and Applications to Stochastic Bifurcation Theory. *Discrete and Continuous Dynamical Systems 4*, 1 (1998), 99–130.

160. SCHMALFUSS, B. Backward Cocycles and Attractors of Stochastic Differential Equations. In *International Seminar on Applied Mathematics — Nonlinear Dynamics: Attractor Approximation and Global Behaviour* (1992), V. Reitmann, T. Riedrich, and N. Koksch, Eds., Technische Universität Dresden, pp. 185–192.

161. SCHMALFUSS, B. The Random Attractor of the Stochastic Lorenz System. *Zeitschrift für Angewandte Mathematik und Physik 48*, 6 (1997), 951–975.

162. SCHMIDT, U. Autonome und nichtautonome dynamische Systeme. Diploma Thesis, University of Frankfurt, 2002. (in German).

163. SCHÖNEFUSS, L. W. Nichtautonome Differenzengleichungen und Kettenoperationen. Mitteilungen aus dem Mathematischen Seminar Gießen 207, Gießen, 1992. (in German).

164. SELGRADE, J. F. Isolated Invariant Sets for Flows on Vector Bundles. *Transactions of the American Mathematical Society 203* (1975), 359–390.

165. SELL, G. R. Nonautonomous Differential Equations and Dynamical Systems. I. The Basic Theory. *Transactions of the American Mathematical Society 127* (1967), 241–262.

166. SELL, G. R. Nonautonomous Differential Equations and Dynamical Systems. II. Limiting Equations. *Transactions of the American Mathematical Society 127* (1967), 263–283.

167. SELL, G. R. *Topological Dynamics and Ordinary Differential Equations.* Van Nostrand Reinhold Mathematical Studies, London, 1971.

168. SELL, G. R. The Structure of a Flow in the Vicinity of an Almost Periodic Motion. *Journal of Differential Equations 27* (1978), 359–393.

169. SELL, G. R. Bifurcation of Higher Dimensional Tori. *Archive for Rational Mechanics and Analysis 69* (1979), 199–230.

170. SHUB, M. *Global Stability of Dynamical Systems.* Springer, Berlin Heidelberg New York, 1987.

171. SIEGMUND, S. *Spektral-Theorie, glatte Faserungen und Normalformen für Differentialgleichungen vom Carathéodory-Typ,* vol. 30 of *Augsburger mathematisch-naturwissenschaftliche Schriften.* Wißner-Verlag, 1999. (in German).

172. SIEGMUND, S. Dichotomy Spectrum for Nonautonomous Differential Equations. *Journal of Dynamics and Differential Equations 14,* 1 (2002), 243–258.

173. SIEGMUND, S. Reducibility of Nonautonomous Linear Differential Equations. *Journal of the London Mathematical Society II 65,* 2 (2002), 397–410.

174. SIENZ, T. Der Attraktorbegriff in dynamischen Systemen. Thesis, University of Augsburg, 2002. (in German).

175. SMALE, S. Differentiable Dynamical Systems. *Bulletin of the American Mathematical Society 73* (1967), 747–817.

176. SRI NAMACHCHIVAYA, N. Stochastic Bifurcation. *Applied Mathematics and Computation 38,* 2 (1990), 101–159.

177. STEINKAMP, M. *Bifurcations of One-Dimensional Stochastic Differential Equations.* Logos Verlag, Berlin, 2000.

178. STRAUSS, A., AND YORKE, J. A. On Asymptotically Autonomous Differential Equations. *Mathematical Systems Theory 1* (1967), 175–182.

179. THIEME, H. Asymptotically Autonomous Differential Equations in the Plane. *Rocky Mountain Journal of Mathematics 24,* 1 (1994), 351–380.

180. VANDERBAUWHEDE, A. Center Manifolds, Normal Forms and Elementary Bifurcations. In *Dynamics Reported,* U. Kirchgraber and H. O. Walther, Eds., vol. 2. Wiley & Sons, B. G. Teubner, Stuttgart, 1989, pp. 89–169.

181. WANNER, T. Invariante Faserbündel und topologische Äquivalenz bei dynamischen Prozessen. Diploma Thesis, Universität Augsburg, 1991. (in German).

182. WIGGINS, S. *Introduction to Applied Nonlinear Dynamical Systems and Chaos,* vol. 2 of *Texts in Applied Mathematics.* Springer, New York, 1990.

183. WIGGINS, S. *Normally Hyperbolic Invariant Manifolds in Dynamical Systems,* vol. 105 of *Applied Mathematical Sciences.* Springer, New York, 1994.

184. YI, Y. A Generalized Integral Manifold Theorem. *Journal of Differential Equations 102,* 1 (1993), 153–187.

185. ZEEMAN, E. C. On the Classification of Dynamical Systems. *Bulletin of the London Mathematical Society 20,* 6 (1988), 545–557.

186. ZEEMAN, E. C. Stability of Dynamical Systems. *Nonlinearity 1,* 1 (1988), 115–155.

# Index

adiabatic solution
 definition, 131
 existence, 131
adiabatic system, 130
all-time attractive
 definition, 21
 linearized attractivity, 126
all-time attractor
 definition, 18
 example, 18, 19
all-time bifurcation
 definition, 43
 dichotomy spectrum of linearization, 125
 example, 46, 47, 137, 144
all-time dichotomy spectrum
 corresponding Morse decomposition, 106
 definition, 98
 example, 102
 spectral manifolds, 105
all-time exponential dichotomy
 definition, 83
 nonhyperbolic, 84
 Roughness Theorem, 112
 uniqueness of the invariant projector, 86
all-time repeller
 definition, 18, 22
 example, 18, 19
all-time repulsive
 definition, 21
 linearized repulsivity, 126

all-time transition
 definition, 43
 example, 144
attractor, 40

center manifold reduction, 48, 179
cocycle, 9

D-bifurcation, 49
domain of
 all-time attraction, 26
 all-time repulsion, 26
 future attraction, 25
 future repulsion, 26
 past attraction, 25
 past repulsion, 25
Duffing-van der Pol oscillator, 190
dynamical system, 9

Euclidean
 norm, 8
 scalar product, 8
exponential dichotomy, 83

forward attractor, 41
future asymptotically autonomous, 153
future attractive
 definition, 21
 linearized attractivity, 126
future attractor
 definition, 16
 example, 18
 existence, 36
 nonuniqueness, 32

# Lecture Notes in Mathematics

For information about earlier volumes
please contact your bookseller or Springer
LNM Online archive: springerlink.com

Vol. 1764: A. Cannas da Silva, Lectures on Symplectic Geometry (2001)

Vol. 1765: T. Kerler, V. V. Lyubashenko, Non-Semisimple Topological Quantum Field Theories for 3-Manifolds with Corners (2001)

Vol. 1766: H. Hennion, L. Hervé, Limit Theorems for Markov Chains and Stochastic Properties of Dynamical Systems by Quasi-Compactness (2001)

Vol. 1767: J. Xiao, Holomorphic Q Classes (2001)

Vol. 1768: M. J. Pflaum, Analytic and Geometric Study of Stratified Spaces (2001)

Vol. 1769: M. Alberich-Carramiñana, Geometry of the Plane Cremona Maps (2002)

Vol. 1770: H. Gluesing-Luerssen, Linear Delay-Differential Systems with Commensurate Delays: An Algebraic Approach (2002)

Vol. 1771: M. Émery, M. Yor (Eds.), Séminaire de Probabilités 1967-1980. A Selection in Martingale Theory (2002)

Vol. 1772: F. Burstall, D. Ferus, K. Leschke, F. Pedit, U. Pinkall, Conformal Geometry of Surfaces in $S^4$ (2002)

Vol. 1773: Z. Arad, M. Muzychuk, Standard Integral Table Algebras Generated by a Non-real Element of Small Degree (2002)

Vol. 1774: V. Runde, Lectures on Amenability (2002)

Vol. 1775: W. H. Meeks, A. Ros, H. Rosenberg, The Global Theory of Minimal Surfaces in Flat Spaces. Martina Franca 1999. Editor: G. P. Pirola (2002)

Vol. 1776: K. Behrend, C. Gomez, V. Tarasov, G. Tian, Quantum Comohology. Cetraro 1997. Editors: P. de Bartolomeis, B. Dubrovin, C. Reina (2002)

Vol. 1777: E. García-Río, D. N. Kupeli, R. Vázquez-Lorenzo, Osserman Manifolds in Semi-Riemannian Geometry (2002)

Vol. 1778: H. Kiechle, Theory of K-Loops (2002)

Vol. 1779: I. Chueshov, Monotone Random Systems (2002)

Vol. 1780: J. H. Bruinier, Borcherds Products on O(2,1) and Chern Classes of Heegner Divisors (2002)

Vol. 1781: E. Bolthausen, E. Perkins, A. van der Vaart, Lectures on Probability Theory and Statistics. Ecole d' Eté de Probabilités de Saint-Flour XXIX-1999. Editor: P. Bernard (2002)

Vol. 1782: C.-H. Chu, A. T.-M. Lau, Harmonic Functions on Groups and Fourier Algebras (2002)

Vol. 1783: L. Grüne, Asymptotic Behavior of Dynamical and Control Systems under Perturbation and Discretization (2002)

Vol. 1784: L. H. Eliasson, S. B. Kuksin, S. Marmi, J.-C. Yoccoz, Dynamical Systems and Small Divisors. Cetraro, Italy 1998. Editors: S. Marmi, J.-C. Yoccoz (2002)

Vol. 1785: J. Arias de Reyna, Pointwise Convergence of Fourier Series (2002)

Vol. 1786: S. D. Cutkosky, Monomialization of Morphisms from 3-Folds to Surfaces (2002)

Vol. 1787: S. Caenepeel, G. Militaru, S. Zhu, Frobenius and Separable Functors for Generalized Module Categories and Nonlinear Equations (2002)

Vol. 1788: A. Vasil'ev, Moduli of Families of Curves for Conformal and Quasiconformal Mappings (2002)

Vol. 1789: Y. Sommerhäuser, Yetter-Drinfel'd Hopf algebras over groups of prime order (2002)

Vol. 1790: X. Zhan, Matrix Inequalities (2002)

Vol. 1791: M. Knebusch, D. Zhang, Manis Valuations and Prüfer Extensions I: A new Chapter in Commutative Algebra (2002)

Vol. 1792: D. D. Ang, R. Gorenflo, V. K. Le, D. D. Trong, Moment Theory and Some Inverse Problems in Potential Theory and Heat Conduction (2002)

Vol. 1793: J. Cortés Monforte, Geometric, Control and Numerical Aspects of Nonholonomic Systems (2002)

Vol. 1794: N. Pytheas Fogg, Substitution in Dynamics, Arithmetics and Combinatorics. Editors: V. Berthé, S. Ferenczi, C. Mauduit, A. Siegel (2002)

Vol. 1795: H. Li, Filtered-Graded Transfer in Using Noncommutative Gröbner Bases (2002)

Vol. 1796: J.M. Melenk, hp-Finite Element Methods for Singular Perturbations (2002)

Vol. 1797: B. Schmidt, Characters and Cyclotomic Fields in Finite Geometry (2002)

Vol. 1798: W.M. Oliva, Geometric Mechanics (2002)

Vol. 1799: H. Pajot, Analytic Capacity, Rectifiability, Menger Curvature and the Cauchy Integral (2002)

Vol. 1800: O. Gabber, L. Ramero, Almost Ring Theory (2003)

Vol. 1801: J. Azéma, M. Émery, M. Ledoux, M. Yor (Eds.), Séminaire de Probabilités XXXVI (2003)

Vol. 1802: V. Capasso, E. Merzbach, B. G. Ivanoff, M. Dozzi, R. Dalang, T. Mountford, Topics in Spatial Stochastic Processes. Martina Franca, Italy 2001. Editor: E. Merzbach (2003)

Vol. 1803: G. Dolzmann, Variational Methods for Crystalline Microstructure – Analysis and Computation (2003)

Vol. 1804: I. Cherednik, Ya. Markov, R. Howe, G. Lusztig, Iwahori-Hecke Algebras and their Representation Theory. Martina Franca, Italy 1999. Editors: V. Baldoni, D. Barbasch (2003)

Vol. 1805: F. Cao, Geometric Curve Evolution and Image Processing (2003)

Vol. 1806: H. Broer, I. Hoveijn. G. Lunther, G. Vegter, Bifurcations in Hamiltonian Systems. Computing Singularities by Gröbner Bases (2003)

Vol. 1807: V. D. Milman, G. Schechtman (Eds.), Geometric Aspects of Functional Analysis. Israel Seminar 2000-2002 (2003)

Vol. 1808: W. Schindler, Measures with Symmetry Properties (2003)

Vol. 1809: O. Steinbach, Stability Estimates for Hybrid Coupled Domain Decomposition Methods (2003)

Vol. 1810: J. Wengenroth, Derived Functors in Functional Analysis (2003)

Vol. 1811: J. Stevens, Deformations of Singularities (2003)

Vol. 1812: L. Ambrosio, K. Deckelnick, G. Dziuk, M. Mimura, V. A. Solonnikov, H. M. Soner, Mathematical Aspects of Evolving Interfaces. Madeira, Funchal, Portugal 2000. Editors: P. Colli, J. F. Rodrigues (2003)

Vol. 1813: L. Ambrosio, L. A. Caffarelli, Y. Brenier, G. Buttazzo, C. Villani, Optimal Transportation and its Applications. Martina Franca, Italy 2001. Editors: L. A. Caffarelli, S. Salsa (2003)

Vol. 1814: P. Bank, F. Baudoin, H. Föllmer, L.C.G. Rogers, M. Soner, N. Touzi, Paris-Princeton Lectures on Mathematical Finance 2002 (2003)

Vol. 1815: A. M. Vershik (Ed.), Asymptotic Combinatorics with Applications to Mathematical Physics. St. Petersburg, Russia 2001 (2003)

Vol. 1816: S. Albeverio, W. Schachermayer, M. Talagrand, Lectures on Probability Theory and Statistics. Ecole d'Eté de Probabilités de Saint-Flour XXX-2000. Editor: P. Bernard (2003)

Vol. 1817: E. Koelink, W. Van Assche (Eds.), Orthogonal Polynomials and Special Functions. Leuven 2002 (2003)

Vol. 1818: M. Bildhauer, Convex Variational Problems with Linear, nearly Linear and/or Anisotropic Growth Conditions (2003)
Vol. 1819: D. Masser, Yu. V. Nesterenko, H. P. Schlickewei, W. M. Schmidt, M. Waldschmidt, Diophantine Approximation. Cetraro, Italy 2000. Editors: F. Amoroso, U. Zannier (2003)
Vol. 1820: F. Hiai, H. Kosaki, Means of Hilbert Space Operators (2003)
Vol. 1821: S. Teufel, Adiabatic Perturbation Theory in Quantum Dynamics (2003)
Vol. 1822: S.-N. Chow, R. Conti, R. Johnson, J. Mallet-Paret, R. Nussbaum, Dynamical Systems. Cetraro, Italy 2000. Editors: J. W. Macki, P. Zecca (2003)
Vol. 1823: A. M. Anile, W. Allegretto, C. Ringhofer, Mathematical Problems in Semiconductor Physics. Cetraro, Italy 1998. Editor: A. M. Anile (2003)
Vol. 1824: J. A. Navarro González, J. B. Sancho de Salas, $\mathscr{C}^{\infty}$ – Differentiable Spaces (2003)
Vol. 1825: J. H. Bramble, A. Cohen, W. Dahmen, Multiscale Problems and Methods in Numerical Simulations, Martina Franca, Italy 2001. Editor: C. Canuto (2003)
Vol. 1826: K. Dohmen, Improved Bonferroni Inequalities via Abstract Tubes. Inequalities and Identities of Inclusion-Exclusion Type. VIII, 113 p, 2003.
Vol. 1827: K. M. Pilgrim, Combinations of Complex Dynamical Systems. IX, 118 p, 2003.
Vol. 1828: D. J. Green, Gröbner Bases and the Computation of Group Cohomology. XII, 138 p, 2003.
Vol. 1829: E. Altman, B. Gaujal, A. Hordijk, Discrete-Event Control of Stochastic Networks: Multimodularity and Regularity. XIV, 313 p, 2003.
Vol. 1830: M. I. Gil', Operator Functions and Localization of Spectra. XIV, 256 p, 2003.
Vol. 1831: A. Connes, J. Cuntz, E. Guentner, N. Higson, J. E. Kaminker, Noncommutative Geometry, Martina Franca, Italy 2002. Editors: S. Doplicher, L. Longo (2004)
Vol. 1832: J. Azéma, M. Émery, M. Ledoux, M. Yor (Eds.), Séminaire de Probabilités XXXVII (2003)
Vol. 1833: D.-Q. Jiang, M. Qian, M.-P. Qian, Mathematical Theory of Nonequilibrium Steady States. On the Frontier of Probability and Dynamical Systems. IX, 280 p, 2004.
Vol. 1834: Yo. Yomdin, G. Comte, Tame Geometry with Application in Smooth Analysis. VIII, 186 p, 2004.
Vol. 1835: O.T. Izhboldin, B. Kahn, N.A. Karpenko, A. Vishik, Geometric Methods in the Algebraic Theory of Quadratic Forms. Summer School, Lens, 2000. Editor: J.-P. Tignol (2004)
Vol. 1836: C. Năstăsescu, F. Van Oystaeyen, Methods of Graded Rings. XIII, 304 p, 2004.
Vol. 1837: S. Tavaré, O. Zeitouni, Lectures on Probability Theory and Statistics. Ecole d'Eté de Probabilités de Saint-Flour XXXI-2001. Editor: J. Picard (2004)
Vol. 1838: A.J. Ganesh, N.W. O'Connell, D.J. Wischik, Big Queues. XII, 254 p, 2004.
Vol. 1839: R. Gohm, Noncommutative Stationary Processes. VIII, 170 p, 2004.
Vol. 1840: B. Tsirelson, W. Werner, Lectures on Probability Theory and Statistics. Ecole d'Eté de Probabilités de Saint-Flour XXXII-2002. Editor: J. Picard (2004)
Vol. 1841: W. Reichel, Uniqueness Theorems for Variational Problems by the Method of Transformation Groups (2004)
Vol. 1842: T. Johnsen, A. L. Knutsen, $K_3$ Projective Models in Scrolls (2004)

Vol. 1843: B. Jefferies, Spectral Properties of Noncommuting Operators (2004)
Vol. 1844: K.F. Siburg, The Principle of Least Action in Geometry and Dynamics (2004)
Vol. 1845: Min Ho Lee, Mixed Automorphic Forms, Torus Bundles, and Jacobi Forms (2004)
Vol. 1846: H. Ammari, H. Kang, Reconstruction of Small Inhomogeneities from Boundary Measurements (2004)
Vol. 1847: T.R. Bielecki, T. Björk, M. Jeanblanc, M. Rutkowski, J.A. Scheinkman, W. Xiong, Paris-Princeton Lectures on Mathematical Finance 2003 (2004)
Vol. 1848: M. Abate, J. E. Fornaess, X. Huang, J. P. Rosay, A. Tumanov, Real Methods in Complex and CR Geometry, Martina Franca, Italy 2002. Editors: D. Zaitsev, G. Zampieri (2004)
Vol. 1849: Martin L. Brown, Heegner Modules and Elliptic Curves (2004)
Vol. 1850: V. D. Milman, G. Schechtman (Eds.), Geometric Aspects of Functional Analysis. Israel Seminar 2002-2003 (2004)
Vol. 1851: O. Catoni, Statistical Learning Theory and Stochastic Optimization (2004)
Vol. 1852: A.S. Kechris, B.D. Miller, Topics in Orbit Equivalence (2004)
Vol. 1853: Ch. Favre, M. Jonsson, The Valuative Tree (2004)
Vol. 1854: O. Saeki, Topology of Singular Fibers of Differential Maps (2004)
Vol. 1855: G. Da Prato, P.C. Kunstmann, I. Lasiecka, A. Lunardi, R. Schnaubelt, L. Weis, Functional Analytic Methods for Evolution Equations. Editors: M. Iannelli, R. Nagel, S. Piazzera (2004)
Vol. 1856: K. Back, T.R. Bielecki, C. Hipp, S. Peng, W. Schachermayer, Stochastic Methods in Finance, Bressanone/Brixen, Italy, 2003. Editors: M. Fritelli, W. Runggaldier (2004)
Vol. 1857: M. Émery, M. Ledoux, M. Yor (Eds.), Séminaire de Probabilités XXXVIII (2005)
Vol. 1858: A.S. Cherny, H.-J. Engelbert, Singular Stochastic Differential Equations (2005)
Vol. 1859: E. Letellier, Fourier Transforms of Invariant Functions on Finite Reductive Lie Algebras (2005)
Vol. 1860: A. Borisyuk, G.B. Ermentrout, A. Friedman, D. Terman, Tutorials in Mathematical Biosciences I. Mathematical Neurosciences (2005)
Vol. 1861: G. Benettin, J. Henrard, S. Kuksin, Hamiltonian Dynamics – Theory and Applications, Cetraro, Italy, 1999. Editor: A. Giorgilli (2005)
Vol. 1862: B. Helffer, F. Nier, Hypoelliptic Estimates and Spectral Theory for Fokker-Planck Operators and Witten Laplacians (2005)
Vol. 1863: H. Führ, Abstract Harmonic Analysis of Continuous Wavelet Transforms (2005)
Vol. 1864: K. Efstathiou, Metamorphoses of Hamiltonian Systems with Symmetries (2005)
Vol. 1865: D. Applebaum, B.V. R. Bhat, J. Kustermans, J. M. Lindsay, Quantum Independent Increment Processes I. From Classical Probability to Quantum Stochastic Calculus. Editors: M. Schürmann, U. Franz (2005)
Vol. 1866: O.E. Barndorff-Nielsen, U. Franz, R. Gohm, B. Kümmerer, S. Thorbjønsen, Quantum Independent Increment Processes II. Structure of Quantum Lévy Processes, Classical Probability, and Physics. Editors: M. Schürmann, U. Franz, (2005)
Vol. 1867: J. Sneyd (Ed.), Tutorials in Mathematical Biosciences II. Mathematical Modeling of Calcium Dynamics and Signal Transduction. (2005)

# Recent Reprints and New Editions

Printed in the United States
By Bookmasters